Johns Hopkins University Press
Director's Circle Book for 2013

Johns Hopkins University Press gratefully acknowledges members of
the 2013 Director's Circle for supporting the publication of works such as
*Chasing Sound: Technology, Culture, and the Art of Studio Recording
from Edison to the LP.*

Anonymous
Dominic and Helen Averza
John and Bonnie Boland
Richard Burgin
Jack Goellner and Barbara Lamb
John T. Irwin
John and Kathleen Keane
Donald Kraybill
Ralph S. O'Connor
Anders Richter

STUDIES IN INDUSTRY AND SOCIETY

Philip B. Scranton, *Series Editor*

Published with the assistance of the Hagley Museum and Library

Chasing Sound

Technology, Culture, and the Art
of Studio Recording from Edison to the LP

SUSAN SCHMIDT HORNING

The Johns Hopkins University Press

Baltimore

The Johns Hopkins University Press
2715 North Charles Street
Baltimore, Maryland 21218-4363
www.press.jhu.edu

Library of Congress Cataloging-in-Publication Data

Schmidt Horning, Susan.
Chasing sound : technology, culture, and the art of studio recording from Edison to the LP /
Susan Schmidt Horning.
pages   cm. — (Studies in industry and society)
Includes bibliographical references and index.
ISBN-13: 978-1-4214-1022-7 (hardcover : alk. paper)
ISBN-13: 978-1-4214-1023-4 (electronic)
ISBN-10: 1-4214-1022-2 (hardcover : alk. paper)
ISBN-10: 1-4214-1023-0 (electronic)
1. Sound recordings—Production and direction—History. 2. Popular music—Production and
direction—History. 3. Sound recording industry—History. 4. Sound—Recording and
reproducing—History. 5. Sound studios—History. I. Title.
ML3790.S346 2013
781.490973—dc23   2013006896

A catalog record for this book is available from the British Library.

*Special discounts are available for bulk purchases of this book. For more information,
please contact Special Sales at 410-516-6936 or specialsales@press.jhu.edu.*

The Johns Hopkins University Press uses environmentally friendly book materials,
including recycled text paper that is composed of at least 30 percent post-consumer waste,
whenever possible.

Contents

Illustrations follow page 108.

Acknowledgments

My interest in the art of sound engineering was sparked by my early experience in a recording studio when my high school rock band, The Poor Girls, made a demo record at Cleveland Recording. That was in the days when you walked out of the studio with a ten-inch lacquer disc, or "acetate," in a paper sleeve, with one song on each side. Watching the engineer work, observing how he operated the controls and cued the tape, how he could hear things we could not, and how he made things sound better by moving a fader or a microphone seemed like magic to me. I did occasional work in other studios over the years, from the garage variety to major independents with my own bands as well as on other artists' recording sessions, but nothing made a more vivid impression than that early session in Cleveland.

Almost thirty years later, I interviewed that engineer, Ken Hamann, for what was to be my first conference paper, a research project that ultimately led to my dissertation and now this book. It was the first of dozens of interviews that have helped me attempt to document the largely undocumented history of sound engineers and their craft, their workplace, and their contributions to the recording art. My deepest gratitude to the men and women whose names appear in the Essay on Sources for generously sharing their time, experiences, and in some cases, lending primary sources that I would otherwise never have seen. Their stories form the backbone of this book, and I regret that so many have not lived to see its publication. Now that I know more about what Ken did, and about the nature of studio recording, I must admit that it still seems like magic.

For support and encouragement from the earliest stages of my research at Case Western Reserve University, I thank my advisor, Carroll Pursell, whose guidance helped me define the project and whose unflagging support helped me complete it, and Alan Rocke, Jonathan Sadowsky, and Mary Davis for helpful suggestions on improving the manuscript. I am deeply indebted to other colleagues and friends who have taken the time to read and comment on chapters, particularly Aaron Alcorn, Molly Berger, Hans-Joachim Braun, Tim Brooks, Chris Butler, Alan Chalmers, Mark Clark, Bryan DeSousa, Michael Devecka, Dave Giovannoni, Mike Gray, Bruce Hensal, David Hochfelder, Paul Israel, Jim Kraft, Bill Klinger, Sarah Lowengard, Alex Magoun, Anna-Maria Manuel, Rebecca McSwain, and Michael Wolfe. Special thanks to Jay McKnight, who read the entire manuscript and saved me from several technical blunders, and to Doug Pomeroy, Richard Riegel, Gene Savory, and Bill Stoddard for helpful feedback and encouragement.

Generous financial assistance from several organizations supported my research and writing. A National Science Foundation Dissertation Improvement Grant (no. 9711127) made it possible for me to conduct the majority of oral interviews, and a National Endowment for the Humanities Fellowship in 2005–6 supported additional research and writing. A short-term publication grant from the American Association of University Women Education Foundation in 2004 helped me complete the article "Engineering the Performance: Recording Engineers, Tacit Knowledge, and the Art of Controlling Sound" for the special issue, "Sound Studies: New Technologies and Music," in *Social Studies of Science*, volume 34, number 5 (December 2004), a theme that became an integral part of this book. Two research grants from the Association for Recorded Sound Collections in 1996 and 2011 helped with the initial stages of research as well as the final stages of manuscript preparation.

As a graduate student, I enjoyed generous support from the Department of History at Case Western Reserve for travel to present my research at conferences around the world, thus providing me with a priceless opportunity to meet scholars interested in music and technology from far and wide and leading to numerous opportunities to publish and present my research in different venues. I am also indebted to the International Committee for the History of Technology (ICOHTEC) and the Society for the History of Technology (SHOT) for frequent travel support, but more important, for providing a nurturing and intellectually stimulating community that welcomes graduate students and advanced scholars alike, and encourages mentoring of younger scholars.

I also want to thank Art Mollela at the Smithsonian's Lemelson Center for the Study of Invention and Innovation for inviting me to take part in the center's "Electrified, Amplified and Deified: The Electric Guitar, Its Makers and Its Players," at the National Museum of American History in Washington, D.C., in November 1996. Checking into the hotel, I met Les Paul, another participant in the event, and that chance meeting led to my interviews with him as well as the title of this book when I heard him describe what he and others were doing in the 1930s with recording experiments as "chasing sound."

For the photographs that appear in this book, I am deeply grateful to David Berger and Holly Maxson, Don Peterson and W. Royal Stokes, Loren Schoenberg and the National Jazz Museum in Harlem, Cynthia Sesso of CTS Images, Sony Music Entertainment, the Western Reserve Historical Society, and Peta Smith for their extraordinary generosity, and Che Williams for help in gathering images from the Sony photograph archives. I also thank Ed Wirth at the Thomas Edison National Historical Park, Anna-Maria Manuel, and Bill Klinger for help locating images of early recording studios, and Nick Bergh for supplying a high-resolution copy of the Western Electric engineer. For research assistance, I thank the many librarians at the St. John's University library in Queens, Kelvin Smith Library at Case Western Reserve University, Bierce Library at the University of Akron, Margaret Smith at New York University's Bobst Library, the New York Public Library, Queens Public Library, Browne Popular Culture Library and the Music Library and Sound Recording Archive of Bowling Green State University, Howard Sanner at the Library of Congress, the Paley Center for Media in New York, and the Archives Center at the Smithsonian Institution's National Museum of American History. I also thank Sara Velez, former librarian at the New York Public Library, for helpful advice on using the Rodgers and Hammerstein Collection.

I am grateful to Jeffrey Fagen, dean of St. John's College of Liberal Arts and Sciences, and to my colleagues in the Department of History for ongoing support and encouragement, especially Dolores Augustine, Mauricio Borrero, Elaine Carey, and Tracey-Anne Cooper. Through teaching seminars, workshops, and writing retreats, Maura Flannery at the Center for Teaching and Learning and Anne Geller at the Institute for Writing Studies have provided fertile ground for scholarly and pedagogical development. I thank them as well as the many colleagues throughout St. John's University for collegial support and stimulating debate.

To several scholars who have provided extraordinary opportunities that helped shape my thinking on sound and helped me mature as a scholar, I owe a special thanks. First, Hans-Joachim Braun, who organized the first of several symposia on music and technology at ICOHTEC's annual meeting in Budapest in 1996, gave me the chance to present my research at the earliest stage as well as the opportunity to play music again with the Email Special, ICOHTEC's "house band." I also thank my fellow musicians Jim Williams, Anthony Stranges, Friedrich Naumann, Jeremy Kinney, and all the pickup drummers who have helped us swing on four continents. I thank Karin Bijsterveld and Trevor Pinch for inviting me to take part in two very stimulating Sound Studies workshops at the University of Maastricht, and other scholars whose work has had a profound influence on mine, especially Mark Katz, James Kraft, Jonathan Sterne, and Emily Thompson. I am deeply grateful to my editor at the Johns Hopkins University Press, Robert J. Brugger, for his patience and belief in this project, and I thank Melissa Solarz for guidance in the final stages, and Andre Barnett for elegant and constructive copyediting. Two reviewers provided tremendous help in strengthening this manuscript, and all problems and mistakes that remain are my own.

I especially want to thank two valued friends who have supported me in one way or another from the beginning of this project: Molly Berger, my dear friend, staunchest supporter, and gentle critic; and Alan Chalmers, who provided expert advice, encouragement, and a shared love of recording technology.

My greatest debt of gratitude is to my family for their love, unwavering support, and reasonable impatience with my progress. I dedicate *Chasing Sound* to my son, Nick Horning, my brother, David Schmidt, and to the memory of my parents, Karl and Marj Schmidt. My dad passed away before I began my research and my mom, the bravest woman I know, did her best to stick around until this book was finished. It is to them I owe thanks for my earliest musical memories, a Christmas gift of Magic Mirror Records that began a lifelong fascination with sound emanating from spinning discs.

Chasing Sound

Introduction

*I*n 1943, accordion player and bandleader Frankie Yankovic booked time in the studios of the Cleveland Recording Company during a two-week furlough before being shipped overseas. In one afternoon, Yankovic and his polka band cut thirty-two sides, about eight times the average number of songs expected in the standard three-hour recording session of that period. The studio, a side business for local radio announcer Fred Wolf, was Cleveland's first professional recording facility, a place where local amateur and touring professional musicians recorded commercial radio spots, popular records for release on local labels, and vanity records for friends and family. Cleveland Recording's equipment, standard for small studios of the era, consisted of a Presto transcription disc recorder, an RCA radio broadcast console, and a few microphones. The band played the tunes from beginning to end as the recording engineer cut the audio signals fed from the microphones through the console to the disc cutter and directly onto lacquer discs. Mistakes were inevitable, retakes expensive, and consequently the initial performances, "clinkers" and all, as Yankovic recalled in his memoir, were recorded for posterity.[1] There were no retakes, no edits, no overdubs—all of which became common after tape recording was introduced just five years later.

Twenty-three years later in Los Angeles, musician and producer Brian Wilson booked seventeen sessions in five different studios, employing dozens of musicians over a nine-month period to complete the recording of a single song: the Beach Boys' "Good Vibrations." For Wilson, the real fun happened in the studio. In 1965, at age twenty-two, he stopped touring with the band, deciding instead to devote his time solely to writing and recording music. Acquiring the tag of "genius" for his songwriting and creative use of the recording studio, Wilson later recalled that he thought of his musical ideas as "toys" and the studio as his "playground."[2] The sessions were not exactly recess, and while Wilson always treated the musicians with respect and courtesy, he worked them hard and was known to spend a whole night on a sixteen-bar phrase until he achieved just the right performance, feel, and sound he was after. Where Yankovic had only an afternoon free, Wilson had the luxury of time and the finances to employ top session musicians for as many hours as he needed them. Of course, there were many other differences between these musicians and their circumstances: Cleveland polka versus California rock, local amateur musician versus internationally famous rock star, but the most significant difference was in the recording equipment at each musician's disposal. The studios where Wilson recorded were equipped with multi-track tape recorders, recording consoles, high-quality German microphones, echo chambers, equalizers, and other electronic devices that did not exist for Yankovic during his marathon session in Cleveland. The idea that music could be edited, overdubbed, layered, and shaped into a final recorded work of art impossible to replicate on stage was simply inconceivable in the early 1940s.

The technology and the different uses to which these two musicians put the recording studio symbolize the transformation in the way records were made before and after the end of World War II. For Yankovic, the studio offered a means to an end, a vehicle by which his music could reach a wider audience. Wilson, by contrast, used the studio as more than a place to document a performance, it was a place where his musical ideas, in concert with an army of musicians, recording engineers, and musical collaborators, and aided by recording technology, came together to make records that were works of art not easily recreated in live performance, if at all. Only the fundamental goal to record music was the same. The contours of how this goal was achieved differed dramatically. Yankovic's records captured a world without retakes; Wilson's ushered in the era of recording studio as creative workshop.

From the early days of acoustical recording up to Yankovic's time, records were meant to document performances; they were "sheet music on shellac," and songs were written and rehearsed before musicians even entered a recording studio.[3] Of course, there were exceptions. Guitarist Les Paul and multi-instrumentalist Sidney Bechet both made multiple-disc recordings in the 1940s, overdubbing instrument after instrument to make a studio creation, but these were still well-rehearsed performances of previously composed music. German composer Paul Hindemith created *Grammophonmusik*, an early experiment in using recording technology as a compositional tool. Like Les Paul, Hindemith altered turntable speeds to make some instruments sound like others, creating what music historian Mark Katz called "a lively study in timbre and polyphony."[4] By the 1950s, magnetic tape recorders made editing and overdubbing easier, and along with stereophonic sound, gave rise to a variety of gimmick records that exploited recording technology to create "audiophile spectacles."[5] By the 1960s, pop music producers were creating new and intriguing sounds with the help of electronic instruments, compressors, tape manipulation, and other studio effects. In 1962, British producer Joe Meek created the first British number one hit in America, "Telstar," a melodic instrumental featuring electric guitar, bass, drums, and an electronic organ known as a Clavioline. What gave this ordinary complement of instruments an extraordinary sound was Meek's use of recording technology: compressors, echo, reverb, delay, triple tracking, and sped-up tape that made single guitar strums sound like angelic harps, a fitting otherworldly touch for a record named after a space satellite, and Meek added sputtering, exploding effects at the beginning and end of the record that mimicked the sound of a spaceship launch.[6] With Meek's innovative production techniques, "Telstar," like its subject, pointed to the future. It became the sound of tomorrow.

Today, it is a given that the studio is where music is created and that engineers and producers are integral to the final product. Much of today's popular music exists only because of studio technology. Without recording, we would have no scratch or sampling, no rap or hip-hop, no techno or electronica, no mash-ups or heavily layered and tracked "songs"—essentially none of the styles of "configurable" popular music predicated on using previously recorded music and manipulating recorded sounds.[7] Beginning in the 1980s, the digital revolution in musical instruments and recording devices gave new meaning to "transformative appropriation," the act of referring to or quoting from old works to create a new work, which, as Joanna Demers points out, had always been a key element in thriving musical cultures.[8] Rather than a brief musical tip of the hat to another composer or recording artist, digital technology made it possible to take a recorded work or any part of it and transform it into another, turning the recorded music of the past into a palette of fungible sounds to be used to create new recorded works. Along with the ability to correct pitch, remove "clinkers," regulate tempo, mimic the sound of any instrument, voice, or style of playing, and with digital musical instruments, create what Paul Théberge aptly described as "any sound you can imagine," recording has wedded technology to music in ways once unimaginable.[9] But the courtship began long ago, as soon as the sound of music was made into a tangible artifact.

Musicians and nonmusicians state unequivocally that the song—historically defined as words and music—is the most important part of any record. Songs, after all, were what the recording industry was founded on, what performers needed and record buyers wanted. Whether on sheet music or shellac, the song powered the commercial revolution in American music.[10] Yet, for over a century, vast sums of money and countless hours of innovation and engineering expertise have been devoted to improving and expanding the technologies of recording and reproduction and, in the process, rendered these technologies integral to the creation of music. The evolution of music recording from the art of capturing a performance to the art of engineering an illusion changed the sound of music, changed the character of the song, and ultimately, reversed the historic relationship between live and recorded music. Through surviving recordings, we can listen to how music changed, but we have little understanding of the *process* by which the emphasis in recording shifted from the song to the sound.[11]

Chasing Sound traces this transformation by charting the development of sound recording technology and practice from the beginning of the commercial recording industry in the late nineteenth century to the rise of multi-track recording in the 1960s and early 1970s. From the earliest efforts to capture sound, to the "quest for fidelity" that was the elusive goal of acoustical-mechanical recording, to the "search for the sound" that multi-track analog tape recording made possible, the first hundred years of recording saw the rise of different concepts of sound recording and different ideas about how the studio should be used.[12] In the early 1890s, the phonograph was still considered a novelty that provided a source of amusement and recreation and the studios in which records were made were nothing more than bare rooms with a partition between the man operating the recording machine—known as the recordist—and the musical performer. Eighty years later the recording industry comprised a system of powerful international entertainment conglomerates from which emerged new skilled professions; new technologies and new companies with large departments to research, to develop, and to market those technologies; new forms of distribution and marketing; and new genres of music and new sounds. The recording studios that generated those sounds ranged from home studios with egg cartons tacked to walls for acoustical treatment to lush, high-tech facilities designed for creature comfort and acoustical perfection.

The recording studio belongs at the center of any analysis because it was there that music began the shift from live performance art to a technologically mediated art. This development continued across all media—radio studios, film sound stages, television studios, not just recording studios. However, it was in the recording studio that the transformation of music from a craft-based endeavor to one reliant on technology for its form and content began. Technology has always been integral to music making, going back to the first musical instruments other than the human voice. Before sound recording, musical instruments were used to make music in real time, with musicians playing alone or together, with one another or before an audience. Sound recording not only changed the relationship of musicians to one another and to their audience but also changed the nature of musical creativity and made nonmusicians essential to the musical art. As recording engineers sought to improve the sound of records and experimented with what could be done in the studio, their improvements gradually changed how musicians and composers conceived the music they wanted to record. Ultimately, this served to change notions of authentic performances, "good" sound, or what constituted music.

In the early years of the recording industry, few considered recording an enjoyable activity. Opera stars found the recording horn intimidating and the whole process nerve-racking. By the early 1950s, however, improvements in recording technology and the sound quality of records transformed studio practice. Cultural critics, audiophiles, and musicians heralded recording as the future of music; one contemporary reviewer declared that the recording studio would become the site where "the extreme compatibility of the artist's *modus operandi* and modern record methods" played out.[13] The gifted musician Glenn Gould left the concert stage to focus on recording, and writing about recording, predicting that technology would enable recording to progress beyond mere archiving of performances to "a higher stage in which technology and technicians would participate in the creative process actively and in their own right."[14] Frank Sinatra devoted as much attention to the details of his recording sessions as he did to any concert performance, declaring in 1961, "I adore making records. I'd rather do that than almost anything else."[15]

In 1966, the Beatles went into the EMI recording studios in London intending to make a record with "new kinds of sounds."[16] Since they had vowed never to tour again, they were not limited by the requirements of live performance, thus giving their imagination free reign. Glenn Gould believed that artists should be able to use recording techniques to "transcend the limitations that performance imposes on the imagination" and that the producer, technicians, and their apparatus should "leave the hallmark of their own art upon the disc or tape."[17] With *Sgt. Pepper's Lonely Hearts Club Band*, the Beatles, their producer, and the recording engineers accomplished just what Gould had proposed. Over the span of just two decades, the recording studio became a site of technological and musical innovation and cultural change, a place where creative uses of technology and generational conflict played out as rules and standards were established and creatively destroyed. In the recording studio more than anywhere else, technological and musical innovation went hand in hand.

From capturing the live performance to engineering sonic illusions of live performances, to creating sounds that have no origin in the natural world, sound recording's history is one of shifting objectives. Although critics debated the relative merits of records in terms of fidelity, authenticity, and more recently the quality of analog sound versus digital sound, from the beginning, recording changed the concept of "original" or "authentic" performance and the definition of "fidelity" was constantly changing. The question became "fidelity to what?" As Jonathan Sterne observed, "every age has its own perfect fidelity," and during the first decades of sound recording, "the idea of 'better' sound reproduction was itself a changing standard."[18] Since then, the history of sound recording has been a series of changing standards as technology transformed recording, leading to new sounds and new methods of producing and disseminating music, which in turn shape listeners' expectations and tastes and influence musicians' and composers' approaches to their art.

It is not news to declare that technology changed music, but how that happened from the perspective of those engaged in the recording process has remained relatively unexamined.[19] This book explores the world of recording professionals—engineers and mixers, record producers, music arrangers, session musicians and contractors, performing artists and songwriters, and independent studio owners and managers—and their collective impact on recording culture during the most formative years of sound recording. The oral history interviews conducted for this study, with recording professionals active in the record industry between the 1930s and 1970s, offer a window into the dynamic nature of studio work and how it changed over time. They suggest that the interplay of technology, engineering practice, musical performance, and creativity that occurs in recording studios involved an ongoing process of negotiation and compromise. They also reveal that the effect of technology on music was neither deterministic nor top-down but rather an ever-changing reciprocal process. Thus, *Chasing Sound* is not only a story of how technology changed music but also a story of how users—musicians as well as technicians—changed technology. In their efforts to improve sound, professionals and amateurs experimented with technique, innovated new devices, found new uses for old technologies, and adapted existing equipment to perform in ways unintended by its makers. From the very beginning of sound recording, users as well as inventors helped to refine technology and practice even as new technologies suggested new applications, a good example of the co-construction of users and technologies.[20]

Through recent works we know much more about the history of sound recording and its effect on musical culture and practices of listening than we did twenty years ago.[21] Musicologists and ethnomusicologists have explored the collaborative nature of recording in contemporary sound studios, from New York to Texas to South Africa.[22] But we have little understanding of the historical development of recording studios or the job of the recording engineer from its origins in the nineteenth-century mechanical engineering tradition to its professionalization in the analog era. In exploring these and other aspects of recording's history, this book differs from previous studies of sound recording in that it favors the production over the reception of music, the creators over the consumers, players over listeners. One challenge of charting this history follows from the lack of written documentation for the work that was done in studios, and the coming and going of studios and record labels over the years has relegated much of the documentation that did exist to dumpsters. Session logs, musicians union contracts, contemporary accounts, where they do exist, reveal only the most basic, and sometimes inaccurate, information about recording sessions; they do not reveal anything about the nature of the work of mixers or what engineers and technicians had to do to solve problems, nor do they reveal the give-and-take of musical ideas and technical limitations that characterize studio recording. The historical record depends on surviving documentation, and while the wealth of recorded music enables us to listen to music history, it is impossible to tell the technical story of recording studios by listening to their musical output alone. Although published interviews with recording and engineering professionals date as far back as the 1960s, for at least the first half of the twentieth century, few people outside of the engineering community and the artists and producers who relied on their expertise either knew or cared what recording engineers did.[23]

One musician who did care was the orchestra conductor Leopold Stokowski. He believed that the engineer operating the controls of his 1931 radio broadcast was the real conductor because he could alter the sound with the turn of a knob, and the maestro insisted that he alone manipulate the controls of his broadcast.[24] In a similar move, jazz guitarist Les Paul made a career out of re-cording his own music in a succession of home studios from the 1930s through the rest of his life, declining to work in the studios of his label, Capitol Records, where he would have had to relinquish control of his recordings to staff engi-neers and producers. Leopold Stokowski and Les Paul were among the very few artists who were fascinated by technology, and until the 1960s, few musicians shared such determination to control the technical aspects of their recordings. For the most part, artists respected the control man's expertise and followed his direction, along with that of the recording director or producer, when in the studio. They knew that successful recordings demanded it; after all, the record-ing engineer had control over the audio content, if not the artistic integrity of the performance as listeners ultimately heard it on record. In the major label studios of Columbia, RCA, Decca, and Capitol, all union closed shops, engineers had sole access to the technology. These men incorporated intuitive knowledge, craft skill, musical sensibility, engineering expertise, and the ability to adapt to unforeseen circumstances. At major record labels, members of the engineering department designed and built equipment to company specifications and to meet the needs of control engineers who worked in the studios. Independent studio owners, at least until the late 1960s, had to build their own recording chain, which required enough technical knowledge to know where to acquire the necessary components, how to assemble them, and how to maintain the equipment. This knowledge was not easily acquired. There were no schools for recording engineers until the 1960s, and few recording engineers possessed de-grees in electrical or any other type of engineering. Their training ground was the recording studio—from the earliest acoustic recording laboratories to the small, independent radio station studios of Middle America, to the major record company studios of Chicago, New York, and Los Angeles. Their work did not fit into existing categories, in part because of its liminal character. Thus, this study also uncovers the origins and growth of a previously unstudied technical profes-sion, one that resided "on the margins" between engineering, craft, and art.[25]

This book also explores the persistence of tacit knowledge as an indispens-able part of the recording engineer's work. From the early experimental meth-ods of acoustical recording to the high-tech, multi-track studios of the 1970s, the increasing complexity of recording technology did not diminish the need for recording engineers to possess certain skills above either technical or artis-tic. As new technologies enabled more creative options and recording became more collaborative between technologists and artists, this symbiotic working relationship required recording engineers to possess significant people skills as well as technical acumen. As one studio manager put it, the engineer was "a technician and a diplomat."[26]

This is also a story of a generational shift in the recording profession that paralleled the broader societal changes of the 1960s. The cohort of recording professionals that emerged from World War II, often out of the signal corps or with some related radio and electronics training, not only had their own musical tastes (primarily big band jazz, rhythm and blues, and classical) but also understood a work culture based on the military chain of command that also characterized white-collar corporate America. As the recording industry flourished, in large part by appealing to a younger audience with the increased desire and ability to consume records, it attracted a younger cohort of independent recording engineers and producers who embraced the music of their generation, rock 'n' roll, and rejected corporate values, forged less formal and reverential relationships with the musicians they recorded, and gradually approached their craft in a different way than did their predecessors. The youth cultural revolution of the 1960s played out in the studio with profound implications for the direction of recording. Whereas the best hands in the profession had developed their expertise through years of experience, rock 'n' roll was about youth, inexperience, and experimentation.

The prominent voices in this study are those of the recording engineers, mixers, record producers, and arrangers. Readers will surely note that very few of these voices belong to women or blacks. For the first century of sound recording, the field of audio engineering and recording studios in particular comprised a profoundly white male-centered culture that reflected corporate culture at large and technical professions in particular. It has its origins in the mechanical engineering background of many of those who were the first operators of recording machines, like the brothers Raymond and Harry Sooy of Victor Talking Machine, and it continued with the radio engineers of the 1920s and 1930s and the generation of World War II veterans who had trained in the signal corps and went on to build the recording studios of the postwar period. Until the 1970s, with very few exceptions, a "woman's place" in recording studios was behind the microphone rather than seated at the control board. Similarly, while African American musicians, singers, and band leaders were involved in countless recording sessions, few were found in the control room until the 1960s, a reflection of the domination of electronics and engineering fields by white men, even though the studio was peopled by a cultural cross section of musicians.[27]

This study focuses on studios devoted primarily to the recording of music for commercial records, including record company facilities and independent studios. There have been many other kinds of sound studios, such as film sound stages and electronic music studios. Although the recording technology is much the same, the goals of film sound differ considerably from those for making commercial records.[28] This book charts technological change and how it altered patterns of work for technicians and artists, but it is not a technical book, nor is it a comprehensive history of recording studios or recording equipment. The coverage of studios and individuals is selective rather than exhaustive but ranges across major label and independent recording studios, record companies, and musical genres. The focus is on popular music because that was where much of the experimentation with pushing boundaries in the studio originated. Finally, although this is not an economic history of the recording industry, it briefly explores the history of several small studio businesses, shedding light on trends in the larger economy and the history of small business enterprise. Recording studios pursued the same winning strategies that other small businesses adopted, targeting niche markets that major record companies initially ignored or marginalized. Because they had less invested in the status quo, these small studios and labels were more likely to take risks, and they were flexible enough to experiment with new methods of production when opportunities arose. Indeed, the business environment, no less than the available technology, determined what was recorded and distributed.

In charting the history of studios and the practice of recording engineers, we start at the beginning, when the old technologies of sound recording were new.[29] Our story begins in the cradle of the early recording industry, New York City and New Jersey, in laboratories, machine shops, and early acoustic recording studios where inventors and mechanical engineers developed the technology and practice of recording through cut-and-try empiricism. The greatest challenge of acoustical recording was to capture the sound of a live performance. By the 1970s, recording had become a musical event in its own right only later to be mimicked live. Where early recording was labor intensive, a more physically demanding kind of work, by the 1970s, recording had become technologically driven, guided by the ability to manipulate performances, to adjust levels, to add effects, and to edit and shape the sound with greater precision. Artists and engineers had at their fingertips a seemingly infinite array of choices and possibilities. The recording console became the engineer's palette. Not surprisingly, the engineer's job became more complex, and his skills all the more important to the record.

Capturing Sound in the Acoustic Era

Recording Professionals and Clever Mechanics

*E*arly phonograph makers and recording companies bore little resemblance to the large entertainment conglomerates they became by the end of the twentieth century. Beginning in small-machine shops and inventors' laboratories, they could be counted among the many specialty manufacturing firms that composed the largest sector of American production in the late nineteenth and early twentieth centuries. Only later would they benefit from the industrial research laboratories that emerged as America grew into a sophisticated industrializing nation.[1] The first decades in the development of sound recording involved a great deal of cut-and-try experimentation and refinement of the different components of the recording chain. Much work went into improving the mechanical apparatus, the sound box and diaphragm, the shape and material of the recording horn, and the composition of the wax recording material, but the studio received little attention. The science of modern acoustics was still in its infancy, and although some recording companies hired scientific consultants, their expertise was devoted to improving the devices of recording and reproduction rather than the design and construction of the studios in which recordings were made. Those studios ranged from the well-equipped Edison Recording Laboratory in West Orange, New Jersey, to office spaces in New York City, to a converted warehouse in Indiana, where recording experts worked to improve the art of capturing sound on record when the art of sound recording was new. Although they described their work as a mixture of art and science and considered their workplaces recording laboratories, sound recording practitioners by necessity and inclination could best be described as systematic tinkerers.[2] Early recordists approached their work empirically, citing experience and knowledge of the art as their only teachers.

Although the recording studio underwent its most dramatic change after World War II, when the booming record industry became both catalyst for and beneficiary of the growth and development of recording technology, an overview of what the early studio was like and how recordings were made in the decades before this period reveal both change and continuity in the technique and practice of sound recording and are essential for a fuller appreciation of the effect of developments during the postwar period. The equipment and processes involved in recording sound compose technological systems that have gradually evolved over time, their form and function directly influenced by the technology and practice that went before. Thus, like other technological systems that Nathan Rosenberg described as path dependent, the recording studio and the technologies it employs retain vestigial features that can best be understood with some knowledge of their evolution.[3] Moreover, a glimpse into the technical, physical, and artistic challenges of acoustical recording and how musicians and recordists adapted their respective skills and talents to meet those challenges reveals the extent to which the technology of recording and musical culture were from the very beginning mutually interdependent: cultural practices in the recording studio evolved from the limitations as well as the possibilities of the technology and the skill and creative ability of those operating the equipment. This period also reveals the varied nature of the recordist's job, which required engineering precision, ingenuity, diplomacy, and a touch of artistic creativity. Tracing the development of studio recording from the perspectives of recordists as well as artists reveals the origins of what became the inexorable link between technology, science, and art that defines sound recording.

The history of the phonograph and the early development of the phonograph industry with its competing technologies, patent wars, and competition between firms have been ably covered by historians and other scholars, and readers may refer to the bibliographic essay for further reading. However, before we enter the acoustical recording studio, a brief description of the devices and methods of sound recording and reproduction is in order.

By 1900, two competing recording formats existed: the cylinder phonograph and the disc gramophone. There were also two methods of inscribing the sound onto wax blanks: vertical or "hill and dale" recording, so named because the stylus traversed the tiny hills and valleys of the groove as it inscribed sound onto the wax surface, and lateral recording in which the stylus moved from side to side, inscribing sound in the groove. Until 1925, recording experts made cylinder and disc records by the acoustical method of recording. The basic difference between acoustical recording and the electrical recording method that replaced it involved the means of capturing and inscribing sound. In the acoustical method, the sounds to be recorded—instrumental and vocal—had to be sung or played into a large recording horn, or several horns of varying sizes, connected to a sound box that housed a diaphragm. The sound waves vibrated the glass or mica diaphragm, which in turn vibrated the cutting stylus attached to it either directly at its center or by a lever. This stylus, moving in a spiral groove pattern, engraved the sound waves on a wax blank cylinder, or disc, rotated by a weight-driven motor running at constant speed. Much work went into refining each of these components, but studios were generally simple rooms with a partition separating the recording equipment from the performers.

The Studios of the Acoustical Era

Early recording studios were anything but glamorous dens of technological marvel and musical creativity. The first studios were actual laboratories where inventors and mechanics experimented with various methods of capturing sound. Before methods of mass production were refined in the late 1890s, cylinders were recorded in "rounds," a process that resembled batch production. Recordists and performers simply made multiple copies, up to ten or more at a time, one after the other, to fill the demand. When Cappa's Seventh Regiment Band recorded "My Country 'Tis of Thee" for the New York Phonograph Company in the early 1890s, they played the number surrounded by ten phonographs. The recording operator stepped before the horn of each phonograph, started the motor, announced the selection to be recorded and the band's name, and followed with "record taken by Charles Marshall, New York City." Marshall then stopped the motor and proceeded to do the same at each recording phonograph until all the ten cylinders had recorded the opening announcement. He then started all ten motors simultaneously; the band played the song until the cylinders were nearly full, at which point Marshall signaled for them to stop at the end of the next phrase—whether or not they had played fast enough to reach the end of the song. The recordist then inserted fresh cylinders in each of the ten machines and began the entire process again. If the band played particularly fast on one round, it might burst into applause in the space left on the cylinder, "shouting and stamping in fervent approbation of its own performance."[4] This went on for three hours until as many as three hundred cylinders had been recorded. In other sessions, the solo artist bore the burden of repeating performances on a single day. *Music Trades Review* reported that America's first black recording artist, George W. Johnson, sang "The Laughing Song" fifty-six times in one day, managing to laugh with "as much merriment . . . at the conclusion as when he started."[5] By the early twentieth century, mass production of cylinders and discs eliminated the need for rounds, but in the nineteenth-century recording session, musicians needed skill as well as stamina; for all concerned, making a record was truly labor intensive.

A different type of recording went on in Thomas Edison's New Jersey laboratory, where his staff undertook extensive experimentation with a wide variety of instruments, recording horns, wax cylinder compounds, and methods of adjusting room acoustics.[6] However, this location proved inconvenient for commercial recording, as most of the talent was located in New York City. By 1906, Edison's National Phonograph Company Recording Department occupied the top floor of the Knickerbocker Building at Fifth Avenue and Sixteenth Street in Manhattan. A visitor to the New York operation in 1906 found a highly professional studio setup, including reception area, manager's private office, as well as a rehearsal and audition room where unknown performers had the opportunity to join the ranks of recording professionals, the men and women who "by means of the Phonograph have become known in the most remote corners of the globe."[7] In addition, there were test rooms, receiving and shipping rooms, and an experimental machine shop. In the largest of two recording studios, equipped with myriad devices, hangings, and other apparatus, large bands, orchestras, and other instrumentalists recorded.

On this particular day, the Edison Military Band was making a record, a process that the visitor described as "arduous and unromantic." Once the record was taken, an assistant removed it from the machine and carried it to the test room where the official critic and his assistant listened and passed judgment. Afterward, improvements were suggested and errors were explained to the musicians. Then another record was taken, and the entire process was repeated until the results were approved, and more masters were made to be sent to the factory for processing. In the smaller recording room, vocalists sang into a horn protruding through a curtained partition in the corner, preventing the artist from seeing the recording machine behind it because "how it is equipped and how it does its work are department secrets that even the artists are not permitted to know."[8] Secrecy permeated early recording culture because recordists and inventors sought to protect their inventions and innovations from being copied by rivals; in such a climate, even musicians apparently could not be trusted.

Not all studios could boast such elaborate operations, but the partition separating the area where the artists performed from the area where the recording expert operated the recording equipment remained fixed in nearly all early studios and became more rigid over time. This distinction between work domains emerged because the technology and the proprietary nature of recording methods demanded it. Recordists moved freely between these realms because the job required it, but they jealously guarded their technique. This secrecy pervaded the competitive early recording business. Both Edison and Eldridge Johnson of Victor Talking Machine protected their innovations that either required no patent or could not be patented, keeping their labs remote and under lock and key.[9]

Recordists at Work in the Studio: Clever Mechanics

Few contemporary descriptions of the recordist's room survive, but a circa 1911 reminiscence by the advertising manager of Columbia's London studio, published in *The Gramophone* in 1940, describes the recordist's "shrine of mystery," which few were permitted to enter, as containing little more than a turntable mounted on a heavy steel base, controlled by gravity weight, a floating arm with its recording diaphragm, a small bench strewn with spare diaphragms, and a heating cupboard where the wax blanks were slightly warmed to soften the recording surface.[10] A sliding glass panel in the partition between the recording room and the studio enabled the recorder (another term for "recordist") to communicate with the artists and conductor. In the early decades of recording, scientific understanding of recording remained scant, so recording experts had to rely on cut-and-try methods and a great deal of ingenuity and experimentation.[11] Each recorder developed and employed highly personal techniques, and many also designed and constructed their own recording apparatus or made these to order for others. Some, such as the brothers Harry and Raymond Sooy of the Victor Talking Machine Company, had machine work experience, but all who took up the recordist's trade entered uncharted professional territory.

In 1918, Henry Seymour, a British maker of recording and reproducing equipment, described the recordist's job as equal parts tenacity and technical acumen, requiring "mechanical finesse," and, because "no written instructions could ever amount to more than a rough and ready guide," considerable resourcefulness. The recording expert, Seymour declared, "is the one man in any record-making establishment who holds the fortunes of the establishment in his hands. Much depends on others, in the allied process, but they are as nothing in comparison to him. It is preeminently *his* work which is judged in the final analysis by the public."[12] While the public may not have been all that discerning about sound quality, at the very least, the recordist not only had to make sure his equipment operated properly but also had to instruct the artists where to stand in relation to the horn and when to change position for certain passages of the song. This positioning represented a major tension between art and technology that defined acoustical recording. Some musicians might get carried away in the emotions of the performance and forget these instructions, putting recordists in the awkward position of manhandling them. Max Hampe, a German recordist for the Gramophone Company in England, was so determined to make sure his recording was done properly that he angered the singer Frieda Hempel by brusquely shoving her forward and pushing her away from the horn during her recording session. Hempel's manager complained that this rough treatment undermined the diva's dignity before the orchestra, but Mr. Hampe claimed only to be conscious "that the lady was being paid 100 pounds a song and that it was up to him to produce results. . . . Successful records meant glory for the manager, but failures the sack for the recorder."[13] This disparity between the recordist's relatively low position in recording hierarchy and the high level of dependence on his ability to work his "magic" in the studio persisted well into the twentieth century. It also justifies Seymour's somewhat grandiose claims about the recordist's importance. Stars got top billing, but the recordist remained anonymous and uncelebrated by all but those in the know.

Recording sessions involving a large number of musicians posed logistical challenges in getting sounds onto disc or cylinder. Because the acoustical recording horn picked up certain sounds effectively but others hardly at all, the necessity of scaling back instrumentation to accommodate the limited capability of the recording horn meant that orchestras and bands for recording purposes and for concert purposes were totally different propositions.[14] Stringed instruments came across particularly weak but by necessity could not occupy the space in front of the recording horn because that was reserved for the singer (see fig. 2). Violins, violas, cellos, and basses sounded like "a pathetic and ghostly murmur," so tubas were often substituted for basses, bassoons for cellos, and after the invention of the Stroh violin in 1901, many ensembles used this instrument in place of, or in combination with, conventional violins or violas.[15] Pianos had to be placed on raised platforms near the horn with their fronts and backs removed to expose the sound. Hawaiian guitars and ukuleles appeared on

record during the vogue for Hawaiian music in the World War I era, but Span-
ish guitars were not played in recording sessions until the early 1920s, and they
were not regularly featured as solo instruments until after the introduction of
electric guitars and amplification in the 1930s.[16] Other instruments such as the
tuba, trombone, and banjo came across so well that they had to be positioned
away from the recording horn or their penetrating sounds might make the recording stylus jump the groove and ruin the take.[17] Drummers were often limited to playing nothing more than a cymbal and wood block, and occasionally
they were replaced by a banjo used as a "tune drum."[18] Because the recording
horn easily picked up high-pitched penetrating tones, sounds emanating from
outside the studio could also wind up on early records. The factory whistle
signaling the changing work shifts at the Victor plant ruined many a recording
made there until it was finally silenced in 1923.[19]

One responsibility of the recording expert, then, was to know how to position the various instruments and voices to get the best recording. Since there
could be no alteration of the recording after it was made, no adjustment to the
levels or tones of individual voices or instruments, a proper balance had to be
achieved in the recording itself solely by proper placement of instruments.
Undoubtedly, this involved compromise and the judgment of the recordist.
In a memoir of his career with the Victor Talking Machine Company and RCA
Victor, recordist Raymond Sooy recalled how he had to place those playing the
cello, oboe, clarinet, cornet, and trombone on high chairs or stools so that they
could concentrate their tones directly toward the recording horns. The proximity made it almost impossible for them to play, Sooy recalled, and "the violinists, while playing, would ofttimes run their bows up the bell of the clarinets
which were being played directly above them or in one of the other musician's
eyes, which would cause a heated argument." At other times, Sooy needed to
cobble together makeshift platforms of precarious stability. While recording
the "Poet and Peasant Overture," the xylophonist, reaching for a high tone,
pushed his instrument so hard it slid off of the flimsy three-foot-high stands,
knocking down the horns and leaving everyone "on the floor in a heap with the
artist underneath the wreckage."[20] Raymond's older brother, Harry Sooy, managed to salvage another recording session through mechanical ingenuity. When
monologist George Graham arrived for a recording date too drunk to stand up
straight, Harry made an iron yoke for his forehead, drove it into the partition
between studio and recordist's room so that it extended to just where he wanted
Graham to stand during recording, then led him up to the horn and let him rest
his head in the yoke "so he could not wobble all about." Since the speaker was

apparently able to talk just as well when he was drunk, "and perhaps a little better than when sober," the recording engagement proceeded.[21] These situations were not common and for the most part, recording sessions were routinized and uneventful, but when challenges arose, whether from temperamental equipment or uncooperative performers, the recordist had to find a way to continue the session.

Artists at Work in the Studio: Recording Professionals

Recording was physically demanding as well as mentally challenging for performers. The successful recording artist had to possess a powerful voice to be picked up by the horn, and this required distinct enunciation because the early phonographs reproduced sibilant sounds poorly, if at all. Billy Murray, one of the early recording stars, described other qualifications, unique to the recording environment, that became equally important: the ability to perform well under pressure, a talent for analyzing one's performance in the studio and making immediate adjustments, and the ability to quickly learn the material and practice the ensemble arrangement.[22] It was not enough to be a popular performer whose records sold well; recording artists who enjoyed the greatest longevity were those who made few mistakes and required few takes during recording sessions. Murray was one of several-dozen professional recording artists who derived most of their livelihood from recording in the first decade of the twentieth century. Studios relied heavily on them, and grooming these artists became essential to the success of the American record industry. Whereas in England and in the rest of the world, stage performers did the recording work, in America, stage celebrities considered the low pay and undignified working conditions of the early commercial studios beneath them. Professional recording artists came to their sessions well prepared, having learned in advance the songs company executives assigned them. Occasionally, they learned a number while in the studio, coached by the house pianist. They might even suggest a song, but the final decision about repertoire rested with company executives.[23]

After the Italian operatic tenor Enrico Caruso recorded for Victor in 1904, recording became both respectable and financially attractive, and more serious artists began making records. With no scenery, lights, or audience, emoting before a large horn was intimidating to performers accustomed to the stage. The Aeolian Company, well-established instrument makers who entered the talking machine industry around 1914, set up their recording facility in a brownstone just around the corner from Aeolian Hall in Manhattan's theater district. The company took pains to provide a warm atmosphere in its facilities to counteract the phonograph fright—the recording equivalent of stage fright—that many performers experienced. From the homey appearance of the building's white front trimmed with red-brick windowsills and flower boxes to the well-appointed reception and rehearsal rooms, replete with "handsome rugs, tapestried furniture and floor-lamps," Aeolian spared no expense in creating a welcoming environment for the artists. But the building's tidy façade and well-decorated artists' room and reception salon contrasted with the studio proper: a bare-walled room filled with musical instruments, risers, straight-backed chairs, stools, and wires hanging from the ceiling to act as music stands, and the recording horn jutting through a wall at the far end of the room. The artists may have calmed their nerves in the plush anterooms, but the studio was a place to work.[24]

Some artists regularly approached the indifferent surroundings of the recording studio with apprehension. Asked by an interviewer how she managed to put so much emotion into her recordings, violinist Maud Powell elaborated the difficulties of conjuring up passion in the throes of nervousness:

> You watch that awful face at the window, waiting for the raising of the eyebrows which tells you to begin, and all the while you are wondering whether you'll be able to find a single note. . . . The artist who makes a record sings into the horn and sees nothing else except a bare wall and the face of the operator at a tiny window. Facing this cold, indifferent prospect, which is ominous in its scientific aloofness, one must dig into one's own soul for the impassioned touch that is afterwards to thrill one's hearers, and—it is something of an ordeal. . . . Does your finger touch by accident two strings of your fiddle when they should touch but one? It will show in the record, and so will every other microscopic accident.[25]

Echoing these sentiments, Ernest L. Stevens, Thomas Edison's personal pianist-arranger and a prolific recording artist in his own right who played on hundreds of commercial and experimental records, felt anxious at the start of each session. Waiting for the red light that signaled the start of recording, Stevens recalled, "I shook like a leaf for the first record, and I did the same thing for the six hundredth."[26] In other studios, a buzzer signaled the beginning of the recording process; all present had to be utterly silent while the recording apparatus cut blank grooves at the head of the record, followed momentarily by a second buzzer signaling that the music may begin. When the selection was finished, the musicians had to maintain complete silence again until the recording expert announced from behind the partition that the recording process was complete.[27] These silences were necessary for technical reasons, but they must have heightened the sense that what was about to be recorded, and what had just been recorded, would become irrevocable documentation of an artist's ability. Those microscopic accidents that Maud Powell referred to, easily passed over in a live performance, became permanent on record. "On the vaudeville stage a false note or a slight slip in your pronunciation makes no difference," declared Ada Jones, but "on the phonograph stage the slightest error is not admissible."[28] Jones found the nervous strain of knowing that "you must not make a single mistake" utterly exhausting. Even Caruso, who loved making records—and reportedly made more money from royalties than any of his contemporaries or from his concert appearances—still "dreaded the recording laboratory more than the most exacting appearance in opera." Why? "Oh, because, because," he explained, "it must all be absolutely perfect, the perfection of the perfect mechanism."[29] Subject to the machine's ability to amplify their human imperfections, many artists never felt comfortable in the recording studio.

Recordist Raymond Sooy found that "no two artists ever face the recording instrument quite alike; some are nervous, some confident; some cannot make records with a spectator in the studio, while others must have someone standing by constantly."[30] Sooy also found that tensions ran high in certain sessions, and mistakes could incite violent outbursts. He saw one artist, enraged at the mistakes made during his session, "pick up his piano accompanist by the neck, shake him like a dog and then throw him bodily out of the studio," after which the artist became so excited, he "beat his own head on the walls of the Recording Studio." One band director bit his fingers and cursed at the musicians when they made mistakes during recording. One day he actually got down from the director's stand, "found the hat of the musician who made the mistake, threw it on the floor and jumped on it," nearly causing a riot.[31] Such tantrums may have been extreme, but the circumstances under which musicians and singers were expected to execute performances that they knew would be etched permanently onto a record did not make for relaxed and inspired performances. "Imagine a vocalist," suggested one early maker of recording machines,

> standing often on a stool or small raised platform with his mouth but a few inches from the bell of a recording horn; all around him, above and below, placed as closely together as possible, the orchestra, consisting sometimes of twenty or more performers, all directing or holding their instruments in such a manner as to get the utmost sound into the horn. Then picture this artist bobbing down under the horn, so that the full effect of the orchestra playing the opening bars may be obtained, and afterwards regaining as quickly as possible his exactly correct posi-

tion to commence his song, and then to keep in mind the whole time that he must not forget to vary his distance from the horn as required by moving away when singing his high or very loud notes, and approaching sometimes very near on the low and quiet ones, and bobbing under the horn again for the closing bars. Can anyone expect any artist to give his best under conditions such as these, and yet this is typical of what frequently occurred. One can only marvel and still further admire the work of both the artists and the recording experts.[32]

The limitations of the acoustical recording apparatus demanded that performers adjust playing style, vocal style, and physical movement to accommodate the available technology. These considerations inhibited spontaneity by forcing the performer to divide his or her concentration between artistic interpretation and recall of the "staging" required before the recording horn. The recording expert instructed, and the artist followed those instructions; those who refused risked making an inferior recording, or none at all. One Hungarian violinist insisted on walking around the Victor studio while he played, despite recordist Raymond Sooy's repeated explanations that this made it impossible to record him. The violinist told Sooy not to interfere, claiming that "it was his artistic temperament," at which point Sooy decided, "that was just about enough for me, so I told him I had one of those 'damn things' myself, and if he didn't stand where I placed him, there would be no records made. I had no more trouble with his artistic temperament after this—he stood exactly where he was placed."[33]

As singers gained experience in making records, they grew accustomed to the horn and how to gauge the appropriate position to accommodate certain notes of their vocal register. In her first recordings for Columbia, opera star Rosa Ponselle marked where she stood during the test recordings with chalk so that she could easily erase and redraw the mark if a subsequent recording from a different distance sounded better. "Being a dramatic soprano," she recalled, "I had to get *way* back" when singing high Cs, then "would run forward for the middle- and lower-register notes. I got plenty of exercise!"[34] Gradually, Ponselle learned where to stand when singing certain notes, but a more troubling restriction of those early recordings was the need to abbreviate the music to fit the four-minute time limitations of the 78 rpm ten-inch disc. Ponselle echoed the sentiments of other dramatic vocalists of her time when she recalled, in 1976, that what she found least satisfying upon hearing her early recordings was the absence of emotion: "I couldn't give as much as I would have liked to give. Emotionally I would have taken more liberties, but what can you do in three or four minutes? Besides, all the cuts they had to make—cut this bar out, and cut that bar out to save a second or two. I knew that I was limited to a short amount of time, and that sort of handicapped me, put a harness on me."[35]

Time constraints and compromised sound quality were major reasons operatic stars avoided recording studios; however, Enrico Caruso's success gave others the confidence to enlarge their audiences through recording and thus expand their careers without damaging their reputations. Of course, proper enunciation, the ability to begin on cue, and proper pacing and tempo were essential in the recording studio, whereas a misstep or botched line of singing on the stage could easily be finessed by a seasoned performer. The live audience was far more forgiving than the mechanical apparatus that picked up and permanently documented one's mistakes. In most sessions, particularly with the major record labels such as Victor and Columbia, several test recordings were made to ascertain balances. With major artists whose schedules were built around live performances, a few hours a year might be all the time they had available for recording, so the recordist had to work quickly and efficiently, maintain the studio in top running order, and elicit the trust and cooperation of the performer. The trust a recordist earned was directly related to his technical skill and ability to work effectively, efficiently, and diplomatically with the recording artist.

Tricks of the Trade (Outsmarting Sound)

From the time that commercial recording began, inventors and recordists sought to achieve more brilliant and lifelike sound, the successful attainment of which relied on the recordist's expertise with the tools of his trade. Selecting the correct thickness of diaphragm, the best means of damping the diaphragm's edges in the sound box, the correct size and shape of the recording horn, the proper design and polishing of the cutting stylus, the proper composition and conditioning of wax—all meant that the recordist had to have experience with mechanical, acoustical, and chemical elements involved in the recording process.[36] No operational manual provided guidance, and the hard-won methods remained highly individual and closely guarded secrets of each recordist, who considered the recording room the sanctum sanctorum few were permitted to enter.[37] One experienced recording expert admitted in *Scientific American* that he knew practically nothing about sound, that each day he encountered some new peculiarity, and that the only way to make good records was through "constantly watching out for the tricks of sound waves and meeting them with counter-tricks."[38]

At the turn of the century, the science of sound was still in its infancy. The work of physicists Wallace Sabine on architectural acoustics and Dayton Miller on sound and the properties of musical and other instruments appears to have had little effect on the recording industry. Miller, professor of physics at the Case School of Applied Science in Cleveland and inventor of the Phonodeik, a device that graphically displayed sound waves, acted as consultant to the Aeolian Company on the design of its phonographs as well as of the Webber piano. Few early recordists had formal scientific or engineering background; rather, they were clever mechanics whose success depended on experimentation, trial and error, and innovative thinking.[39]

To illustrate the type of mechanical and acoustic variables the recordist contended with, the recording horn and diaphragm provide two ideal examples. First, since the acoustical system relied on sound wave pressure to activate the diaphragm and the needle to cut the wax, every step along the recording chain involved a loss of that acoustical power, leading to a loss of most overtones, harmonics, and low fundamental notes that impart richness to musical sound, thus resulting in the thin, tinny, or metallic quality on acoustical recordings. Moreover, sympathetic vibrations occurred when a particular note resonated with the horn's natural resonant frequency, resulting in distortion.[40] Because the horn's particular density and resilience played a part, papier-mâché horns proved less susceptible to the influence of sympathetic vibrations, whereas brass horns had upper harmonics that tended to produce blast. Recordists often wrapped the outside of metal horns with electrician's tape to dampen these "wolf-tones."[41]

Recording horns came in many shapes and sizes, and each captured a particular range of sound best; the ideal horn for recording a soprano voice, for example, would not be right for an orchestra. Theoretically, each instrument required a different-sized horn to be recorded optimally, but while it was possible to connect several horns to one recording machine, multiple horns reduced the overall volume and required a more sensitive diaphragm.[42] The recordist had to have considerable experience with the various sizes and shapes of horn to know which would best capture the particular grouping of instruments and voices to be recorded. Like the horn, the diaphragm also possessed a natural vibratory frequency, which meant that it responded optimally to one note, less well to a wider range of notes.[43] Since the glass diaphragm also possessed weight that would make it less responsive to the acoustical sound waves, the thinner the diaphragm, the more sensitive it would be to the sound pressure, which it then transmitted to the cutting stylus. The larger the recording horn, the lower the note it could capture, and the lower the note, the greater the sound pressure and the more likely that the diaphragm would drive the stylus to travel across grooves or jump the surface of the wax blank.

The recordist, then, had to choose the various components involved in recording based on a compromise between achieving the greatest sensitivity with the fewest undesirable effects. Still, there was no guarantee that blast would not occur when the records were played back on a home phonograph. Lack of standards in recording and playback systems meant that a record that reproduced optimally on the studio equipment utilizing an ideal sound box for that instrumentation might reproduce poorly on the playback equipment of the record buyer, which utilized one sound box for a range of records available to the consumer.[44] Moreover, the average phonograph owner in the early twentieth century knew little about his or her machine, particularly the significance of choosing the right needle. Consequently, many home phonographs lacked the tonal qualities intended by the instrument's manufacturer, because the conditions of use were less than ideal.[45] Thus, even the most experienced recordist could never be sure that his recording would sound on the home phonograph as it did in the studio.

The recordist also had to possess considerable skill in dealing with people. While the recording artist may have been able to assess the aesthetic quality of his or her own performance, recordists had to steer clear of matters of taste, limiting their critical judgment to the technical quality of the recording. One of the key "Requirements Necessary for a Good Recorder," according to Raymond Sooy, was to gain the artists' confidence and to take care "never to reveal his disappointment whether the voice he is recording be good or bad, but to try under all circumstances to get the best record possible." Sooy suggested that five years of experience was essential, and because recording was a line of work entirely different from any other line of business, "in a class by itself," its practitioners required a special talent. Like the recording artist, the recordist either had it, or didn't; even experienced recorders might lack what Sooy called "the necessary knack of recording." That knack involved quick thinking, patience with often temperamental people, and the ability to remain unflappable in the face of trying situations, particularly while on location. Once Victor completed its first portable recording machine in late 1902, the company began sending recordists to do "export recording" in Mexico, Cuba, South America, England, and throughout Europe.[46] In unfamiliar makeshift studios, ranging "from a 'grass hut' to a 'palace,'" the recordists needed to know how to drape the space to make it acoustically optimal for recording and had to possess enough mechanical or electrical ability to repair the equipment should anything go wrong. Moreover, he had to be financially responsible for large sums of company money when he was on the road, because he not only had to pay artists on the date but also had to furnish the instrumentation used on the records.[47]

Finally, the recording engineer needed to understand the value of sound vibrations and know how the grooves should look and when their appearance might indicate a problem. Each instrument, Sooy noted, has a different "character of sound vibration," or frequency, measured in cycles per second, and experienced recordists could identify sound vibrations by examining the waxes with a magnifying glass after a recording was made. They knew from visual inspection, not necessarily by ear, which vibrations would not reproduce properly in the manufacturing stage and could then identify which musicians to rearrange in the studio to eliminate the problem. In the end, such experience saved money, cautioned Sooy, for "failure to detect the problem in recording might mean that the entire engagement would be lost." Hence, the kind of tacit knowledge learned by experience and intuition would be highly valued in the acoustic era. As Henry Seymour noted, "The great paucity of really good records and the plethora of the commonplace is evidence sufficient that really good recorders are extremely few, and are not to be picked up at random."[48]

Training and Tacit Knowledge: What the Recordist Knew and How He Knew It

As a completely new line of work requiring new skills, recording constituted an autonomous body of technical knowledge separate from science. Technology, as Walter Vincenti demonstrated in his study of aeronautical engineering, may *apply* science, but it is not the same as or entirely *applied* science. The creative, constructive knowledge of the engineer is the knowledge needed to implement the art of the engineer, and this was particularly true for the early recordists.[49] The Sooy brothers' memoirs, despite having been written years later from what appear to have been lab notebooks or daybooks, provide a window into the recordist's career path, training, and working life. What do they tell us? Both Harry and Raymond began as mechanics in the late nineteenth century. After being laid off from his job at the Atlantic Tool Works in Philadelphia during an acute slump in machine work in the first half of 1898, Harry, the older brother, sought employment at Eldridge R. Johnson's machine shop in Camden, New Jersey. Johnson and his small staff were at that time making Berliner Gramophones and doing general machine work. Harry Sooy began by making lathe turning tubes for the Berliner machines and moved his way up to the recording laboratory, becoming a full-fledged recorder in 1900. That same year the Berliner Gramophone Company went into receivership and E. R. Johnson formed the Victor Talking Machine Company. In 1903, Harry recommended his brother Raymond to recording manager C. G. Child when an extra hand was needed in the recording laboratory, then located in Philadelphia. Raymond got his first chance at operating the recording machine a mere three months after joining the company, thinking himself "some pumpkins" to have been given the chance to record tenor Harry MacDonough, a veteran of the early cylinder days and one of Victor's best-selling artists, singing "In the Good Old Summer Time." In February 1904, Raymond assisted on the Carnegie Hall recording sessions of Enrico Caruso, but it was not until September 1908 that he became a full-fledged recorder and was immediately sent to Mexico, the first of many location recording trips that took Raymond Sooy, frequently with Mrs. Sooy, around the world, towing as many as ten trunks in recording equipment.[50]

As recordists for the world's largest record company, the Sooy brothers' gained experience and on-the-job training with Victor Talking Machine Company and RCA Victor that had to be considered among the broadest and most expert in the trade. Not all recordists, however, had the opportunity to acquire such professional skills and worldly experience, particularly in the studios of the many independent record labels that emerged after 1914. Up until then, Edison, Columbia, and Victor had been the main patent holders and manufacturers of phonographs as well as recorded cylinders and discs. Between 1914 and 1916, some sixty new phonograph makers entered the trade, and many also established their own record labels.[51] The most successful were musical instrument manufacturers, furniture makers, and other companies with established retail distribution networks. In addition to independent record shops, records were also sold in general stores and in furniture shops with phonographs. The phonograph, like the parlor piano, was considered as much an item of furniture as a source of entertainment.[52] Even the demonstration rooms and listening booths that talking machine dealers used featured elaborate designs and period décor. The Unit Construction Company promoted its Unico System listening rooms as "sales builders" and offered six standard styles, including Louis XVI, Colonial, and Modern French interiors.[53] Just as these ancillary businesses appealed to the dealer's desire to increase sales, so the companies now entering the recording trade needed to establish legitimacy to promote their product. When it introduced the Vocalion label, Aeolian boasted that its technical staff included musicians, artisans, and noted scientists. The Otto Heineman Phonograph Supply Company introduced the Okeh label by announcing the addition of two seasoned musical and technical personnel, Charles L. Hibbard and Fred W. Hager.[54] As relative newcomers to the recording business, these companies could not expect to compete technically with Victor and Edison. Instead, these labels made a different contribution to phonographic history, recording vernacular music and regional styles that the established labels initially shunned.

With the emergence of big business firms in the mid-nineteenth century in fields where technologies permitted economies of scale in production and distribution of goods, these big businesses came to dominate certain sectors of the economy. Small businesses, which had been the norm in America before the 1880s, did not disappear; they persisted in the industrial sector by developing market niches ignored by large manufacturers or by becoming intermediate goods suppliers to large industrial firms.[55] Although the recording industry was not in the same league as the railroad or steel industries, it had grown tremendously since the 1890s and quickly spread internationally with the establishment of the Gramophone Company in Europe and subsequent moves into South America, Asia, Africa, and Australia.[56] Victor had more than $33 million in assets by 1917, over ten times its assets in 1902.[57] Victor had built its reputation on high culture, devoting three times the advertising space in the trades to its Red Seal artists than it gave to its popular artists on the Black Label, while openly admitting that sales were "just the other way, and more, too; but there is good advertising in Grand Opera."[58] Like other small businesses in America, the new record labels such as Vocalion, Okeh, and Brunswick developed niche markets by responding to a demand for popular musical genres, particularly jazz and blues.

Gennett Records of Richmond, Indiana, was one such small record label.[59] Started in 1915 by the Gennett family, owners of the Starr Piano Company and the newly established Starr Phonographs, the Starr label (as it was originally called) recorded the popular repertoire of the day—marching bands, orchestras, and stage performers—in a studio in the piano company's New York City office. The Gennett family soon constructed a phonograph-manufacturing and record-pressing facility in Starr Valley, the Richmond, Indiana, headquarters of the Starr Piano factory and the family home. They also converted a shed on the premises into a second recording studio even more primitive than the Manhattan office studio. Sandwiched between the noisy Whitewater River and a rail spur used by the Chesapeake and Ohio Railroad to deliver freight to and from the manufacturing plant, the 125-foot by 30-foot recording room was sound-proofed with sawdust between the interior and exterior walls and deadened by floor-to-ceiling monk's cloth draperies.[60] It was so acoustically dead that the musicians frequently complained they could not hear one another as they played but were nevertheless drowned out by the occasional screeching of the locomotive cars outside that interrupted many a session. Moreover, because the company apparently did not have the warming cabinets used by other studios to keep the wax discs at a constant temperature, the studio was kept uncomfortably warm year round to keep the wax discs soft enough to receive the cut of the recording stylus. Musicians found the horrendous climate in the studio a difficult environment in which to play, and the small electric fans situated on wall shelves flanking the recording horn did little to alleviate their suffering.[61]

Yet this was the studio in which Gennett Records, as the label was renamed after World War I, recorded some of the most historically important records of this century because it opened its doors to anyone, recording early jazz, blues, and "Old Time" country music before the established recording companies saw the potential market for what they considered lowbrow musical forms.[62] The family did not hire music or recording professionals for its recording studio, preferring to recruit from the nearby piano factory. Since the studio staff did not possess technical know-how, they merely captured the performances on record, although often under less than optimum conditions. The recordists from Gennett's New York recording operation frequently complained about the quality of discs recorded at the Richmond studio.[63] Throughout the 1920s, chief recordist Ezra Wickenmeyer made, in some instances, dozens of test recordings before the actual recording session began, repositioning the instruments each time until he had achieved the correct sound balance. An instruction sheet handed to King Oliver's band before they began their historic April 1923 sessions in Richmond explained that they would make an initial test recording to "judge positions and tone and arrangement of music and everything that is necessary to make a good record."[64] After fixing whatever faults existed they would make another wax, and even if it took "one hundred test waxes in a date although this would be almost a physical impossibility," they would do so until they played back a wax that sounded perfect, and then they would begin to make a master, cutting three masters for each number tested. Apparently, the Gennett recording pledge was put to the test that day as perhaps it had never been before. When the band began to play, according to Frederick Ramsey's account of the session, they "nearly smashed the recording machinery on the first test." The recordists determined fairly quickly that the real threats to their machinery were Joe Oliver and Louis Armstrong, two very strong cornetists. They moved the brass players twenty feet from the recording horn, and at that distance, Louis Armstrong recorded a solo for the first time in his life.[65]

In addition to Oliver's band, Gennett recorded the New Orleans Rhythm Kings, Bix Beiderbecke and His Rhythm Jugglers, Blind Lemon Jefferson, Big Bill Broonzy, Jelly Roll Morton, and Hoagy Carmichael. Because the recording staff did not possess the professional musical expertise to interfere with the artists' natural presentation of the music, they allowed musicians to express themselves freely. Thus, Gennett recordings have preserved an unadulterated account of early jazz and old-time country music that recording directors at major labels would have attempted to polish and refine.[66] Gennett Records is one of the earliest examples of small independent recording studios whose contribution to America's musical legacy surpassed its technological prowess.

The Gennett family had no intention of changing musical history. They were not musical visionaries but hard-nosed manufacturers and retailers who sold records to boost sales of their phonographs, but their customers comprised the black and rural consumers largely neglected by the dominant labels.[67] Perhaps more intentionally, if not altruistically, Gennett left an equally valuable legacy in challenging the Victor-Columbia patent monopoly on lateral recording. Having initially entered the record trade with vertical-cut "hill and dale" discs, which did not sell as well as the lateral-cut discs, Gennett decided to defy the patent and make their own lateral-cut records without paying Victor's licensing fee. These could then be played on the Victor and Columbia machines found in so many American homes, thus widening the market for Gennett records. Although Victor brought suit for patent infringement, Gennett and the other recording companies supporting its challenge—General Phonograph, Aeolian-Vocalion, and the Canadian Compo Company—ultimately became the victors when the US Circuit Court of Appeals invalidated the Victor patent and declared the process in the public domain. This ultimately opened up competition for new independent record companies, widening the consumer market for their releases and promoting improvements in recording processes during the 1920s.[68]

Conclusion

The early decades of sound recording were a period of cut-and-try experimentation and adaptation on the part of performers and recordists. Acoustical recording was an inventive and demanding enterprise. Professional recording experts did not purchase ready-made recording machines with instructions, rather the equipment remained highly individual and unique. Recording studios ranged from the elaborately equipped Edison Recording Laboratory in West Orange,

New Jersey, with more than a hundred different recording horns of various sizes and shapes (see fig. 1), to Gennett's Indiana operation, a single-story shed nestled among factory buildings and railroad tracks, equipped with comparatively rudimentary recording technology and expertise. Though recording studios varied in size, equipment, and technical capability, the quality of their records depended largely on the skill of the operator. Those who became recording experts had a special talent for their craft and the drive and tenacity to overcome the obstacles to capturing sound waves by acoustical-mechanical means. The recordist learned through experience what worked and what didn't; as one recordist observed, "sound waves may follow and be governed by established laws of nature . . . but unknown laws always intervene, and many are the problems that must be decided and acted upon as the result of long training and good judgment."[69]

In the first three decades of commercial recording, roughly the 1890s to the mid-1920s, the limitations of acoustical recording made it impossible for weak, soft, or subtle vocalists to become successful recording artists, indeed to become recording artists at all. Nevertheless, commercial records in these years reflected considerable diversity. As early as 1893, the Columbia Phonograph Company's recorded repertoire included the United States Marine Band, marches, polkas, gallops, yorkes, waltzes, schottisches, airs, sacred music, artistic whistling, whistling with piano, songs, piano and solo instrument, dialogues, and a series of auctioneer records.[70] Between 1895 and 1925, popular styles included Tin Pan Alley songs, Broadway show tunes, ragtime, "coon" songs, "darky dialect" stories, novelty numbers, quartet arrangements, parlor ballads, early jazz (originally called "jass"), blues, dance music, hymns, and early country, or Old Time, music.[71] In addition to other styles of music (opera, band, orchestral), the recording phonograph was put to other uses by amateurs and professionals alike, making records of everything from sermons to smut and strange songs that artist R. Crumb described as "audial surrealism."[72] Almost from the moment the phonograph was introduced, amateur ethnographers began to employ portable cylinder recorders to study and document other cultures, and recording companies made many location trips to record foreign performers.[73] The popularity of "race" records, pioneered by small labels such as Okeh, Gennett, and Vocalion in the 1920s, led Columbia and Victor to undertake location recording in the American South, recording in churches, hotel rooms, warehouses—anywhere the company could set up its portable facilities.[74]

One reason for this diffusion of recording activity was the wider availability of recording equipment in the 1930s, and the greater ease with which even the novice could make a record on these new machines. The first step in that process was the successful development of the electrical process of recording, a goal pursued by recordists, inventors, and recording companies dating back to Edison and Berliner. But, as the next chapter reveals, the successful endeavor was made by a team of researchers with no prior connection to the commercial record business.

2

The Studio Electrifies

Radio, Recording, and the Birth of the Small Studio Business

*F*ueled by the dance craze of the 1910s that introduced new steps such as the turkey trot and the foxtrot, the record industry, by World War I, had entered what some historians referred to as its first "golden age."[1] By the early 1920s, the spectrum of recorded music had broadened to include more jazz, blues, gospel, and hillbilly music, and new record labels such as Gennett, Okeh, Paramount, Black Swan, Ajax, and Vocalion focused on these niche markets. In addition, a huge demand for foreign music records emerged as America's growing immigrant communities, struggling with assimilation, sought familiar music from the homeland.[2] The blues craze that began with Okeh's 1920 release "Crazy Blues" by Mamie Smith and her Jazz Hounds prompted Columbia to begin releasing blues records, and by 1923, the label had signed blues and jazz artists, including Bessie Smith, Fletcher Henderson, Clarence Williams, and Ethel Waters, while Victor concentrated on selling its Red Seal catalog to the African American market.[3] The recorded output of the 1920s embodied what one musicologist aptly dubbed "the motley fabric of popular music" during the Jazz Age.[4] Record sales reached $106 million in 1921, but as records became the most popular form of home entertainment, a competing entertainment technology, radio, introduced a new way for the listening public to hear music.[5] By 1922, a radio boom had swept the country and record sales began a three-year decline. Radio has been blamed for the recording industry's reversal of fortunes, but radio ultimately became the catalyst for sweeping changes in sound recording. The communication technologies developed during World War I to improve and expand wireless transmission and later developments in early radio broad-casting ushered in a transformation of the recording studio as engineers em-ployed microphones and amplification devices. Electrifying the studio became the most revolutionary improvement in sound recording to date and the first step in transforming recording from the art of capturing sound to the engineer-ing of an illusion.

Electrical recording transformed the recording studio, the work of recordists and musicians, and the record industry during the interwar period. First, more than any other innovation up to that time electrical recording dramatically im-proved the sound of records. Replacing the recording horn with microphones also eased cramped conditions in the studio. It became possible to record larger-sized ensembles more effectively and thus a broader range of music. The change from acoustical to electrical recording signified more than a simple im-provement in the sound of records; it marked a radical shift with long-term con-sequences for recording. It transformed the recordist's craft from cut-and-try empiricism to mathematical and scientific control and measurability. Although it did not eliminate experimentation in the studio, it changed and expanded what recordists needed to know. Electrical recording also made possible the de-velopment of instantaneous recording methods, which encouraged a new group of amateur enthusiasts to enter the recording field, many of whom ultimately made recording their profession. Finally, it gave rise to electrical transcriptions and syndicated radio program services—the offspring of the union of radio and recording and an arena in which many fledgling recording engineers began their careers. These developments in turn gave rise to small independent "air check" studios not affiliated with the marketing, sale, or distribution of records. From the early 1930s on, transcription and production studios mushroomed in New York, Chicago, and Los Angeles, as well as in cities outside these key entertain-ment centers, such as Cleveland, Seattle, San Francisco, and Atlanta.[6] This first revolution in recording technology produced not only better-sounding records but also more of them and more individuals involved in making them.

From Recording Experts to Engineers: Transforming the Craft of Recording

On January 1, 1925, Bell Telephone Laboratories incorporated as a wholly owned subsidiary of American Telephone & Telegraph and Western Electric, thereby officially becoming the second-largest industrial research laboratory in the United States after General Electric. AT&T promoted Bell Labs as a national research laboratory in which the benefits of pure research could be carried over to different scientific, technical, and economic interests.[7] Among the first businesses to benefit from the work of Bell System scientists under this broad research agenda were the recording companies and film industries. The story of how Bell Labs and Western Electric, the manufacturing and licensing arm of AT&T, became involved in the entertainment field has been told elsewhere, but it is worth recounting the outlines of the story to illustrate how a new principle was applied to the talking machine art, thus revolutionizing the work of the recording studio.[8]

Since the late nineteenth century, record companies, inventors, and individual experimenters had sought ways to improve the sound of records. The first few decades of sound recording focused on improving the design of records and playback equipment, and a good deal of empirical knowledge about studio recording had been accumulated through trial and error. The major drawback of the mechanical-acoustical recording method rested in the weak power of sound waves to drive the cutting stylus sufficiently to engrave sound into the record grooves. Every step along the recording chain involved a loss of acoustic power and a consequent degradation of sound quality from increased noise levels and reduced frequency response. In 1905, Sir Charles A. Parsons, inventor of the steam turbine, devised a talking machine that employed air pressure, adjustable valves, and a reed-like sound box to intensify sound reproduction. His invention, the Auxetophone, was intended for public use and on a calm day was said to be audible for miles. Victor Talking Machine acquired control of the patent in the United States and produced five hundred units, but the device required more routine mechanical attention than expected and initial enthusiasm quickly waned.[9] Any attempts to acoustically amplify the sound in playback still could not make up for the inadequate volume and limited range of frequencies captured in the initial recording. Many companies and individuals experimented, but no systematic development occurred until after World War I when research into acoustical theory and advances in communication and amplification led to the era of scientific electroacoustics.[10]

In the fall of 1919, a team of Bell Labs engineers under the leadership of Joseph P. Maxfield began work on developing electrical recording. Equipped with new tools—condenser microphones, vacuum tube amplifiers, electrical filters, loudspeakers—as well as experience dating back to about 1912 in using acoustic recording and reproducing equipment to analyze speech in the development of the telephone, Bell Labs was uniquely poised to undertake the project.[11] Theirs was neither the first nor the only effort to come up with such a method. More than a dozen others were known to have experimented with electrical recording between 1919 and 1925. The most successful was Orlando Marsh of Chicago, but others never publicized their efforts because of either competitive secrecy or failure to achieve satisfactory results.[12] The Bell team succeeded where others failed for several reasons. Like other big businesses Bell had the synergistic muscle of a major research and development laboratory behind it, which provided not only the most extensive body of experience in the field of sound transmission but also ample experimental resources. Between 1912 and 1916, Western Electric engineers not only developed the high vacuum tube (from Lee DeForest's Audion, which the Bell System purchased in 1913) and applied it to long-distance telephone service but also recorded sound on film, constructed improved electrical reproducers for the phonograph, developed the condenser microphone, and invented public-address systems used in World War I military aircraft.[13] Most important for the development of electrical recording, by applying the laws of physics and mathematics to the study of sound reproduction, they were able to overcome the impasse that frustrated empirical cut-and-try methods.

Analyzing the problem before them, of "taking sound from the air, storing it in some permanent way and reproducing it again without appreciable distortion," the Bell engineers examined the existing acoustical methods of both recording and reproduction.[14] They studied every aspect of the recording process: the studio design and acoustical characteristics, the desired frequency range to achieve a more faithful reproduction of speech and all types of music, and the difficulties associated with the existing acoustical-mechanical recording method, from the problems confronting the musicians in crowding around the horn to those confronting the recordist in achieving balanced sound and sufficient volume. By using condenser microphones and vacuum-tube amplifiers, and devising an electromagnetic rubberline recorder and reproducer based on the principle of mechanical analogs of electrical filters that had been developed for telephone systems, they eliminated most—but not all—of the major problems associated with acoustical recording and reproduction.[15] The lack of low tones on acoustical records led to a metallic sound in reproduction, and the loss of higher harmonics created a muffled quality and the inability to hear sibilants.[16] To create what Maxfield called "the illusion of the presence of the artist," he and his team knew they needed to control the acoustics of the room in which the record was made; reproduce weaker sounds of performances, often carried by higher frequencies; and reproduce low bass notes.[17] The key breakthrough in their analysis also became an important milestone in the application of electrical theory to mechanical transmission problems. Maxfield noted that in teaching electricity it had been customary to apply analogous mechanical principles. His colleague Henry C. Harrison devised an electromechanical recorder by doing the reverse: using electrical-network analogs for the mechanical components of the acoustic recorder.[18] Thus, instead of the weak sound waves driving the diaphragm and cutting stylus, the electromechanical recorder transformed sound into electrical signals. When amplified these signals drove a coil that drove a moving armature attached to the stylus that cut the grooves, thus matching the level and frequency content of the original sound waves.

The new electrical recording chain reconfigured the studio and recording room and gave the recordist—still a few years from being referred to as the recording engineer—increased control. Instead of a horn protruding through the wall or dominating the center of the room, performers now sang into a microphone connected by wire to the recording room that housed the amplifier and electromagnetic recorder. Together, these composed what Maxfield and Harrison called the recording amplifier system, the amplified equivalent of the acoustic recording horn and sound box. The recording horn, diaphragm, and cutting stylus of the acoustical system had been replaced by the condenser microphone, vacuum tube amplifier, and electromagnetically powered cutting stylus.[19]

A more revolutionary change involved the introduction of completely new components to the system: a volume indicator for measuring the power delivered to the recorder and an audible monitoring system, enabling the operator to listen during recording. Instead of only listening and determining proper balance by ear, or by inspecting the grooves after recording, the recordist could now see the signal level being recorded and adjust accordingly (see fig. 3). This ability to measure the volume quantitatively and to monitor the sound audibly and visually during recording had never been possible before. The acoustical recordist could only adjust volume by altering the placement of musicians around the horn or by changing the diaphragm in the sound box, a judgment only arrived at after a test. Now the recordist could adjust the sound level from the control room during a test. In acoustical recording, the only way to know how the recording would sound was to play back a test wax master, thus destroying that particular recorded performance. If the level sounded right, the recording had to be taken again exactly as played on the test. With the electrical process, monitoring the performance as the recording was under way enabled the recordist to experiment with adjusting levels in the course of a single take, thus saving time and wax, and relieving artist and recordist alike of the tedium of repeated tests until the proper balance had been struck. Although monitoring did not eliminate the need for test records, it reduced the number of tests necessary to achieve the desired balance.

As described by its inventors, the Maxfield–Harrison recording method heralded three major improvements immediately apparent to the listener. First and most noticeable was the sudden presence of bass and higher harmonics, giving the music, in Maxfield's words, "body and weight" as well as "definition or detail."[20] Now the dynamic range of large orchestras and choruses could be recorded and more fully reproduced. Second, the records had "atmosphere," or "room tone," giving the listener the sense that the music had been played in a room or a hall. This was possible in part because the instruments were no longer crowded around the horn and because the sensitive microphone, unlike the acoustical horn, could pick up room acoustics.[21] Finally, new records seemed louder because of the wider range of recorded and reproduced notes. The inventors stressed that this apparent loudness was in fact due to the presence of added notes either weakly reproduced or not at all audible in the older system, not by increasing overall volume.[22] The overall impression, as one listener described the experience of hearing an early electrical recording, was "as if the doors of my machine were a window opening on to the great hall in which the concert was held. If it produces any less perfect result in your hands," the writer asserted, "blame your reproducing apparatus and not the record."[23]

This was no small matter, for records require two discrete systems—the system of recording and the system of reproduction. At different points in the development of recording and playback technology, the capabilities of one invariably exceeded those of the other. Thus, although acoustical *reproduction* could extend to about 4,000 cycles per second, the frequency response of acoustical *recording* was limited to approximately 200 to 3,000 cycles per second.[24] With the new electrical recording method, frequency response extended from 100 to just more than 5,000 cycles per second, thus capturing low fundamental notes and higher overtones of speech and music. However, Maxfield and Harrison had only electrified *recording*—not the phonograph, the system of *reproduction*. To accommodate the disparity, they designed an improved acoustical-mechanical phonograph and reproducer based on the electrical principles of resistance and matched impedance, thus ensuring that the wider frequency range captured in the recording would be transmitted in playback.[25] The results of Maxfield and Harrison's research, the Orthophonic Victrola, became the first consumer phonograph specifically designed to play electrically recorded discs and reproduce the wider frequency response they offered. However, the key to the audibly improved records, explained Maxfield, lay in the new methods of recording.[26]

Diffusion of Electrical Recording into Record Company Studios

The Bell Labs work had been kept under wraps until Maxfield and Harrison determined the commercial feasibility of the process, at which point Western Electric, AT&T's manufacturing arm, undertook licensing agreements with the two leading US record companies. Rather than sell the system outright, they leased the equipment for a onetime fee of $50,000 and a royalty on each record made. In 1924, they approached the Victor Talking Machine Company and the Columbia Phonograph Company about acquiring rights to the new recording process. Victor executives hesitated, and Columbia, which had gone into receivership in October 1923 after debilitating losses following overproduction and the 1921 business depression, was not in a position to license the system.[27] The other major US record label, Brunswick-Balke-Collender, did not receive an offer from Western Electric but worked with General Electric to develop its own, albeit short-lived, method of electrical recording. Named the "Light Ray," Brunswick's system was a complex method that employed a minute crystal mirror and an electromagnetic cell rather than a traditional microphone.[28]

In the meantime, Sir Louis Sterling of Columbia Graphophone Company in England, which had been actively involved in electrical recording experiments of its own since 1919, learned of the Bell Labs development and hoped to acquire the Western Electric system for British Columbia. Because Western Electric would only license the system to foreign companies through an American affiliate, Sterling could acquire the technology only through purchasing a controlling interest in the ailing Columbia Phonograph Company, thereby gaining the rights to the new electrical process for his Columbia Graphophone Company in England. Sterling's faith in the superiority and ultimate commercial potential of electrical recording can be gauged by his investment: $2.5 million for Columbia Phonograph and $50,000 for the Western Electric licensing fee, plus a royalty on each record made by the process.[29]

Conversely, Victor's sluggish response to the Western Electric offer and to electrical recording in general has been attributed in the standard phonograph histories to a range of causes: from Victor president Eldridge Reeves Johnson's illness to the company's general antagonism to radio and anything connected with it. According to bandleader Paul Whiteman, a year before Western Electric made its offer, Victor had rejected an electrical recording system devised by an English inventor. Whiteman, one of Victor's major artists, met the inventor when his band made its first trip abroad. Whiteman was impressed with the recordings he heard and paid the inventor's passage to the United States. To-gether, the inventor and the bandleader, then riding high on the popularity of his recordings, tried to sell the process to Victor, only to be met with resistance at "the lower executive echelons." Victor had its own research department, and Whiteman was told that the invention did not have a chance.[30] Although White-man does not provide the exact date of the encounter with Victor executives, it may have coincided with experimental work in electrical recording carried out by Victor employee Albertis Hewitt.[31] More than likely, their rejection of the idea can be attributed to the corporate not-invented-here ethos. As phonograph his-torians Oliver Read and Walter Welch observed, Victor "had been accustomed for years to giving aspiring inventors from outside the realm a courteous but firm brush-off."[32]

Victor's hesitance about licensing the Western Electric system, despite its superior quality and commercial feasibility over any other method, can be un-derstood in light of its significant investment in equipment and inventory and thus in maintaining the status quo. However, it also suggests a cultural resis-tance to new technology as much as any corporate rejection of outside inven-tion. According to Joseph Maxfield, some of the Victor executives initially rejected the system, because although the new equipment sounded "more natu-ral," it, nevertheless, "didn't sound like a phonograph"; they believed the public would not buy it. A listening test conducted by Victor's vice president at home with his neighbors acting as the unwitting test audience revealed that they were wrong. When the host apologized for what he referred to as "experimental" re-cordings, telling them they could resume listening to the older acoustical re-cords, "his neighbors made it clear that they had no further interest in listening to the old records on the old machine, and they insisted on playing through each of the new records several times."[33] The remaining Victor skeptics conducted similar listening experiments in their own homes, until finally, as Maxfield later recalled, the last hold-outs "shook their wise old white heads and said, 'well, we don't understand it, but if that's what the public wants, let's give it to them.'"[34] Believing that record buyers would reject the unfamiliar, they underestimated the public's ability to recognize and embrace something unquestionably better sounding than the status quo. Indeed, consumer demand would be the primary motivation, in the minds of most record executives at any point in time, to in-vest in improving the sound of records. Only for recording engineers did the in-trinsic value of improved sound quality supersede commercial considerations; listeners and musicians ranged somewhere in between.

Why did these two leading record companies respond so differently? One explanation may be found in their respective leadership. Columbia's Louis Sterling was a self-made businessman who had worked his way up from New York's Lower East Side to become a manager of record companies—first an English cylinder company and then Columbia. He had an innate business sense but was not an inventor and apparently had no allegiance to a particular embodiment of recording technology. Victor president Eldridge Johnson, by contrast, had built up his company from the beginning with technology of his own making. He believed firmly in Victor's expertise with acoustic recording and, although keenly aware of its shortcomings, Victor engineers and recordists sought to improve the existing method (acoustical recording) over adopting a radically new one (electrical recording). This commitment, coupled with ever-increasing record sales as its competitors suffered declines, explained Victor's reluctance to change. They had too much invested in the status quo, and the status quo worked. By September 1924, however, expected sales did not materialize, and by winter, any remaining resistance to electrical recording had apparently ended. Ultimately, corporate survival dictated the decision to pursue technological change.

At Victor and Columbia, the transition to the new system proceeded swiftly. On January 3, 1925, Harry Sooy's proposed trip to the West Coast to visit with Victor dealers was abruptly cancelled so that he could do work for Eldridge Johnson related to electrical recording, the first reference to the process in the elder Sooy's memoir. The sudden postponement, Sooy wrote, "was brought about by some three or four records submitted by the Western Electric Company, which showed great possibilities in the art."[35] Within weeks, Sooy was manufacturing waxes submitted by Western Electric, and by the end of January 1925, Joseph Maxfield of Bell Telephone Laboratories went to Camden, New Jersey, to inspect the layout for wiring and locating boxes to demonstrate the electrical recording system in the Victor recording lab. Bell Labs sent the recording equipment by truck a week later, and on February 9, Raymond Sooy reported that the Camden studios made their first experimental electrical recording, a piano solo by Mr. Watkins from the Bell Labs.[36] On March 11, they began work on the Victor catalog, with Olga Samaroff as the first Victor artist to make a commercial electrical recording.[37] Not until the end of July did Sooy and an assistant make their first electrical recordings of the Jack Shilkret Orchestra in the New York Studio, then located on West Forty-Fourth Street.

On March 31, 1925, Columbia staged a dramatic public experiment to demonstrate the new capabilities of electrical recording equipment. Suspending a single microphone from the ceiling of the Metropolitan Opera House, they recorded a concert arranged by the Associated Glee Clubs of America, featuring fifteen glee clubs comprising 850 voices, thirteen different conductors—one each for the thirteen numbers performed—and a finale in which the entire audience was invited to join in and sing. Columbia quickly processed and released its first electrical record, a twelve-inch disc of "John Peel" sung by the glee clubs, and "Adeste Fidelis," which featured the glee clubs and the audience, amounting to thousands.[38] The results were as impressive as the event. Columbia claimed that this record alone brought back customers who had not purchased records for months. A New Zealander reported that the record, "though not faultless, had such wonderful power and definition that those who had previously regarded chorus records as the ugly ducklings of the gramophone library said at once:—'Here, indeed, is something new.'"[39]

In smaller studios with fewer resources, such as Parlophone Records studio in London, the transition took longer because existing studios could not be converted to electrical recording, and a new location had to be found.[40] Some small record companies simply could not afford to convert to electrical recording. Others, such as Columbia's budget-priced Harmony label, used the acoustical facilities of the parent company into the late 1920s. At first, companies did not advertise the new process for fear that existing stocks of acoustical records would become worthless, but eventually, all boasted how their records were made with the words "Electric" or "Electrical Process," even though the recording quality varied dramatically.

The New Studio

The layout of the new electrical recording studio differed only in degree from that of the acoustical studio. The recordist and recording equipment still occupied a separate room adjacent to the studio in which the artists performed, but the separation had begun to grow more distinct, with triple-pane glass and soundproof partitions between control room and studio. The recordist no longer had to do double duty, running the machine and positioning people around the horn. Now, he had control through knobs at his fingertips and the ability to communicate with the studio through an intercom system like that used in radio broadcast studios. The necessary separation not only provided protection for the recording equipment, which was sensitive to vibrations, extraneous noise, and environmental conditions, but also provided what amounted to a barrier between artist and technician, a boundary between realms, and a stronger distinction between their respective roles. In the 1920s, this division was a symbol of greater efficiency and control for the engineer, but over time, it led to dramatically different listening environments in the control room and in the studio.

From the performer's perspective, the microphone allowed more freedom of movement and lessened the tedious and demanding conditions of recording. Recording artist Franklyn Baur, who enjoyed a successful career before and after the introduction of electrical recording, commented just two years after the first electrical recordings had been released that in addition to incomparably better sound, recording time had been cut by two-thirds, and it was no longer necessary to "nearly crack our throats singing into that hated horn."[41] Baur's recollection describes three major benefits of electrical recording: (1) better-sounding records, (2) less strain on the performer, and (3) less time to make a record. However, the new process also posed certain challenges. Nathaniel Shilkret, musical director for Victor, pointed out the difficulty of recording tenor voices, which sounded either thick or hollow, until recordists figured out how to make the necessary adjustments.[42] The microphone, a more sensitive instrument, not only picked up a wider frequency range but also reacted to environmental conditions; heat, humidity, and vibrations plagued electroacoustical devices even more than they affected acoustical equipment.[43] However, the acoustical recordist had dealt for years with the "tricks of sound waves," and the microphone's idiosyncrasies simply posed different sound problems to solve.

From the recordist's perspective, the key benefits of electrical recording were greater control and measurability. Sound quality could now be quantified. Sound waves, now that they had been transformed from acoustical power into electrical signals, could be better controlled. However, this did not eliminate the need to consider acoustical parameters. On the contrary, room acoustics became more important as the microphone's increased sensitivity required more control of the acoustical properties of the recording studio.

Before microphones, studios were kept free of absorbent materials that would deaden the sound. Some recording experts even tried collecting reflected sound to increase the weak bass response of acoustic records; others suspended ceiling wires at right angles to the recording horn, believing this would enhance "sonorous reflection by sympathetic vibration."[44] Henry Seymour described the ideal 1918 recording studio as a bare room with a domed or arched ceiling, painted or wood walls, preferably well varnished to aid sonorous reflection.[45] One notable exception to this rule was the Edison Columbia Street studio in West Orange, New Jersey, where the sidewalls, floors, and ceiling, even the piano and piano bench were padded with cow hair. This was used for experimental purposes, whereas the New York studio, where commercial recording took place, was not so deadened and more pleasant for musicians.[46] However, the "sonorous reflections" that served to enhance acoustical recordings proved difficult to control when using the more sensitive condenser microphone with an omnidirectional pattern; whereas the acoustical horn collected only those sounds directed toward it, the condenser microphone picked up virtually all sounds indiscriminately.[47]

In an enclosed space, the duration of the sound after the source stops is known as the room's reverberation time and is considered the most important single factor in auditorium acoustics.[48] Because recordings were meant to capture the live performance, it was believed that reverberation was also essential in the recording studio. In the late nineteenth century, Harvard physicist Wallace Sabine conducted numerous experiments in acoustics and sound-absorbing materials and deduced a formula for predicting a room's reverberation time based on the room size and total absorption of the materials and surfaces therein.[49] Sabine's work concerned auditorium acoustics, but the principles surely applied in recording studios. Although his experiments coincided with the birth of the commercial recording industry, few if any recording experts may have been aware of Sabine's discoveries. As scientists, however, Maxfield and Harrison were well acquainted with his work and relied on it in their electrical recording experiments. They determined that, in electrical recording, however, the time of reverberation should be somewhat less than that of the concert hall or the room in which an imaginary listener might hear the music being performed, because the phonograph only afforded single-channel reproduction, rather than binaural (two-ear, or what came to be known as stereo) listening. Under such conditions, they explained, "whether for radio or the phonograph, the ability of the listener to separate reverberation from the direct music by means of the sense of direction is completely removed and there is thrust upon his attention an apparently excessive amount of room echo."[50] Because our ability to determine spatial relationships by our sense of hearing relies on both ears, with only one ear we cannot properly decode the sounds we hear, rendering their spatial relationships unintelligible.

In radio broadcasting in which the first commercialized condenser microphones were used, engineers deadened the studio for years with burlap and then used rugs and heavy draperies over the walls to minimize reverberation, which challenged musicians and singers to stay in tune.[51] The same acoustical treatment was attempted in the recording studio when condenser microphones first came into use. Bessie Smith's first electrical recording session offers an example of the awkwardness of attempts to adapt the studios to electrical recording. The session took place in Columbia's New York studio, located at Broadway and Fifty-Ninth Street, on May 5, 1925, a month after the company's first electrical recording "experiment" with the chorus of thousands at the Metropolitan Opera House. On hand were the Columbia recordists and several Western Electric engineers, in addition to recording director Frank Walker, Smith, and her accompanists, the Fletcher Henderson Hot Six. The Western Electric engineers determined that the large studio required some acoustical damping, so they rigged a tent of monk's cloth from the ceiling to cover the singer, the musicians, Walker, and the engineers. During the session, the ceiling wires holding the conical tent snapped, "reducing everyone to a mass of bobbing bubbles under a sea of monk's cloth," said producer George Avakian. "That was the end of the session—and the tent theory of electrical recording." According to Smith biographer Chris Albertson, the singer emerged from the collapsed tent muttering, "'I ain't never *heard* of such shit!'"[52] The experience must have been frustrating, but the recordings they managed to complete, "Cake Walking Babies" and "Yellow Dog Blues," revealed the remarkable difference in the sound of electrically recorded music—fuller, crisper, clearer, and deeper than acoustical recordings. Bessie Smith's powerful voice had been ideally suited to the acoustical horn, but the microphone picked up her more subtle inflections and made it possible for her to record with a full band for the first time, rather than with the customary trio of musicians.

Such experiments with acoustical treatment were common in the early days of using microphones on film sound stages and in radio and recording studios.[53] By 1927, however, record company studios had changed their approach, moving away from the absolutely dead studio to more live rooms, making it easier for performers to sing.[54] In 1932 radio stations began to adopt a "live end–dead end" approach to studio acoustics. The "live" end of the broadcasting studio, free of absorbent materials, provided localized reverberation that facilitated musical performance, while the "dead" end of the room kept the effective reverberation at the microphone to the desired low value. Radio station WCAU in Philadelphia, the first purpose-built broadcasting facility in the United States, became the first to incorporate this approach. Its striking Deco exterior of glass, steel, and bronze, all lighted by the bluish hue of General Electric's new low-voltage mercury-vapor lamps, was matched by the acoustically engineered studios and control rooms within, incorporating the latest in materials and soundproofing techniques. Musical considerations evidently figured prominently in the station's planning, which included a specially designed laboratory and workshop for Philadelphia Orchestra conductor Leopold Stokowski.[55]

The microphone afforded more intimacy, capturing more subtleties in the vocalist's performance. Beginning in 1925, singers such as "Crooning Trouba-dor" Nick Lucas, "Whispering" Jack Smith, and Gene Austin had begun to capi-talize on the microphone's sensitivity, singing in ways that would never have gained them entry to an acoustic recording studio. In his 1954 study of changing musical styles over the first half-century of recording, Hughson F. Mooney rec-ognized that radio and electrical recording, as well as the public-address system and film sound track, collectively gave rise to new styles of vocalization and or-chestration during the 1920s and 1930s. Where once "songs had been popular-ized by lung power unaided by electrical amplification," Mooney noted,

> Now, since the microphone and amplifier were sensitive to the slightest breath, and could convey the tiniest modulations, singers no longer needed stentorian voices in order to "carry" to the farthest corner of the hall or vibrate the stylus in the old mechanical recording horn. Radio ... not only obviated the necessity of vi-tality in singing, it helped make vitality unpopular for two decades ... [and] stim-ulated a demand for the languid, caressing vocals facilitated by the microphone. ... The electronics revolution thus furthered the evolution of the slow ballads which, reiterated by the radio during the stay-at-home depression and war years, enthralled us for two decades.[56]

Other media scholars have noted the profound effects of aural technologies on the musical culture of the interwar period.[57] By the 1930s, Rudy Vallee and Bing Crosby had begun to recognize the microphone for what it was: an instrument through which they could express themselves rather than merely a receptacle, a conduit through which their voices traveled to the record only if belted out loudly enough. Electrical recording and the microphone, then, ultimately cre-ated not only a greater intimacy between the listener and the performer but also a more intimate relationship between the performer and the technology. Recording technology became a more refined tool—for the performer and for the recordist.

Recording circa 1927: New Freedom in the Studio

Concerns about overcoming physical, psychological, technical, and artistic challenges dominated firsthand accounts of working in the acoustical recording studio. After the introduction of electrical recording, such accounts were fewer and less detailed and overall indicate that musicians paid less attention to the technical process and more to the interaction with fellow musicians. Once the initial adjustments were made, the technology of electrical recording was far less imposing than that of the acoustical studio, and the greater fidelity of the process favored musicianship and style over physical stamina. Most important, the new method enabled musicians to play more freely, an essential component of the emerging hot jazz style. Between the two world wars, jazz emerged as the most popular form of vernacular music in the United States and spread rapidly to Europe. The freedom expressed in jazz, its unrestrained solo improvisation, blurred distinctions between composer and performer and brought performer and audience closer. More intimate than formal music, its evolution paralleled the changing technology of recording.

On December 9, 1927, guitarist Eddie Condon and a group of musicians assembled at the Okeh Records studio in Chicago, "a barnlike place at the corner of Washington and Wells," as he described it in his jazz memoir, *We Called It Music*.[58] As the band arrived and Gene Krupa began to set up his drums, the recording director said, "You can't do that. . . . You'll ruin our equipment. All we've ever used on records are snare drums and cymbals." Since the drums were the backbone of the band, Condon pleaded for at least one take with them. The recording director acquiesced but expressed concern that the bass drum and tom-toms would "knock the needle off the wax and out into the street."[59] Fears about the effects of powerful players such as saxophonist Sidney Bechet, who had already made the needle jump off the wax in the Okeh studio even before the introduction of the microphone, made recordists apprehensive about the effect such playing would have on their equipment.[60] Obviously, after electrical recording had been introduced, recordists had even more reason to fear the effect of some instruments on sensitive recording equipment. Gene Krupa's drums ultimately posed no danger to the equipment, and according to Condon's account, not only did the records sell well, but the drums, which had rarely appeared on record up to that point, became a sensation among musicians.[61]

Condon's description of the date suggests that once they settled the issue of instrumentation, little about the technique of recording intruded on the musicians' enjoyment of the performance, and that any tension that may have been felt, if not relieved by the Prohibition era bootleg, actually served to heighten the experience. Condon's recollection of the session relates the excitement and his own sense of drama:

> The rules and regulations were explained to us; one white light—get ready, one red light—play. We were to run through the numbers and hear them played back to us so we could iron out the rough spots; then the master records would be cut. We warmed up and told the engineer we were ready. Rockwell and McKenzie went

into the control room. I could hear the boys fidgeting behind me. Someone muttered, 'Damn!' Somebody else whispered, 'Where did you put the bottle?' The white light flashed. I swallowed. The red light came on. I gave the boys the beat and we jumped into *China Boy*. We opened on the nose, all playing, with everyone knitting from his own ball of yarn. The nights and years of playing in cellars and saloons and ballrooms, of practicing separately and together, of listening to Louis and Joe Oliver and Jimmy Noone and Leon Rappolo, of losing sleep and breathing bad air and drinking licorice gin, paid off. We were together and apart at the same time, tying up a package with six different strings. Krupa's drums went through us like a triple bourbon. Joe Sullivan took a chorus and all the good things he had learned from Earl Hines came out in his left hand. MacPartland followed him for half a chorus. Tesch finished it; then we went into ensemble, followed by Bud on tenor saxophone. Lannigan took a release of eight bars and we finished on an ensemble, with the tom-toms coming through strong. Quietly we waited for the playback. When it came, pounding out through the big speaker, we listened stiffly for a moment. We had never been an audience for ourselves. Then Joe's piano chorus started and smiles began to sprout. MacPartland, Tesch, Bud, Lannigan—as each heard himself he relaxed. At the finish we were all laughing and pounding each other on the back. We were the happiest kids since the founding of Fort Dearborn. Rockwell came out of the control room smiling. 'We'll have to get some more of this,' he said. 'Can you boys come again next week?' 'We could come tomorrow,' I said. 'How were the drums?' Rockwell nodded toward Krupa. 'Didn't bother the equipment at all,' he said. 'I think we've got something.'"[62]

Bearing in mind that Condon's colorful account, written two decades after the event, may well have embellished the more exciting details at the expense of the tedious, it nevertheless provides one of the few detailed descriptions of a recording session from this period and presents a striking contrast to the recollections of acoustical sessions.[63] It shows that recording in 1927 could be as enjoyable as a live performance, with just enough tension to inspire playing, as Condon put it, "together and apart at the same time," followed by tremendous relief on hearing a good playback. Condon's account also underscores the importance jazz musicians placed on listening to other musicians—live and on record—as they honed their craft, and the subsequent exhilaration they felt on hearing themselves on record for the first time. Although jazz records had only existed for a decade by the time the Condon-McKenzie Chicagoans did their Okeh date, recorded music by then had become, in Eric Barnouw's words, "the great school for a rising generation of musicians."[64]

The greatest benefit of electrical recording from the artist's perspective was its easing of the cramped conditions of playing into a horn. From a musical perspective, it also introduced the possibility of recording large ensembles with better results than had been achieved in acoustical recording. Both Columbia and Victor attempted to acoustically record orchestras but with mixed results because so much sonic information was invariably lost. Bandleader Paul Whiteman recalled an attempt to use twenty-five musicians instead of his customary nine the first time he recorded George Gershwin's *Rhapsody in Blue*, a year before Victor licensed the electrical method from Western Electric:

> We filed into a tiny, dusky room, barren, crowded and uncomfortable. The acoustics were far from perfect in the studios of those days. In the middle of the room stood a tower, made up of four ladder-like supports tapering to a narrow point. This pylon was about eight feet high. Four recording horns, which looked like megaphones, were attached to the pylon in the form of a four-leaf clover. Four or five of us gathered around each horn so that we'd be close enough for the stylus (recording needle) to pick up the sounds we made. My boys had to be athletes. When a solo passage was to be played, the musician would move up close to the horn and play directly into it. Then he had to back out in a hurry, dodging out of the way of the next man who was hurrying toward the pylon.[65]

Whiteman's account illustrates the continued challenges of acoustically dead rooms and the need for fast footwork before the acoustical horn, but it also reveals a growing determination by musicians and engineers to make better sounding records. Clearly undeterred by his unsuccessful effort to convince Victor to purchase the English inventor's electrical method, Whiteman and his band "enjoyed experimenting [with] a group of engineers at Victor who shared our zest for new ways to improve recording," which sometimes involved, according to Raymond Sooy, also making as many as twenty-two takes to achieve a satisfactory master.[66] Not all musicians would have been accorded such extensive time and resources in the studio, but record sales justified the attention. Whiteman led one of the most popular dance orchestras of the 1920s, selling more than a million copies of the 1920 recording of "Whispering," which remained in the Victor catalog for at least a decade. Whiteman's account of studio experiences also suggests that he took special interest in the recording process, and he was acutely aware of the benefits of electrical recording for the musicians as well as the music. Writing just a year after Victor's conversion to the Western Electric system, he described its effect on the composition of his orchestra: "The saxophone, for instance, had always had a shadow or understudy. A third saxophone now was added and in time the orchestra developed the full Wagnerian quartette of instruments in this group. The one trumpet was reinforced by a second and the now popular combination 'straight' and 'comedy' trumpets came into existence. The banjo, instead of just marking time began to make new excursions into the realms of rhythm."[67]

As a bandleader, Whiteman surely reveled in the expanded instrumental lineup. The single saxophone expanded to a horn section, and the banjo, no longer "marking time," developed as a rhythm instrument. The transition also posed challenges as it required rescoring, rehearsing, and "consultations not always free of the heat of argument." Although careful not to reveal any secrets of the Victor engineers' techniques, calling the work of the recording studio "a star chamber matter," Whiteman admitted that some of the early electrical records were spoiled "by men swearing softly at themselves before they learned the new adroitness which the delicate mechanism of the recording room required."[68] Just as recordists had to learn new techniques, so musicians needed to adjust their playing to accommodate the more sensitive microphone. As the Condon session illustrates, within two years of the adoption of electrical recording, most problems associated with instrumentation had been solved, and musicians enjoyed a new freedom in the recording studio.

Electrical recording transformed the sound of records and the conditions for performers and recordists in the studio. For the industry, electrically cut records pleased the public, and record sales that had declined after the introduction of radio rose during the years 1926–1929. But the boom was short-lived. After the stock market crash of 1929, record sales declined sharply during the 1930s, reaching an industry low of $6 million in 1933.[69] Plummeting sales led to cutbacks in production, corporate consolidation, and the disappearance of some small labels. On June 16, 1931, Victor's Raymond Sooy issued a memorandum limiting sessions to processing one wax master per song, unless the wax became defective, and Victor began to curtail field recording activities and increase the use of the New York studio.[70] Warner Brothers Pictures purchased the record division of the pool table giant Brunswick-Balke-Collender in 1930 and moved the headquarters of Brunswick Records, then the second-largest American record company, from Chicago to New York. Within a year, Warner Brothers sold the Brunswick Record Corporation to Consolidated Film Industries, which also owned the American Record Company. Louis Sterling sold Columbia Phonograph in 1931 for a bargain price to Grigsby-Grunow, manufacturers of Majestic radios, washing machines, and refrigerators.[71] By 1934, Consolidated owned Columbia, as well as the independent labels Vocalion, Cameo, Banner, Perfect, and Okeh.[72]

Thomas Edison ceased to make records and phonographs after the stock market crash, thus leaving the industry his invention spawned.[73] Eldridge Johnson sold Victor Talking Machine to financiers in 1926, and in 1929, Victor became the RCA Victor division of David Sarnoff's Radio Corporation of America.[74] None of the original inventors remained in the recording business by 1930. Many improvements to the sound of records or advancements in the technology and practice of studio recording during the 1920s emerged from outside the record industry. However, new developments appeared by the late 1920s, emanating from the prosperous field of radio broadcasting.

Electrical Transcriptions, Instantaneous Recording, and the Birth of the Independent Recording Studio Business

Several innovations, made possible by the advent of electrical recording, led to significant growth of the technology and culture of the recording studio, even during the bleak years of the Great Depression. Until the 1930s, most recording studios were associated with record companies. Although there were a few independent recording studios, such as Marsh Laboratories in Chicago, the expense of making recordings, processing the wax master, making shellac pressings, and acquiring expertise for quality recording, meant that most recording studios were affiliated with companies that made and distributed the records.[75] During the 1920s and early 1930s, demand for recording services from different industries and broader participation by amateurs and aspiring professional recordists escalated.

First, and concurrent with electrifying the studio, came movie sound. The introduction of the microphone, amplification, and loudspeaker made clear that potential uses for recorded sound were not limited to the phonograph. As they developed electrical recording for record companies, Western Electric and Bell Labs also worked on similar systems for sound motion pictures—one based on disc recording, the other on sound-on-film.[76] Although Western Electric and Bell offered the disc system to several large film companies, they all rejected it.[77] In 1925, Warner Brothers became the first company to adopt the Western Electric disc recording system to make sound films, and in August 1926, the company premiered its first sound movie, *Don Juan*, under the trademarked Vitaphone name. Until they moved operations to the West Coast, Warner Brothers had contracted Victor Talking Machine to process and press records used for Vitaphone motion pictures.[78] By mid-1928, the demand for commercial recording laboratories as a result of sound recording for motion pictures had grown to necessitate hiring sixty new employees for Victor's recording department.[79] The film industry did not continue to use the disc-recording method, and by the 1930s, most movie companies had converted to the RCA Photophone sound-on-film process.[80] Moreover, Hollywood sound engineers, many of whom had originally come from radio and the record industry in the East, soon learned to reconceptualize their idea of "good sound," which differed markedly from that of the record industry, to hew to a different "aesthetic norm" in their professional work.[81] This did not end increased demand for disc-recording services. The film industry's abandonment of disc-recording technology aided the rise of another recording business: prerecorded programs on electrical transcriptions for broadcasting.

The first completely recorded radio program had been the late 1920s brainchild of Raymond Soat, program director of a small radio station. Seeking a more streamlined method of coordinating his one-man operation of selecting records, announcing commercials, operating the turntable, and engineering the broadcast, Soat prerecorded entire programs. He petitioned the Federal Radio Commission to call his recordings "electrical transcriptions" to differentiate them from "mechanically reproduced" commercial records, which were normally used in radio but not highly regarded by broadcasters because of poor sound quality.[82] Until radio stations began to run recorded music through control amplifiers, broadcasting music was simply a matter of placing a microphone before a Victrola horn. For the most part, music and other programming were broadcast live either from the studio or through a wired connection to a remote source, such as a ballroom or a concert hall, and the phonograph was used primarily to fill unexpected gaps in program routines.[83]

One radio entrepreneur saw the potential for mass producing Soat's idea. Percy Deutsch founded the World Broadcasting System in 1929, and with Charles Gains and A. G. Sambrook, created the first music library service in 1933.[84] Deutsch licensed the Western Electric sound-on-disc systems developed for the movie business by Electrical Research Products Inc. (ERPI).[85] Unlike the method employed for commercial music records, the ERPI recording technology used a sixteen-inch wax disc rather than the standard twelve-inch discs, cut these using vertical rather than lateral modulation, and operated at $33^1/_3$ rpm rather than the standard 78 rpm for home phonographs.[86] These recordings thus differed from regular commercially produced records in two important respects: first, they contained a full fifteen minutes of recorded music or other program material per side (as opposed to three to four minutes); and second, they were of higher sound quality because of increased volume and wider frequency response. Because electrical transcriptions offered large and small broadcasters numerous advantages—the records could be made in studios designed for the best acoustical effects, errors in performances could be eliminated by rerecording, the program could be played or replayed at any time, and stations could eliminate the expense of wire line connections—they very quickly became adopted throughout the industry.[87] In 1929, radio station WOR in New York became one of the first to use electrical transcriptions regularly in its broadcasts and later became the first New York radio station with commercial recording studio facilities of its own.[88] Radio transcription studios flourished in the 1930s, providing employment and training for a new group of recording enthusiasts.

In December 1930, *Electronics* magazine introduced readers to the wonders of electronics in music and a radio studio that revealed the marriage that had taken place between recording and broadcasting.[89] Several electronic musical instruments were pictured: the electronic organ of M. Coupleaux of Paris, the American theremin with keyboard, and several antenna-controlled theremins that resemble elaborate pogo sticks, designed for stringed instrument players.[90] Also featured was a spacious and elegant recording studio, its monitoring room and the disc-cutting equipment. This was no record company operation, but rather the electrical transcription recording facilities of the Judson Radio Program Corporation in New York City. In 1926, Arthur Judson, concert manager of the Philadelphia Orchestra and the New York Philharmonic, conceived the idea to supply radio stations with talent and programs while giving his concert artists a place in the broadcasting boom. Judson ultimately founded what became the Columbia Broadcasting System, the second major broadcasting network after Radio Corporation of America.[91] The Judson monitoring room, where engineers maintained proper balancing of microphones and correct volume in recording, contained a bank of amplifiers, dials, and switches like those used in any radio station; and the studio, free of the risers and overhead wires to hold music stands that filled acoustic studios, contained two grand pianos, tympani drums, vibraphones, chimes, a single floor microphone, and the windowed control room off to the side. Here, Judson recorded electrical transcriptions of studio performances for later broadcasting.

By the mid-1930s, Judson and the World Broadcasting System had been joined by other syndicated program services: Standard Program Service, C. P. MacGregor, NBC-Thesaurus, Associated Program Service, and Lang-Worth Feature Programs. These companies had studios of their own, or they used record company or independent recording studios. By 1935, *Broadcasting Yearbook* published a "Directory of Transcription and Recording Producers," which listed forty-one companies with their own manufacturing plants or direct tie-ups to manufacturing, only five of which were record company studios (American, Columbia, Decca, Gennett, and RCA Victor).[92] Small recording studios had cropped up across the nation, mostly in New York, Chicago, and Hollywood, but there were others in Cleveland, Washington, DC, San Francisco, Atlanta, Minneapolis, and Seattle. In 1936, more than 350 radio stations made yearly contracts with World, Standard, MacGregor, or NBC; three years later, more than 575 stations subscribed to at least one of these services, and nearly half of them used two or more.[93] By 1938, the broadcast industry reference source *Radio Annual* listed ninety-three companies that offered transcription services, including recordings, air checks, scripts, production services, radio package shows, slide film production, original musical jingles, sound for mechanical displays, sound effects and mood music, dubbings, audition studios, and some, such as Kasper-Gordon, Inc., in Boston, even manufactured portable transcription playback machines.[94]

To keep up with demand, recording equipment manufacturers grew considerably during the 1930s. Not all transcription studios used the Western Electric system, nor did all electrical transcriptions emerge from sleek production facilities. Complaints about the poor quality of some of these, especially the "dubbings," or rerecordings of commercial phonograph records for broadcast, surfaced within the first few years.[95] Since the early days of the phonograph, home recording units and other portable recorders had been available, but these were of poor sound quality and difficult to operate. Electrical recording led to improved portable recording units and attachments to radio-phonograph combinations that used aluminum, zinc, plastic, and other blank or pregrooved discs.[96] Electro-Vox recording studio, founded in Los Angeles in May 1932 by Bert Gottschalk, used the "Gottschalk Process," which recorded on bare aluminum.[97] Other companies, such as Speak-O-Phone, manufactured the equipment and discs and promoted their products by touting recording studios as small business ventures.[98] Presto Recording Corporation pioneered high-quality disc recorders and introduced the lacquer-coated aluminum disc in 1934, which revolutionized recording practice.[99] One of Presto's owners, George Saliba, published numerous "how-to" guides aimed at radio technicians, experimenters, and home recording buffs during the early 1930s.[100] By 1936, the lacquer disc was widely used by radio stations, amateur home recording buffs, and even in some recording studios and had a huge effect on network installations, and small radio stations. NBC installed six Presto instantaneous recording lathes in its New York studios to record news and other programs beginning in 1936.[101]

With the rise of network radio, demand for recording services increased from artists, advertisers, program producers, and the stations themselves to form a permanent library of broadcasts. Radio station WOR started its own recording studio after borrowing a Presto machine to record a program in response to accusations by bandleader Sammy Kaye that its transmission quality was poor. Only an air check, or recording of the broadcast as it came over the air, could prove otherwise. When Kaye heard the quality of the discs, he declared, "I could buy that!" WOR's chief engineer J. R. Poppele, responded, "Fine, it will cost you ten dollars," and that was the first money WOR ever made on a recording session, inspiring Poppele to propose a permanent recording facility.[102]

In addition to cultivating a broader culture of recording among radio amateurs and small-business entrepreneurs, these transcription studios provided a training ground for some who went on to professional engineering in the postwar record industry. Frank Laico began at World Broadcasting, working his way up from mimeographing scripts to operating the recording lathes. He then became an engineer with Columbia Records after World War II.[103] Bill Savory worked as an occasional recordist and maintenance engineer at several small New York recording studios, collecting air checks of jazz and swing bands, small groups, and soloists before working for Columbia, Angel, and Capitol Records.[104] Moreover, advances during the decade that followed the introduction of electrical recording—improvements in record materials, lateral groove cutting, turntables and drive mechanisms, and recording styli—ultimately benefited the commercial record industry.[105]

Conclusion

The first revolution in recording technology came from outside the commercial record business, but electrical recording had manifold consequences that affected a range of musical, engineering, and business developments. In music, electrical recording improved the sound and volume of records, captured a wider range of frequencies, and consequently opened up many possibilities for wider instrumentation, larger ensembles, and different styles of musical expression. Because the microphone and amplification obviated the need for lung power, softer singing styles and thus new and different recording artists emerged. Because the microphone eliminated crowding, a necessity with the acoustic horn, musicians were liberated from cramped conditions and could play more freely, respond more spontaneously to one another as in the live performance.

The mechanical engineers who operated the recording studios of the acoustic era developed highly individualized technique on the job. With the introduction of electrical recording, the studio recordist acquired a new knowledge base, new techniques, and new understanding of the behavior of sound as it was transformed into electrical signals and back into acoustical energy. Electrical recording transformed the recordist's job from a craft-based endeavor reliant on empirical knowledge to an engineering skill, although it did not do away with the need for tacit knowledge and artistic sensibility.

Did the Bell Labs team succeed because of or despite the fact that they were engineers rather than recordists? In the eighteenth and nineteenth centuries, scientific investigations into the nature and behavior of electricity and of sound had laid the basis for the inventions and systematic tinkering of Edison, Emile Berliner, and others associated with the phonograph and sound recording. Then, however, the development of the art focused on mechanical rather than on electrical engineering. Despite several efforts to electrify recording, the keys to successful electrical recording were to be found in the tools of scientific electroacoustics: the vacuum-tube amplifier, condenser microphone, electromechanical pickup, and mechanical-acoustical impedance. The entry of the industrial research lab into the record business turned the focus away from the independent tinkerer-recording expert. It served to raise the scientific standards of what had been a craft-based endeavor, considered by practitioners as "an art and a science." It did not remove the amateur recordist from the scene, and even as the new technological system challenged veteran recordists to learn new tricks, it opened up the field of recording to amateur radio operators and younger electronics enthusiasts—a new generation of recording engineers, many of whom began by "dropping the needle" in the electrical transcription studio, recording air checks of radio programs for advertisers, or dubbing big band music off the air using an attachment on their home radio-phonograph combination.

The radio boom of the 1920s may have caused a temporary decline of the record business, but the marriage of radio and recording in the 1930s, through the electrical transcription studio, improvements to recording technology, the development of instantaneous recording and the growth of independent recording studios, ultimately fueled the growth of the record industry, which by 1939 had entered its own boom period.

A Passion for Sound

Amateur Recordists, the Audio Engineering Society, and the Evolution of a Profession

*A*mateurs played a considerable role in the growth and development of recording technology. Just as early twentieth-century wireless experimenters, mostly white, middle- to upper-class men and boys, found a resolution to the contradictions of modern life in mechanical and electrical tinkering, reclaiming a sense of mastery, even masculinity, through control of technology, so too did amateur recordists, fascinated with the ability to harness sound waves, seek to capture a moment in time or a musical performance on a blank cylinder or disc, thereby making experience repeatable.[1] Guitarist Les Paul, who parlayed his boyhood fascination with the phonograph into a successful recording career, was among many who experimented with home recording in the 1930s, learning from failed experiments as well as successes. "All of us," he recounted in 1996, "we were just chasing sound."[2] Whereas wireless operators experienced the excitement and mystery of communicating over vast distances through the ether, the thrill of becoming the heroes of naval disasters, or playing practical jokes over the airwaves, amateur recordists found excitement in technical tinkering and, much later, the sheer exhilaration of high-fidelity sound.

The transformation of the recording industry after World War II resulted from a series of developments both inside and outside the recording studio, including technological innovation, economic prosperity, increased leisure time, and a growing demand for entertainment and consumer goods. Critical to the industry's success was a new generation of audio engineers, many of whom gained their technical training as signal corpsmen during the war and after the war pursued careers in radio, television, film, and recording. Their training and war experience provided a sense of mastery as well as camaraderie and cooperation, quite a departure from the era of trade secrets and competition. In the postwar recording industry, management may have sought proprietary secrecy, but the rank and file more often than not readily shared technical know-how, a necessary step in the formation of a professional organization of audio engineers from "a cottage industry of whacky inventors," who had long been part of an imagined community of audio lovers.[3]

To trace the emergence of this community of professional recording engineers, this chapter charts the growth of amateur recording activity from the early twentieth century to the formation of a professional society and publication of the first technical journal devoted exclusively to audio engineering and sound recording. In its early years, the Audio Engineering Society sponsored annual "audio fairs" open to the public, at which the latest audio technologies both for professional and home use were demonstrated, which helped to foster interest in audio recording and reproduction and thus encouraged a new generation of audio amateurs.

Capturing the Moment, Falling in Love with the Phonograph: Amateurs and the Growth of Interest in Recording

Cylinder phonographs and records became available to the public in the early 1890s, and all were equipped with a recorder and some blank records.[4] In 1900, Thomas Edison's National Phonograph Company encouraged a variety of uses for its recording phonographs through an instructive booklet entitled *The Phonograph and How to Use It* (see fig. 6).[5] Divided into three parts, the book acquainted the reader with the history of the phonograph and the operation of various Edison models and concluded with a testimonial on the benefits and many possible uses of phonograph recording for the amateur. In one chapter, a woman recounted how she and her sister gave a "phonograph party," recording their guests and playing back the results to the delight and dismay of the attendees, who quickly discovered why, as the book warned, "the amateur will do well to avoid the sorrow that is almost inevitable in attempting to make a record of a high tenor, a soprano or a violin."[6] In 1905, the trade publication *Talking Machine World* ran a story about how discouraging it is to hear oneself for the first time. A musician who visited a fair where a talking machine company had its latest

phonographs on display agreed to allow the company agent to record him playing his flute. After hearing the playback, he asked whether it was indeed an exact
reproduction of his performance. The attendant assured him it was and asked
whether he wished to purchase the machine. Having now heard himself for the
first time as others heard him, the man replied, "No, but I'll sell the flute."[7] That
story, however apocryphal it may have been, illustrates that, even under the
best conditions, few wanted to hear themselves as the talking machine made
them sound—tinny, high-pitched, and barely audible; it took powerful voices
to rise above the technical limitations of the acoustical record. Hence, the National Phonograph Company's efforts to educate the public was part of boosting
the idea of home recording and, thus, sales of its machines to America's newly
emerging leisure class.

Historian T. J. Jackson Lears argues that turn-of-the-century middle- and
upper-class Americans attempted to counteract the negative physical and psychological effects of industrialization through an antimodernist quest for experience rooted in the past. One manifestation was the emergence of the Arts
and Crafts movement and the popularity of handicrafts.[8] Other scholars have
explored more closely the rise of amateur and hobbyist activity during the 1880s
and 1890s as increased leisure time invited creative work in the private sphere
for self-renewal apart from the rigors of the workplace while bridging the worlds
of work and home.[9] In her study of amateur film, Patricia Zimmerman traced
the rise of a social concept of amateurism to this period as a result of urbaniza-
tion and increased leisure time.[10] Amateur publications and societies became
increasingly popular, especially in sports, art, and engineering where amateur
activity—doing something for pleasure rather than for financial gain—proliferated. The phonograph, then, came at the historical moment when its use as a
recreational and creative tool could fill a sociocultural need.

Early amateur recording activity focused on preservation and documentation—on capturing the moment like a snapshot in sound. Scientists were among
the first to perceive the usefulness of recording devices for studying the nature
and evolution of language, and ethnographers quickly recognized the phonograph as a tool for capturing and preserving Native American languages and
other folklore.[11] Presaging what would become standard practice decades later,
musicologists encouraged musicians to record their original compositions and
improvisations rather than risk losing spontaneous inspiration through inability to notate quickly the ideas as they occurred to them.[12] These users rarely, if
ever, ventured beyond recording for documenting, studying, or analyzing some

event or expression. For them, the recording phonograph was a tool for study rather than a source of entertainment or an object for tinkering.

Once the development of the vacuum tube made it possible to amplify radio signals, others engaged in recording wireless transmissions. A New Jersey wireless operator named Charles Apgar achieved fame as the "pioneer home-recorder" when he recorded his first radio transmission in 1913 using a wax cylinder machine and a crude electrical amplifier. Two years later he turned amateur spy when his recording of suspicious code messages emanating from a Long Island wireless station revealed that the code was being transmitted to a station in Nauen, Germany.[13] A decade later Frank L. Capps, an employee of Thomas Edison and later of Bell Laboratories in Chicago, conducted home experiments, recording Woodrow Wilson's Armistice Day broadcast on November 10, 1923.[14] These and countless other recordists employed crude homemade amplification devices to improve the sound quality. These actions marked them as a different breed of home recordist from those who simply sought to document events using the recording cylinder phonograph. Driven by the desire to explore the limits of existing technology, these experimenters modified commercial phonographs or built their own devices from scratch. Capps went on to invent a burnishing stylus for lacquer disc recording and then started a company in 1929 that became the major supplier of professional recording styli to recording studios. Not all of these audio tinkerers made their avocation into an occupation, but whether they thought of themselves as hobbyists, tinkerers, or amateurs, rarely were their efforts merely casual.

At the same time, phonograph manufacturers began to introduce different types of instantaneous disc-recording machines. In 1920, Pathé introduced its Voicewriter, a home recording system that employed embossed aluminum discs, and several other companies soon introduced pregrooved discs as well. RCA introduced a recording attachment for its radio-phonograph combination, the Radiola 86, which involved a switch enabling the owner to record on pregrooved plastic discs. While not of high quality, these machines promoted the idea of home disc recording to those less inclined to tinker. Well aware of its product's shortcomings, the company made a special record that instructed its salesmen on how to sell the system, emphasizing the fun of home recording, and discouraging any attempt to be "too serious about it." Home recordists should not expect professional results, and Victor openly distinguished between "home recordings made by amateurs under haphazard conditions, and with very simple equipment, . . . [and] Victor records made by the best talent in the world in laboratories costing hundreds of thousands of dollars with recording experts and costly equipment."[15] Although it was certainly in Victor's interest to promote the superiority of its artists' recordings, this distinction between amateur and professional recording emphasized the importance of both recording expertise and expensive technology.

Interest in sound recording, public address, reproduction, all things audio had grown tremendously during the 1920s and 1930s as radio broadcasting flourished, offering opportunities to enterprising radio servicemen and ham radio enthusiasts who wanted to turn a hobby into a part-time business. When the Depression cast millions out of work, leisure, so desirable in the Victorian era, became a liability in the 1930s.[16] Hobbies took on greater significance because they could be therapeutic for the unemployed and financially remunerative. Scores of home hobbyist/do-it-yourself magazines suggested ways to make and save precious cash.[17] Monthly publications devoted to technical tinkering, including *Popular Mechanics* and *Modern Mechanix*, were filled with how-to articles on auto mechanics and small-engine repair, and a number of specialized periodicals geared toward the radio "ham" and electronics enthusiast appeared. Beginning with his first publication in 1908, *Modern Electrics* publisher and electronics entrepreneur Hugo Gernsback had become an avid promoter of amateur radio and home recording.[18] His next periodical, *The Electrical Experimenter*, was the first to report Charles Apgar's achievement in 1915, and in July 1929, he began publishing *Radio-Craft* magazine, aimed at the serviceman, dealer, and "radiotrician." The December 1930 issue featured the first of many articles on home recording as "the latest adjunct to radio."[19]

Home recording grew in popularity as companies offered kits and radio-phonograph combinations with home-recording attachments, including the Presto home recording kit, the Radiola 86, the Sentinel Chromatrola, and the Audak Musichrome. Hobbyists as well as manufacturers suggested myriad uses for recording letters, ideas, and company circulars and promoted these as being modern.[20] That recording might be considered forward thinking and "modern" was evidently not just a sales ploy. In his 1936 film *Modern Times*, Charlie Chaplin criticized the monotony and standardization that machines had created in people's lives and mocked the many contradictions of scientific and technological advancements. The first words uttered in this predominantly silent film came over a microphone as the president of Electro Steel Corp. barked an order to the factory floor: "Section Five. Speed 'er up. 4-0-1!" The next words in the film emanated from a portable phonograph. Attempting to sell the president on the efficiency potential of the Bellowe's Feeding Machine, the salesman opened the phonograph case, cranked up the turntable, placed the tonearm on the disc, and let the "mechanical salesman" do the sales pitch. Here was yet another example of "technological unemployment—new machines displacing men from their jobs," which Chaplin believed was the root of America's economic woes.[21] For Chaplin and other critics of the seeming stranglehold of science and technology on everyday life, the microphone, loudspeaker, and phonograph—no less than the assembly line and machinery of mass production—were guilty of restricting freedoms and robbing human agency, all delivered with the promise of greater efficiency.

But for others, the new disc-recording phonograph was an exciting tool with myriad possibilities. With instantaneous lacquer-coated discs, recordings could be made and played back with no need for the subsequent processing required to make commercial records. In June 1931, *Radio-Craft* began publishing a series of a dozen articles on various aspects of instantaneous recording. The author, George Saliba, vigorously promoted the art of recording to radio technicians and experimenters, suggesting that "wherever there is an audio amplifier available a recording system is waiting to be exploited."[22] Saliba and his partner Morris Gruber co-founded the Presto Recording Corporation, building on the invention and subsequent development of the cellulose nitrate lacquer-coated recording disc and associated disc-recording equipment for the amateur and professional markets. Thus, Saliba had a vested interest in proselytizing the practice of recording. The response to his articles revealed that a significant community of recording enthusiasts was hungry for more, as evidenced by increasing correspondence from readers.[23] Amateur recording had become so popular that in 1932 *Radio-Craft* published a booklet of Saliba's technical advice, entitled *Home Recording and All about It.*[24] Reminiscent of the National Phonograph Company's booklet of 1900, *Home Recording* was an in-depth technical manual for the serious amateur recordist or radio serviceman, and Saliba's sixty-two pages of detailed instructions about various methods of recording, equipment needed, and advice on how to obtain optimal results revealed that the technology had improved considerably over previous systems. "With a little patience and experimenting," Saliba assured the reader, "the home recordist can achieve results that will be almost on a par with commercially-pressed records."[25] Indeed, instantaneous recording gave rise to a recording studio business by making quality recording possible without the need for elaborate and expensive processing facilities (see fig. 7).

Increasingly, radio publications and equipment companies promoted home recording as a moneymaking venture, blurring the line between hobby and profession. Arthur Heine of the Speak-O-Phone company promoted his firm's product, a portable disc recorder, not as an expensive hobby but as a business opportunity: "A profitable business medium has been found in 'recording studios,' which the radio man may establish at little expense." The Depression had fueled the entertainment industry, particularly radio and motion pictures, leading to a demand for new talent as the country sought more forms of amusement. Moreover, Heine pointed out, these new entertainment media demanded constant change and improvement, as well as new influx of talent. Vaudeville artists once repeated the same program around the country for a year or two, but now the public wanted a new radio program every week, even every day, and this required a complete change of script or new performers. The local movie house might show four new features each week, he went on, but hundreds of broadcasting stations required a new program every fifteen minutes, eighteen hours a day. Prospective customers, Heine assured, were everywhere, "in every hamlet of the country, thousands of men and women, boys and girls are searching their innermost selves to discover if they can fill a niche in this fertile field of occupation." Since there were thousands of uses for recordings, Heine noted, an ideal business would be the small recording studio, where prospective artists can test their abilities by making what the company advertised as "A Snapshot of Your Voice."[26]

For many amateurs, recording remained a part-time pursuit, but for those who sought to make it full time, opportunities abounded. By 1935, the radio professional's reference source, *Broadcasting Yearbook*, listed forty transcription and production companies and more than sixty equipment manufacturers, although some of these supplied only broadcasting and not recording equipment.[27] Over the following decade, the number of transcription and production companies had increased nearly nine times to over 350, and the number of equipment manufacturers and suppliers had doubled.[28] Few of these companies were large enough to branch out into all types of equipment; most specialized in a single or related group of technologies. Audio Devices, for example, specialized in instantaneous recording blanks. Frank L. Capps Company dealt in professional recording styli. Presto started out producing recording heads for embossing aluminum blanks, before later expanding into studio and portable recorders, blank discs, and eventually tape recorders. Presto, Speak-O-Phone, and other companies that made instantaneous disc-recording equipment geared for the home or semiprofessional market encouraged many future recording engineers to get their start. The major record companies used professional quality recording lathes made by Western Electric and Scully or designed and built by the company's own engineers. But many of the lower-priced recording devices were also capable of high-quality performance, and these appealed to the serious amateur. In addition, by the late 1940s, an increasing number of instructive articles, books, and pamphlets on the art of recording helped open up the recording field to more participants.[29]

William Savory, a self-taught engineer who worked for Columbia Records in the 1940s and early 1950s, recalled that a number of small studios opened in New York City during the mid-1930s with just such recording equipment, rim-driven turntables sold by Presto or Allied Radio. Audio entrepreneurs rented small two- or three-room suites in office buildings, recording programs and music off the air for artists, or recording live in their studio for song-pluggers, advertising agencies, and singers or musicians who wanted to make demonstration records of their talents.[30] Savory, who had embarked on what he described as a "self-education regimen" during the depths of the Depression by enrolling in a tutorial in electrical engineering, walked into one of these studios one day in the mid-1930s to make a piano demonstration record of his own. Savory had already attempted to build a recording device and knew quite a bit about the technology. As he recalled, something in the studio wasn't working that day, so he fixed it and promptly was asked, " 'Hey, can you do this on a regular basis?' And I said, 'Well, sure, okay.' So they gave me an hourly figure, and [I] came in there late at night on a scheduled basis when everything had calmed down, and fixed the place. There was no restriction on how many places I could fix so I went to all of them eventually, on a regular maintenance schedule, as well as an occasional recordist on a fill-in basis."[31]

Savory found ample work during the 1930s doing studio maintenance and recording air checks of jazz and swing bands, small groups, and soloists. In 1938, he worked on a system to electronically reproduce original wax cylinders for the National Vocarium, the studio established in the basement of Rockefeller Center by G. Robert Vincent in 1935 to preserve his large collection of historic cylinder and disc recordings.[32] Savory assisted Vincent in making more voice recordings of contemporary orators, such as William Lyon Phelps, and Vincent recalled Savory climbing up a phone pole "in the chilly Jersey countryside" while Vincent ran the equipment to record Benny Goodman remotes.[33] Savory joined Columbia Recording Corporation in 1940, assisting in maintaining and operating the company's new facility in Chicago, before enlisting in the Navy. After a year at the US Naval Research Laboratory in Washington, DC, Savory returned to New York and enrolled in physics and math courses at Columbia University. He returned to Columbia Records as a member of the team of engineers who worked under William Bachman, developing the 33⅓ long-playing record, making the first transfers from shellac disc to tape to LP master.[34] In 1950, Benny Goodman heard some of Savory's earlier line-check recordings of his band and was so impressed with the quality, he convinced Columbia they should be made available. In 1952, Savory's recordings were released as Benny Goodman, 1937–38 Jazz Concert No. 2 on Columbia Masterworks (see fig. 8).[35]

Moses Asch started as a radio repairman in 1926 but always considered himself a tinkerer and inventor. When the Depression severely cut into his business, he went into partnership with an electronics firm, Radio Laboratories in Brooklyn, New York. In 1938, one of its clients, radio station WEVD, required recording services for some of its programming, so Asch set up a small disc-cutting studio in a section of Radio Laboratories, by then located in the station's building at 117 West Forty-Sixth Street in Manhattan. There he recorded the Jewish program material—orchestral theater music, Yiddish songs, cantorials—that the station needed for its broadcasts. After the partnership dissolved in 1940, Asch advertised as Asch Recording Studios. In addition to transcription services, Asch continued to record ethnic and folk music genres, niche markets not well served by the major record labels during the Depression but which nevertheless continued to have wide audiences. This music became the basis for what would grow into Folkways Records, a label that profoundly influenced a rising generation of folk musicians who found inspiration in the old-time music Asch documented.[36]

Frank Laico came from a family of seven and was working in a Manhattan produce market in 1938, training to be a butcher, when a customer asked him whether he intended to be a butcher all his life. When Laico definitively replied, "No," the customer, who was the treasurer for World Broadcasting, the largest transcription program service, offered him a job. Laico, who unlike Savory and Asch had no previous interest in electronics or recording, quickly accepted the offer, but soon tired of his entry-level position running the mimeograph machines to copy scripts. Walking to different floors during his free time, one day he followed the sound of music, pushed open a heavy door and found himself in the recording room. "I found that so exciting," Laico recalled, "I went in very quietly and started talking to the people in there, and I got fascinated, so I kept coming back."[37] Deciding that he would prefer this work to running the mimeograph machine, Laico asked the engineer in charge, Harold Lester, how he might transfer to the recording department. Lester told Laico to "talk to the Big Boss," Charles Lauda, chief engineer and recording industry veteran who began his career with the Aeolian Company in the 1920s.[38] Lauda, who no doubt wanted

engineers with more experience, initially rejected the request but eventually offered Laico a position as an apprentice, dropping cutters on sixteen-inch wax
platters for seventeen dollars a week.[39]

Within two years, Laico was running one of World's three studios and earning a substantial raise. In 1943, he was drafted. The day after he left for basic
training, an Army Signal Corps officer visited World Broadcasting and informed
chief engineer Charles Lauda that they were taking over part of the facility and
would need one of his key engineers. Laico later learned that Lauda told him,
"Well, the one you want is at Camp Upton, he went there yesterday, and if you
can get him out, this way we could both be using him."[40] Laico was soon reas-
signed to the Signal Corps operation at Bell Labs on West Street in New York
City, where he found himself working on one of the most high-level security
projects of World War II, the secure speech transmission work of Bell Labs.
Until about 1943, anyone determined to break through radio signals and own-
ing the right equipment could compromise national security. "Project X," as it
was known at Bell Labs, focused on creating a true secrecy system for speech
and became one of the most closely guarded projects of World War II and for
years after.[41] The system involved having a key with six random levels. Laico
worked on the key production and synchronization, recording the key on high-
quality phonograph records. The project's success hinged on the precision of
the recording process and thus demanded very accurately driven turntables
used for making the records as well as for playing them back at the terminals.
Laico's Signal Corps work later also included advanced electronics training at
the Pentagon and part-time work at U.S. Recording Company in Washington,
DC, another transcription and air check studio. In 1946, Laico returned to New
York, and after working briefly at Reeves Sound and at the United Nations, he
accepted a job offer from Vincent Liebler, chief engineer of Columbia Records.
The two had known each other since Laico's days at World Broadcasting, where
he mastered some of the Columbia discs in those days before the record company had its own studio.

Unlike Savory, Asch, or Laico, Donald Plunkett's interest in recording began with his discovery of a Columbia Graphophone, a cylinder phonograph, in
his grandmother's attic in the mid-1930s, and he immediately "fell in love with
the phonograph." Soon after, he heard his first Victor phonograph at someone's
home, and as he recalled, "those two things started me off, and I guess I've been
in love with it ever since." He began by taking the machines apart and since
people were happy to unload their old-fashioned Victrolas, he acquired a steady
supply of phonographs to disassemble and examine. But Plunkett grew more

interested in broadcasting than tinkering, because radio in the early 1930s, unlike the record industry, was alive and healthy. Plunkett used his contact with a friend's father to obtain a job as a messenger boy at NBC and then entered an apprenticeship and was sent to the RCA Institute for training.[42]

The RCA Institutes, located in New York and Chicago, succeeded the Marconi School of Instruction sometime after formation of the Radio Corporation of America on October 17, 1919.[43] Originally organized for the training of marine operators, the Institutes became the pre-eminent school for the training of radio engineers, not only RCA personnel but anyone involved in radio. By the late 1930s, the Institutes' General Course was especially suited to those interested in the radio engineering profession and included training in Morse Code, theoretical and applied radio, college-level physics and calculus, and broadcast operating and radio service work.[44] Although they offered no hands-on training in recording or disc cutting, the RCA Institutes provided a comprehensive general background in electronics, which was essential for NBC personnel who worked in the network's vast new recording department, by the 1930s occupying half of the fifth floor at 30 Rockefeller Plaza.

The early career paths of William Savory, Moses Asch, Frank Laico, and Donald Plunkett illustrate the multiple opportunities for audio enthusiasts who got their start in the 1930s when the record industry was just beginning to rebound from the effects of the Depression. With the exception of Asch, who remained an independent recording engineer and label owner, each of these men enjoyed careers with major recording companies. All but one, Laico, had been fascinated with the phonograph or electronics as an adolescent, and each one entered a profession for which he had little or no training or, as in Savory's case, was self-taught. Each began as an amateur and became part of an emerging profession. As such, they represent hundreds more across the country for whom recording was both a fascination and a potential career.

As early as September 1941, record sales had begun to approach pre-radio levels, but supplies of key recording and manufacturing materials such as aluminum, copper, and shellac were dwindling or threatened.[45] In January 1942, a new trade publication devoted to sound recording died a quick death because anticipated advertising did not materialize, and in May, *Business Week* reported that manufacturers were drastically cutting the number of their releases, so that an orchestra leader who may have recorded forty sides in 1941 would be lucky to record ten in 1942, and new artists or those of marginal popularity were unlikely to record at all until after the war.[46] The effects of such cutbacks became apparent at war's end when independent record labels sprang up all over the country, many of them now recording those artists passed over during wartime.

Wartime cutbacks did not mean that amateur recording came to a standstill. It became part of the war effort. At USO locations in New York, Washington, DC, and San Francisco, servicemen and women were able to record audio letters for their loved ones, and companies such as Sonora Radio explicitly advertised their radio-phonograph-recording units as a way for servicemen and families to stay in touch. Promoting audio letters through sentimental appeals to families separated by war, one Sonora advertisement depicted a Navy seaman holding a microphone, recording, "Sweet Music for Mom." Although the music was simply an audio letter to the folks at home, the emphasis was on records, because wartime shortages restricted availability of radios. At the bottom of the ad was the assurance, "Records today, Radios tomorrow."[47] A corresponding advertisement shows the family gathered around the home unit, listening to "the record that money can't buy. . . . It seemed that Bob was right there in the room, so faithfully did the SONORA Radio-Phonograph reproduce his words."[48] Such advertisements promoted the idea of recording as a means of communication more immediate and personal than letter writing by emphasizing its ease of operation and the quality of sound. By the early 1940s, home recording devices had become more accessible to the whole family, not just the tinkering father or son.

When "ham" radio operators were ordered off the air during World War II, many turned to recording as an alternate hobby. The Astatic Company of Youngstown, Ohio, suggested to radio hams who had been silenced for reasons of national security that they "satisfy their natural urge to work with electronic gadgetry" by taking up recording as a second hobby.[49] A number of hams had already heeded this advice. In 1948, Columbia Records released *I Can Hear It Now*, a collection of recordings documenting historic events from the Depression to the Japanese surrender ending World War II, many of which had been made by amateur recordists whose efforts made it possible for American record buyers to become "ear-witnesses" to world history in the making.[50]

From Competition to Cooperation: The Sapphire Club

With so much recording activity, both professional and amateur, why did it take over half a century for a society of recording engineers to professionalize? The radio engineers initiated their own organization in 1912, motion picture engineers formed theirs in 1916, yet these industries did not date as far back as the commercial recording industry, nor were the technologies on which they were based any older than the phonograph.[51] One major obstacle to interaction among recording specialists was the shroud of proprietary secrecy surrounding their art. Moreover, the record industry was not as highly profitable before World War II as either radio broadcasting or motion pictures. The recording industry had just begun to benefit from a healthy resurgence in phonograph record sales beginning in the late 1930s, broadly heralded as a comeback by 1941, only to be curtailed by wartime restrictions. In addition, a musicians' union recording ban during World War II put a halt to any new commercial recording work by union musicians from August 1942 until November 1944.[52]

But the war brought unintended benefits for the recording industry and studio recording in other ways. Technological research and development related to the war effort later yielded benefits for sound engineering, the shellac shortage drove companies to investigate alternative materials, which resulted in an improved record made of Vinylite, and the exigencies of material shortages forced record company engineers to begin communicating and cooperating with one another to share scarce resources. The last development proved to be significant as it laid the groundwork for the later formation of the Audio Engineering Society.

The Sapphire Club represented the first attempt of recording professionals to openly communicate on a regular basis. It comprised one hundred members in New York and fifty in Hollywood, all associated with sound recording. The Sapphire Club broke the decades-old ban on intercompany exchange of know-how.[53]

The early recording business developed in a climate of keen competition in which recordists and inventors guarded technical secrets to protect their innovations and maintain a business advantage.[54] Security became a concern as recordists sought devious means, such as assuming the guise of a meter reader, to gain entry to a competitor's studio.[55] The Sooy brothers of Victor Talking Machine jealously guarded their cutting heads, according to bandleader Paul Whiteman, keeping them in "little leather boxes which never left their possession, day or night."[56] The Sooys appear to have protected not only their technology but also their technique. In 1914, Victor contracted Lubin and Company of Philadelphia to film the plant and its working departments to show to talking machine dealers attending their annual convention. On the day the photographers came to the recording rooms, they filmed the artists performing the "Sextette from Lucia" in the studio, and according to Harry Sooy, when the photographer came to the "operating room" to film "what he and many others thought was the actual procedure of recording a record," Sooy admitted in his memoirs that "this mechanical work was arranged specially for the motion pictures, with Yours Truly at the machine."[57] Even during the 1920s, inventors' and manufacturers' consultant Harry Gaydon had advertised his services as a "specialist in Sound Recording and Reproducing Devices" with the assurance, "absolute secrecy guaranteed."[58]

With this prevailing attitude, it is little wonder that recordists had neither the inclination nor opportunity to organize at a time when so many other professional societies were forming in the United States.[59] This situation changed gradually during the 1930s as electrical transcription studios and equipment manufacturers mushroomed to meet the needs of broadcasting. By the early 1940s, recordists had advanced from the relatively isolated "recording experts" of the acoustic era, highly protective of their technique, to "recording engineers" and members of a larger audio engineering community in which exchange of technological information was not only permissible but encouraged. A report of the first Hollywood Sapphire Group meeting indicated that many involved in this recording community believed such an organization was long overdue.

The Sapphire Group was first conceived in New York City in the early 1940s during informal luncheon meetings of a group of recording engineers and Wally Rose of the Frank L. Capps Company, makers of professional recording styli.[60] Rose met for lunch regularly with some of his major clients: G. E. Stewart of the NBC Recording Division; Vincent J. Liebler, chief recording engineer of Columbia Recording Corporation; and Albert Pulley, Liebler's counterpart at RCA Victor. Donald Plunkett, a Sapphire Group member and later a founding member of the Audio Engineering Society, recalled that these lunches led to a kind of informal bartering and trade group, as the record companies began to feel the pinch of rationing during World War II and their stock of needed supplies dwindled. The monthly meetings afforded them not only the opportunity to commiserate but to share and trade surplus of lacquer recording discs, vacuum tubes, or any of the important (and rationed) supplies needed for the recording industry.[61] According to Plunkett, this situation brought about the end of the corporate closed shop, but the doors had already begun to open. Certainly, the Sapphire Clubs on both coasts signaled the increasing awareness of recording professionals that sharing information was essential to survival.[62]

The group assumed the name Sapphire Club in honor of its use and dependence on sapphire cutting styli, the chief product of the Capps Company.[63] As the membership expanded, the lunches became increasingly lengthy, with conversation running well past the lunch hour, prompting the members to change the venue to a monthly dinner. By 1942, the Sapphire Club had a group pin (a sapphire on an oval representation of a phonograph disc) and a printed membership list and held their monthly meetings at the New York Athletic Club. Membership, by invitation, was limited to one hundred individuals, all engaged professionally in some aspect of sound recording. Industry representatives visiting New York from the West Coast were invited into the club, and by 1945, at least eight members of the New York Sapphire Group lived in Hollywood. Robert Callen of NBC and Chester Boggs of Columbia Recording Corporation in Hollywood began to organize a West Coast branch, and by early 1946, the Hollywood Sapphire Group held its first meeting. A little more than half of the thirty invited sound recordists attended the meeting, representing record companies, film studios, radio networks, independent recording studios, and equipment manufacturers. Each member spoke after the dinner, and the talks, quite appropriately, were recorded on sixteen-inch transcription discs and forwarded to the New York Sapphire Group.[64]

The Hollywood Sapphire Group grew rapidly, with participation from all fields of recording. The dynamic mix of film sound and recording industry professionals provided fertile ground for discussion of common terminology and tools and standards of practice. The Hollywood Sapphire Group focused on technical matters, particularly the problem of standardization, more than did the New York group.[65] By the second anniversary meeting in March 1948, the Hollywood Sapphire Group had expanded to fifty members, had formed a Recording Standards Committee and three subcommittees, and had drawn up a "Proposed List of Preferred Terms for Disc Recording."[66] The New York group must have shared the concern about standards, but according to most accounts, it was strictly a social group. However, inventor and New York Sapphire Club member Norman Pickering recalled increasing talk about audio technicalities, and a suggestion by one member that they open these discussions to everyone, including nonprofessionals with a serious interest in audio. An announcement in the trade brought an overwhelming response, revealing an undercurrent of serious interest, both professional and amateur, in the engineering of sound.[67] The time had come for this widely scattered community of audio enthusiasts to organize.

The Audio Engineering Society, Audio Fairs, and Audiophiles

The Sapphire Groups were founded by recording professionals—manufacturers of equipment and engineers affiliated with record companies, broadcasting stations, film sound studios, independent recording studios, and transcription services—but the field of audio encompassed all forms of sound engineering. By 1947, this included not only broadcasting and recording on disc, wire, and tape, but also public address, transmitter and receiver manufacturing, industrial sound, and acoustics. The term *audio engineering* was a relatively new one in the postwar period, so new that there was no publication devoted to it until 1947. Early that year, *Radio* magazine announced that it would be changing its name and focus to *Audio Engineering*, starting with the May issue. "Because there has been no technical magazine devoted solely to this field," explained editor John H. Potts, "all engineers interested in audio engineering have had to gather piecemeal, from a large number of sources, such information on the subject as is published."[68] Potts, an electronics pioneer, author of technical articles, and an engineer for RCA, General Electric, Westinghouse, and Sperry, knew well what literature was available. He knew that recording received little attention in the professional technical literature, and he vowed to place special editorial emphasis on it in the pages of *Audio Engineering*, securing the most prominent chief engineers in recording and broadcasting for his editorial advisory board.[69] The need for standardization had become particularly acute, with different record labels employing different crossover frequencies, different degrees of pre-emphasis at the higher frequencies, varying groove depth, and other factors that affected reproduction. Even the best reproducing equipment could not play all records satisfactorily. By providing a forum for the interchange of ideas, Potts hoped "to be able to contribute in some measure to eventual standardization of these varying techniques."[70]

Cutting and mastering techniques had long been highly individualistic. Recording engineers did their best to mask the surface noise on records by boosting frequencies in the high ranges as the records were mastered (pre-emphasis); likewise, they rolled off the bass response to limit groove width. Each record and transcription company established its own "recording curve," or "recording characteristic," which meant that commercial phonograph records then had widely varying levels of volume and tone. A further complication was that each company made its records sound best on what they considered to be the best phonographs—their own if they were also a phonograph manufacturer—and models were changing rapidly. This caused a problem for broadcasters and jukebox operators who sought more uniformity in the sound of the records they played.[71] Some broadcasters were forced to employ up to ten different equalizing networks to accommodate the disparities between transcription discs.

In the fall of 1941, the National Association of Broadcasters (NAB) adopted sixteen technical standards and good engineering practices for electrical transcriptions and recording for broadcasting, including standard recording characteristics for vertical transcriptions and for lateral transcriptions.[72] Most companies complied, but when the NAB Committee on Recording and Reproducing Standards met again in 1947, the group had to consider establishing new standards to accommodate the changes adopted during and after World War II.[73] Moreover, standardization ultimately affected the consumer of audio products and thus concerned the companies hoping to satisfy consumer demand.[74]

At this time audio engineers, and especially recording engineers, maintained a kind of liminal professional identity. Some were members of the Institute of Radio Engineers (IRE), the American Institute of Electrical Engineers (AIEE), the Society of Motion Picture Engineers (SMPE), or the Acoustical Society of America, but their specific technical interests only occasionally overlapped with broadcast and electrical engineers, film sound recordists, or acousticians. C. J. LeBel spearheaded the drive for a separate society devoted to audio after becoming fed up with the Institute of Radio Engineers' refusal to give much attention to the audio field at conferences or in publications. As early as 1945, he suggested the possibility of breaking away to form a society devoted strictly to the interests of audio engineers. Just as the IRE had broken away from its parent organization, the AIEE, in 1912, so the audio engineering community was on the verge of declaring a separate identity in the dawn of its own takeoff decade.[75]

How could recording professionals and other audio-related industries arrive at standards if they had no recognized umbrella organization within which to effectively discuss and resolve these problems? The Sapphire Club had been a beginning, but now it was time to take it a step further because so much had happened since the end of the war. Decca Records in England, for example, had just introduced full frequency range recording (ffrr), which opened up a world of new sonic possibilities. The enhanced high-fidelity recording technique was developed by British Decca recording engineer Arthur J. Haddy, who during the war had been asked by the Royal Air Force to solve the problem of detecting submarine propeller noises by recording a frequency range much wider than existing recording equipment had been capable of capturing. He succeeded in recording a frequency range of 80–15,000 Hz, and after the war, this work resulted in "ffrr," then heralded as a "major advancement in the science of sound recording."[76] Professional magnetic tape recording was on the horizon, as were the $33^{1}/_{3}$ microgroove long-playing record and the germanium transistor, and microphone quality had also benefited by war research. If technical standards had been difficult to establish during the prewar period, the emerging technologies for recording sound posed an even greater need for an organized approach.

The immediate postwar period held great potential for the recording industry and promised challenging but inspiring work for the recording engineers and equipment manufacturers, if only they could organize, standardize, and move the technology forward. At least that was the sincere belief of a relatively small but dedicated group of people, such as LeBel, Pickering, and others on both coasts, who were concerned with these developments in audio and felt that existing organizations had too long neglected the field in favor of radio and electronics. At the IRE's winter 1947 convention, only three papers dealt with record engineering, yet a significant number of high-quality recording units were featured in the exhibits.[77] Evidence suggests IRE marginalization of the audio contingent stemmed from the belief that audio lacked legitimacy as an engineering discipline.[78] Since the IRE was concerned largely with measurements, components, and theory, some members viewed the audio crowd as hobbyists interested only in improving the sound of records.[79] Many within the IRE, such as H. H. Scott, manufacturer of audio tuners, straddled both worlds. They recognized that the nontechnical listener judged sound reproduction subjectively and that the listener's ear did not always agree with the engineer's measurement results.[80] From the days of acoustical recording, recording professionals and serious amateurs had learned that the ear was often a more reliable measure than any scientific instrument, and in a demanding recording situation, as LeBel once put it, "the ear complains faster than the eye viewing a meter dial."[81] Balancing listening with reading the meters is another example of the tacit knowledge recording engineers needed in their work.

By the end of 1947, a letter appeared in *Audio Engineering* seeking all those interested in forming an "Audio Association." Since they now had a magazine serving the neglected field of audio, was it not time to form an association similar in function and purpose to the IRE and the SMPE in their respective fields?[82] A reply to the letter, printed in the following issue, stated that "a group of us, long active in broadcasting and recording, feel the same way. Audio engineering will be unhampered only when it has a society devoted exclusively to its needs—controlled by, and only to benefit, the audio engineer."[83] The author was none other than C. J. LeBel. His reply also announced the forthcoming organizational meeting of "an audio engineering society" and invited interested individuals to send him their name, address, company affiliation, and title or nature of work. At the first meeting on January 8, 1948, a small group of attendees elected a steering committee.[84] Word spread quickly and, by the next meeting held on the evening of February 17, 137 people crowded into the RCA Victor recording studios on East Twenty-Fourth Street in New York City. Among those present were recording engineers, inventors, musicians, equipment manufacturers, audio specialists, and a formidable number of audio amateurs, an eclectic assemblage with a shared interest in recording. The group included millionaire inventor and entrepreneur Sherman M. Fairchild, record reviewer Edward Tatnall Canby, stylus manufacturer Isabel Capps, lathe manufacturers Lawrence Scully and George Saliba, inventor Emory Cook, musicians Les Paul and Fred Van Eps, and recording engineers representing every major label, including Al Pulley of RCA, Bob Fine of Mercury, Clair Krepps from Capitol, and Vincent Liebler from Columbia.[85] Collectively, they composed the first association dedicated to filling a desperate need for "the promulgation of engineering information and standards in the field of recording, transmission, and reproduction of sound."[86] At the third meeting, held in the same location on March 11 and considered the first official meeting of the Audio Engineering Society, renowned acoustical engineer and inventor Dr. Harry Olson of RCA gave a loudspeaker demonstration and technical paper on "Some Problems of High Fidelity Reproduction."[87] This was the first of many technical meetings that immediately became a regular feature of the AES. In September, Vin Liebler of Columbia Records demonstrated the newly launched LP microgroove record, which had been reported in *Audio Engineering* the month before.[88] These demonstrations accomplished two goals: they not only drew together interested observers but also lay the foundation of a forum for technical exchange and education, which was critical to improving the art and technology of sound recording and reproduction.

Shortly after AES's formation, it appeared that even the IRE had a significant membership of audio enthusiasts, and, indeed, many retained membership in both societies. On June 2, 1948, the Professional Group on Audio (PGA) became the first of many technical groups to form within the IRE Professional Group System; its purview, the "technology of communication at audio frequencies and with the audio-frequency portion of radio systems."[89] These subgroups within IRE recognized that the time had passed when "all members of the organization had an essentially equal interest in the activities of all others."[90] A decade before, *Communications* magazine was founded on the conviction that many branches of the communications industries, including broadcasting, electrical transcriptions, equipment manufacture, telephony and telegraphy, and radio and facsimile transmission were becoming inescapably linked.[91] By the end of World War II, it was difficult to stay abreast of new developments in each field. These industries were growing, and the flood of information—the "information explosion"—was spreading.[92]

In October 1949, the AES, which by then had quintupled its membership, held its first convention at the Hotel New Yorker. Part of this meeting included a manufacturers' display of sound recording and reproducing equipment, a professional exhibit that was open to the public and came to be known as the Audio Fair.[93] The fairs became an annual event, always announced and reviewed in the *New York Times*, drawing increasingly larger crowds of audio engineers, sound experts, high-fidelity enthusiasts, and curious listeners to view and hear the latest audio gadgets, recordings, turntables, amplifiers, loudspeakers, tape recorders, and other apparatus in the high-end sound field.[94] By the early 1950s, there were Audio Fairs in Chicago and Los Angeles, sponsored by two of the many local AES chapters sprouting up across the United States. Edward Tatnall Canby, an audio enthusiast, teacher, arts critic, and founding AES member, routinely covered these Audio Fairs for *Saturday Review*. His record reviews for that popular literary magazine included critical assessment of the engineering, as well as the performance and content of the records he reviewed. Canby, who was knowledgeable about the science and art of recording technology, epitomized the burgeoning community of technically educated reviewers in the era of high fidelity.[95] The dual nature of the early Audio Engineering Society as both a technical professional group and a booster club for consumer audio, with members straddling the worlds of professional audio, music, audiophiles, and journalism, ensured public dissemination of high-quality (if sometimes baffling) technical information in mass circulation newspapers as well as special-interest periodicals and books.[96]

The first issue of the *Journal of the Audio Engineering Society* in January 1953 contained thirty articles that revealed how long the AES had been waiting for this opportunity to publish papers, even though *Audio Engineering* regularly published AES news as well as some technical papers. McProud, now confident that the technical crowd would have its publication vehicle, began to turn the editorial focus of *Audio Engineering* toward a lay audience, and by March 1954, *Audio Engineering* had dropped the second word from its title and shifted its focus slightly away from the professional and serious amateur to the audio newcomer.[97] Beginning with the March 1954 issue, the magazine was titled simply *Audio*, and in the April issue, the editor once again stated: "Readers will have noted rather more material directed to the newcomer to our ranks—and possibly less that was intended for the scientist. Whatever reputation *Audio Engi-*

neering enjoyed was built solidly on its technical and practical articles, and we continue to assure all our readers that we shall continue to serve along those lines. . . . We trust we may continue to please everyone."[98]

This editorial shift illustrated that a division between the "lover of audio" and the technological enthusiast or hobbyist/experimenter had taken place. A major catalyst for this bifurcation of interests occurred with the professionalization of the audio field through the formation of the Audio Engineering Society, with the division becoming more definitive once the society began publishing its own journal. In addition, as the phonograph and hi-fi components industries began turning out more and more consumer goods in response to growing demand, the publishers of *Audio* saw a far more lucrative market in the hi-fi buff, the home consumer, rather than the professional or the do-it-yourselfer.[99] What began as amateurism in its truest sense—a love of the art—became a fad by the time the term *hi-fi* came into common usage, when it transitioned from a hobby to an industry.

Conclusion

The formation of the Audio Engineering Society, its annual meetings, and its technical displays profoundly affected the growth of both consumer and professional audio and the development of the recording industry. The organization acted as a forum for knowledge trading, established standards, boosted the manufacturing of recording as well as playback equipment, and most important encouraged would-be engineers, some of whom might never have chosen their careers were it not for a local AES field trip to a recording facility. All of this is impossible to measure, but even today, young engineers who may not be inclined to join the society eagerly attend the biannual conventions where the latest audio equipment is demonstrated.

The professionalization and popularization of recording was a mutually reinforcing process in which manufacturers responded to consumer demand even as they fueled consumer interest by offering more products. In this invention push–market pull scenario, the recording studio and recording engineer were a unique combination of consumer and producer. Although they produced a product in the form of a sound recording, the studio and its engineers also consumed many different technological tools in the process of creating that product, and as these tools became commercially available, more people got involved in the business. The society to which many of these recording professionals belonged acted as a means to disseminate information about products, processes, technical developments, employment opportunities, and not a little industry gossip. The annual audio fairs and conventions provided a forum for information trading, socializing, networking, and, especially, selling products through initial public demonstration.

The formation of a professional organization devoted to advancing the cause of sound engineering was an important factor in the proliferation of recording studios and recording technology. By bringing together a dynamic but disparate group of serious amateurs and professionals all concerned with developing the quality and potential of sound recording and reproduction, the AES catalyzed the growth of postwar recording studios and the increase of sound recording and reproducing technologies. Like all professional societies, the AES fostered standards of practice, educational presentations, technical meetings, and publications. It also promoted audio engineering as an engineering field in its own right, with a specific knowledge base. However, unlike most professional engineering societies, the Audio Engineering Society included many members who did not make a living by practicing within the field on which the organization was founded. Moreover, its membership traversed social ranking and cultural pedigree. Unlike the electricians of the late nineteenth century, these audio devotees did not exclude nonprofessionals or seek prestige and professional standing to assuage occupational and status anxieties. Instead, they sought to know each other and to establish a forum for discussion and interchange of ideas about the rapid changes taking place in the audio field.[100] This, at least, was the initial impetus for the organization. The broad and expanding field of audio had grown tremendously during the 1930s and especially during World War II. Phonograph records became part of the war effort in multiple ways, from entertainment and morale boosting through the Armed Forces Radio Service and the V-Disc program, to sonar research and high-level security communications.[101] In the postwar era, this work paid dividends for the emerging recording industry. While record company management may not have liked the idea of sharing technological know-how with competitors, recording professionals actively sought to interact with one another and with those in related fields outside of the record business. These interactions provided fertile ground for the technological growth so crucial to the development of sound recording practice.

When High Fidelity Was New

The Studio as Instrument

Record sales rebounded in the years leading up to World War II, and with consumer demand, companies paid renewed attention to the sound of records and to the acoustical properties of the studios in which they were made. While listeners enjoyed natural reverberation in live symphonic or big band performances, records could not convey room ambience in the same way because of the extraneous noise of shellac discs and the limited frequency range phonographs were capable of reproducing. Throughout the 1920s, uncontrollable reverberations and echoes posed repeated problems in recording that could only be solved by damping wall surfaces and eliminating pockets in the corners of studios.[1] Consequently, most recording studios followed the model of the broadcasting studio, adopting an acoustically dead environment to minimize reverberation. In the early 1930s, radio stations began using "live end" and "dead end" studios as well as echo chambers to enhance the broadcasting of music and speech.[2] Soon, preexisting halls and churches with more reverberant acoustics became the preferred type of recording environment, and the best of these acquired superb reputations among those who used them, inviting comparisons to fine musical instruments. A studio's acoustical signature, what recording engineers and producers referred to as its particular "sound," became such an identifiable quality that engineers devised ways of recreating such acoustics in smaller spaces. Reverberation, once engineered out of existence in both public spaces and motion picture sound studios, became an important asset in music recording not only for singers challenged to sing in tune but also for its ability to make records sound louder.[3]

The key to exploiting reverberation successfully was in the ability to control it. Although some believed that recording during this period came closer than ever to recreating the live musical experience, achieving a realistic sound required skillful manipulation by the recording engineer of the recording equipment as well as the vocalists and instruments around the microphones. Some record producers chose to go beyond recreating the live musical performance. They believed that records should offer the listener something more. This new emphasis on record production marked a shift in the significance of studio technology and design and a dramatic increase in the importance of the producer, the engineer, and the studio in the outcome of a recording session.

Record Boom: The Record Industry, Broadcasting Studios, and Recording Studios

The upswing in record sales began in 1934 when RCA revived its long-dormant recording division by advertising phonograph records in radio trade magazines, and a new company, Decca Records, introduced the thirty-five-cent record that featured jazz and popular music by name artists such as Bing Crosby, Chick Webb, Ella Fitzgerald, and the Mills Brothers.[4] Improvements in the frequency response of records, increasingly referred to as "high fidelity," not only boosted record sales but also created demand for better home reproducing equipment. In 1938, Lafayette introduced the "modernized" home phonograph with "bass boosting" and "high fidelity" response. This phonograph was marketed as more capable of faithfully reproducing the improved quality of records.[5] In the workplace, employers recognized that production, efficiency, and employee morale benefited from recorded music piped into factories.[6] The popularity of jukeboxes and swing bands and the continued improvement in the quality of records and phonographs further fueled record sales, which, by 1939, had reached an annual gross figure of $36 million industry-wide, six times the sales figure of 1933, the industry's worst year.[7]

By 1940, twenty-seven new small labels emerged, and records once again became more widely available, appearing at newsstands and in retail stores such as Woolworth's. Both Columbia and RCA Victor poured more than one million dollars each into advertising, most of it aimed at consumers.[8] On the eve of America's entry into World War II, record sales surpassed the previous peak year of 1921, despite serious shortages of key record manufacturing materials: aluminum, copper, and shellac.[9] Moreover, consumers had grown more musically and technically sophisticated; even young listeners, according to one *New York Times* report, debated the relative merits of pickups, amplification, and needles "with the glibness of scientists' shoptalk."[10] Fanatic record collectors, once considered mere eccentrics, gained respectability as connoisseurs of the phonographic art.[11] For those to whom records meant more than casual entertainment, the quality of reproduction and recording assumed increasing importance in their purchasing decisions. Books to guide the uninitiated record collector in assembling a sound library critiqued everything from artists' performances to the material quality of recording and pressing.[12] By the time the United States entered World War II, the recording industry's healthy growth spurt was only temporarily halted by wartime rationing and the first musicians' union recording ban of 1942–44.[13] Even during the ban, which forbade union musicians from making commercial recordings or transcriptions, American Federation of Musicians president James C. Petrillo permitted musicians to play on recording sessions for the Armed Forces Radio Service and the V-Disc program, which ran from October 1943 to May 1949. V-Disc recording kept musicians and a handful of engineers and producers working during the war years when record companies were forced to cut back because of matériel shortages. More important, the V-Disc program was a major morale booster, bridging the gap between the home front and millions of GIs stationed overseas by providing a steady supply of jazz and popular records, thus also spreading interest in American popular music abroad, a trend that escalated during the Cold War.[14] After the war, returning soldiers formed a ready labor pool trained in electronics and capable of building a new recording industry ready to respond to pent-up consumer demand for records, phonographs, and live entertainment.

One might expect that the recording studios in which these increasingly popular records were made should reflect the prosperity they generated for the major record labels—RCA Victor, Columbia, and Decca—who sold most of the phonograph records. However, despite the gains made by the recording industry during the 1930s, the commercialization of radio had made broadcasting a more profitable enterprise.[15] Consequently, the most advanced studio facilities were found in the radio networks—those of the National Broadcasting Company and the Columbia Broadcasting System, parent organizations of RCA Victor Records and Columbia Recording Corporation, respectively.[16] The mid-1930s, a decade of technical development in broadcasting, including improved amplifiers, transmitters, microphones, and loudspeakers, necessitated progress in the design of the "point of program origin," the broadcast studio.[17] These are worthy of consideration as they became the model the phonograph recording studio first emulated and later rejected.

The Modern Broadcast Studio: NBC Radio City and CBS

In New York, NBC completed its massive Radio City studio complex in 1933. The structure cost five million dollars, more than half of which was used for sound equipment and acoustic treatment.[18] The primary concern was to prevent outside sound from leaking into the studios. The designers incorporated elaborate means of achieving sound isolation, such as "floating construction" of walls, triple-pane plate-glass observation windows, and many types of acoustical treatment—rock wool, perforated asbestos tiles, and proprietary materials such as Rockoustile, Acoustone, and Acousti-Celotex—on the walls and even within the building's air ducts.[19] In describing these studios at a meeting of the Acoustical Society of America, NBC engineers stressed that their design criteria were set not by "the critical judgment of comparatively few people" but by "years of operating experience and the collective judgment and opinion of thousands of listeners." In other words, while broadcast studio design was "no longer a matter of guesswork or 'trial and error' methods," neither was it determined solely by theoretical analysis but also by what sounded good to the listening public.[20] This kind of empirically driven decision-making process about what constituted "good sound" would become even more important in the selection and design of recording studios.

In 1940, CBS remodeled a former school of music located across the street from its East Fifty-Second Street headquarters for its new broadcasting studios. The designers incorporated innovations in room design and acoustical treatment "intended to result in studios more 'live,' or brilliant, than any built up to the present time."[21] Employing nonparallel opposite surfaces and serrated walls and ceilings allowed designers to eliminate dead spots, slaps, and unwanted echoes. The two large studios featured movable panels called "Acoustivanes," which the audio engineer operated from the control console by push button, thus enabling him to determine how much resonance to add at certain frequencies or how much of the absorbent material behind the vanes to expose. Goodyear Tire and Rubber Company supplied the rubber flooring in the studios and control rooms, and Johns-Manville supplied Transite, the acoustical wall treatment that minimized noise and maximized sound isolation and control. Although the new CBS studios did not approach the scale of NBC's Radio City complex, the plain stucco façade with a single row of windows on the seventh floor and neon CBS sign over the main entrance lent the building a starkly modern appearance. *Architectural Forum* considered the studios in this monolithic-looking building to be "the last word in broadcasting studio design and equipment."[22]

Both the NBC Radio City studio complex and the CBS Fifty-Second Street studios set standards of aesthetic and acoustic excellence in broadcast studio design. They were, of course, also showplaces for the public, who "trooped by thousands day after day" to Radio City, not only to observe the broadcasts as members of the studio audience but also to take guided tours led by smartly uniformed pages.[23] As places designed for technical performance as well as public display, the broadcast studios differed in purpose from the recording studios, which were off-limits to the public. There was also a difference between the acoustical criteria for radio programs, which involved speech as well as music, and the most desirable acoustical conditions for phonograph recording, which involved primarily music. Only after improvements in recording and reproduction made the ambient sound of the studio more important was this difference in acoustical criteria fully recognized. Consequently, recording studios in the 1930s often incorporated some of the acoustic treatment of the broadcast studio. Yet compared with the neoteric facilities of the broadcasting companies, the recording studios were architecturally, if not technologically, primitive.

Recording Studios: RCA Victor, Decca, Columbia, Capitol

RCA Victor New York studios, after decades of moving from one location to another, finally settled at 155 East Twenty-Fourth Street in 1936, where they would remain until the company built new state-of-the-art studios at Radio City in 1969. The Twenty-Fourth Street location, far from the bustle of Midtown Manhattan, was nestled between Miller's Harness Store and a stable on the block between Lexington and Third Avenues.[24] Mitchell William Miller, a classically trained oboist and recent graduate of the Eastman School of Music in Rochester, remembered his first recording sessions in this studio, accompanying Elisabeth Schumann on Bach's "Wedding Cantata." The studio had padded walls and rugs on the floor, a control room with a simple RCA recording console, and since the recording was made on hot wax, the musicians could not hear a playback until they received test pressings a week later.[25] Miller's account describes the basic technical components of recording in the 1930s' studio of what was then the oldest, most-established American record label. Even after the introduction of the lacquer-coated instantaneous disc in 1934, original recordings continued to be made on either solid wax discs or flowed-wax platters until the 1940s when they were supplanted by lacquer.[26] Because these would be destroyed if played, they had to be processed and copies pressed on shellac before anyone could hear a playback. The recording consoles employed in recording studios had evolved from the original control panels of the Western Electric electrical recording system. RCA Victor and Columbia Recording used consoles designed and built by their respective engineering departments. Smaller recording studios used broadcast consoles manufactured by RCA, Raytheon, Western Electric, and Gates Radio. The engineer could adjust the volume of the input from the microphones and output to the disc cutter, and the balance between instruments, but not much else. The most popular microphones were RCA ribbon microphones, either the diamond-shaped RCA 44 bidirectional or the bullet-shaped RCA 77 unidirectional, both of which were developed in the early 1930s in the RCA laboratories under Dr. Harry Olson. These became the most widely used studio microphones during the 1930s and 1940s.[27]

Similar conditions prevailed in the recording studios at Decca and Columbia. George Avakian's first recording session took place at Decca Records in 1939. Then a twenty-year-old college student and a self-confessed jazz fanatic, Avakian organized and produced a six-disc album entitled *Chicago Jazz*, which Decca released in 1940.[28] Because of the interest in 1920s "hot jazz" stimulated by the popularity of swing music, some companies began reissuing hard-to-find 78s by such musicians as Bix Beiderbecke, Bessie Smith, Louis Armstrong, and Duke Ellington as multi-disc albums in the 1930s. Avakian's *Chicago Jazz* was the first jazz album with original (not reissued) recordings and featured musicians associated with the Chicago school of jazz—notably Eddie Condon, Pee Wee Russell, Bud Freeman, and Jimmy McPartland, among others. Avakian vividly remembered Decca's recording lathe as being regulated by gravity rather than by electricity:

I worked in the period of 78 rpm recordings which were made with so many problems of a technical nature in terms of running the equipment at a consistent speed that the first session that I ever did had the turntables running not by electricity but by a device of pulleys and ropes which were attached at one end to a large stone that slowly descended from the ceiling of the studio, and at the other end of the system the turntable was rotated through a series of gears at a very consistent speed. Gravity was the driving force behind the operation of the turntable, and it produced steadier speed than electrical power did at the time.[29]

Electrical current was subject to line loss and current variations, so the mechanical governor driven by weight provided a constant rate of speed more precise than spring or electric motors; it also eliminated vibrations and smoothed out the cogging effect of the rotating motor (see fig. 5).[30] Decca's so-called gravity-fed recording lathe remained in use even after the war. Tom Dowd, who in 1947 began his recording career at Carl Fischer Recording Studios and later became the chief engineer of Atlantic Records, recalled Decca's small studio at 50 West Fifty-Seventh Street as "an amazing place—they had a lathe in there that was run by sandbags."[31] Others said the studio was "small, stuffy," and even into the 1950s, "pretty barren."[32] A nontechnical feature of the Decca studio even more memorable than the recording lathe and indicative of the label founder's basic philosophy was the picture over the control room window of an Indian, hand cupped over brow, scanning the distance. The caption read, "Where's the melody?"—the standard by which Jack Kapp judged any song to be recorded for his label. From the beginning, recalled Don Plunkett, Kapp's attitude was, "if you can't whistle it, don't bother to record it."[33]

Melody may have been a prerequisite for a hit record, but it was not the only consideration in the recording studio as sound and technological sophistication became increasingly important to recording engineers and record company A&R men. When CBS bought the American Record Company in 1938, it inherited offices and studios at 1776 and 1780 Broadway that producer John Hammond, Columbia Recording Company's first new employee, considered completely inadequate.[34] Even with the expertise of American Record's chief engineer, Vincent J. Liebler, a seasoned veteran of the recording industry since 1928 who became director of recording operations for Columbia, the American Record Company studio facilities could not compete with those of its chief competitors, RCA Victor and Decca. Victor had the East Twenty-Fourth Street studio, but in April 1927, they began leasing the grand ballroom of Liederkranz Hall on East Fifty-Eighth Street for some of their sessions. When Jack Kapp started Decca Records in 1934, he used the studio of the old Brunswick-Balke-Collender Building at 799 Seventh Avenue.[35] After Decca moved to 50 West Fifty-Seventh Street in 1938, Columbia moved into 799 Seventh Avenue, but Columbia president Edward Wallerstein considered these studios inadequate to serve Columbia's artists, so he struck a deal with World Broadcasting to use its studios at 711 Fifth Avenue. Relocation of studios was nothing new, and repurposing existing structures was the way labels set up their recording operations until the 1950s when the Capitol Records Tower in Hollywood became the first purpose-built studios in the American recording industry.[36]

In 1942, songwriter Johnny Mercer, music store owner Glenn Wallichs, and film producer Buddy DeSylva started a new label, Capitol Records, which became the first Los Angeles–based label to achieve success commensurate with major status and by 1945 had claimed sales just behind Decca.[37] Until 1949, the label had no studio of its own, so Capitol artists recorded at C. P. MacGregor Recording Studio during the first year, later moving to Radio Recorders, at that time the largest independent studio and record-processing facility on the West Coast.[38] Most of Capitol's records were made at Radio Recorders until the record company moved to its own studios on Melrose Avenue. By 1946, sales were so strong that the company decided to establish a New York studio to record and make master discs nearer to the pressing plant in Scranton, Pennsylvania. Clair Krepps, who had been a radio and radar specialist in the Navy during World War II, was designing radios for Westinghouse in 1946 when he received a Sunday night phone call from his former naval commander, Warren Birkenhead, asking Krepps if he would like to get into the record business. Birkenhead had just been hired as Capitol's chief engineer by Johnny Mercer, and his first assignment was to set up the New York studio. Birkenhead contacted Krepps and Lieutenant Oliver Summerlin, both of whom had been under his command during the war, thus as Krepps saw it, transferring the Navy's chain of command to the record business.[39] Both Krepps and Summerlin had extensive electronics experience, but like almost everyone who entered the recording business after the war, no experience making records.

Nevertheless, in 1946, Krepps and Summerlin set up Capitol's first New York studio, initially just a mastering room with a Scully recording lathe, amplifiers, and speakers, located in an office space on Broadway and Fifty-Seventh Street. Capitol in Los Angeles sent recordings to New York to be copied or remastered and then shipped to the pressing plant in Scranton, Pennsylvania. A quick turnaround was important in the event of a rejected pressing, which was a common occurrence at the time, and the Scranton Record Manufacturing Company, the oldest and one of the largest pressing plants in the country, had a substantial stock interest in Capitol and therefore devoted much of its production capabilities to the label, giving the company an edge in getting its product to market.[40] Krepps engineered live recording sessions via telephone line hookup. The artist performed in another studio or hall; his or her performance was miked and transmitted over equalized telephone lines to the mastering studio where Krepps cut the disc on the Scully lathe. This was how Nat "King" Cole, singing in the studio of radio station WMCA at 1657 Broadway, recorded "The Christmas Song," on June 14, 1946, one of the most popular records in recording history and one of the first recordings Krepps made.[41] Later, Capitol negotiated a deal to use the Pathé studios on Park Avenue and 106th Street, and the MGM studios on Fifth Avenue. Not until after 1949 did they move permanently into their own space at 151 West Forty-Sixth Street, the studios formerly occupied by Muzak.[42]

Columbia, Decca, RCA, and Capitol, along with other independent studios, continued the redesign and reuse of existing structures for recording into the late 1960s when RCA Victor built its massive recording complex at Rockefeller Center. It was not easy to find a location that fit the demands of a recording studio, particularly the need to prevent outside noise or vibrations from leaking in and, similarly, to make noise at all hours without disturbing neighbors. New York's skyrocketing real-estate values offered companies little alternative but to exploit the city's supply of old, solidly built, abandoned structures that proved to be ideal venues for recording with natural reverberation.

Temples, Churches, and Dance Halls: The Big "Natural" Sound

None of the labels' in-house studios were large enough to accommodate orchestras or the big bands of the era, so each company regularly used alternate venues for their large ensemble recording during the 1940s and 1950s. Decca used the Pythian Temple, an elaborate Egyptian-themed structure on West Seventieth Street built by the Knights of Pythias in 1926.[43] The windowless building held two ballrooms, the largest of which Decca and its subsidiary labels, Coral and Brunswick, used exclusively throughout the 1950s, recording everything from Bill Haley and the Comets' "Rock around the Clock" to Leroy Anderson's "Sleigh Ride." RCA Victor used Manhattan Center on West Thirty-Fourth Street, the 1906 opera house built by Oscar Hammerstein and later purchased by the Masons who added an acoustically exquisite ballroom, and Webster Hall on East Eleventh Street, a dance hall and meeting hall that had been the site of bohemian costume balls, society weddings, and a speakeasy during Prohibition.[44] These venues were ideal for symphonic and other large-ensemble recording sessions but required a certain amount of refitting. RCA engineers built a small control room off to the side of Webster Hall's mirrored ballroom for recording sessions during the week but closed this off on the weekends when it was an active dance hall. The use of these existing structures as recording studios had benefits and certain drawbacks. Some of these large rooms had natural acoustics conducive to recording symphonic music and big band jazz, but because they were not designed to be soundproof, engineers had to devise ways to minimize the effects of extraneous sound picked up by the recording equipment. RCA engineer Ray

Hall recalled that cooing pigeons caused problems at Manhattan Center and the sound of rain hitting the roof at Webster Hall could stop sessions in progress there. At Carnegie Hall, the rumble of the nearby Seventh Avenue subway line forced engineers to either halt the session or try to filter out low tones by using high-pass filters.[45]

One site appears to have been ideal from the beginning, and it came to symbolize the pinnacle of recording studio acoustics during the big band era. Liederkranz Hall on East Fifty-Eighth Street, the nineteenth-century home of the German singing society, the Liederkranz Club Chorus, was transformed during the 1920s into a recording and broadcasting studio used by both RCA Victor and Columbia.[46] Acoustically, Liederkranz earned a reputation as one of the finest recording spaces to exist in New York and became the most desirable recording room throughout the 1940s.[47] According to George Avakian, who after graduation from Yale and after four years in the Army headed A&R for Columbia's Popular Albums and International Departments, Liederkranz Hall signified the new emphasis on the sound of the studio, not just the music being recorded:

> By the time the music business really started to get big in terms of sales, the sound on recordings became important. The first really great studio for sound was the Liederkranz Hall . . . an L-shaped room inside a building owned by the Liederkranz Society which was a German singing society formed in the nineteenth century. . . . This particular room turned out to be marvelous for recording because it was old, solid wood, and its natural sound was quite terrific. Victor used it, but Columbia really made the greatest use of it and early 78 rpm recordings of the Columbia pop artists of the late 1930s and 1940s established that hall as the industry's standard for sound.[48]

Conductor Andre Kostelanetz made his first recordings there for RCA in the mid-1930s. Although he was initially unimpressed with the room's appearance, Kostelanetz soon recognized its superior sound. "To have played there is to be spoiled forever as far as acoustical standards are concerned," he recalled in his memoir. "One mike picked up everything." Kostelanetz conducted the Coca-Cola radio program from Liederkranz Hall for five years beginning in 1938 and claimed that his ear had grown so sensitive to the acoustical perfection of the room that he could tell just "by how the orchestra sounded on a given morning whether the floor had been swept the night before."[49] Don Plunkett recalled that Liederkranz "had such *wonderful* sound," in part because the spaciousness of the room, and the reverberant quality of the wood had the effect of giving a record more presence and seemingly more volume. Greater volume had become particularly desirable because companies wanted their popular records to stand out when they were played on the radio and on jukeboxes, often located in noisy public places. A record with presence would grab the listener's attention and thus be more likely to sell. Don Plunkett cited Les Brown and his Orchestra's "Joltin' Joe DiMaggio," recorded at Liederkranz Hall in 1941, as an example.[50] The record actually demonstrates both dead room and live room ambience since the lead vocalist sounds dry compared to the instruments, as if she sang in a vocal booth. But the brass section, trombone solo, drummer's rim shots mimicking the sound of a bat, and the male chorus, shouting "Joe, Joe DiMaggio, we want you on our side!" all have tremendous presence and reveal the spaciousness of the room. Plunkett pointed out that in those days it was important to achieve volume without exceeding groove dimensions.[51] The louder the record, the greater the excursion of the cutting needle and that meant wider grooves and less playing time. Consequently, a studio that could lend that quality of apparent loudness became extremely valued, both aesthetically and commercially, in the big band era.

By 1947, the success of the "Liederkranz sound" and the cumulative experience gathered in various recording studios, concert halls, and music rooms demonstrated that recording studio acoustics no longer followed the model of the broadcast studio. Although each studio had its highly individualistic setting, it seemed clear that the reverberation time of the studio should increase with the volume of music. This could be conceived as a simple formula: the bigger the band, the bigger the sound, the bigger the room needed to successfully convey it on a record. However, so many factors went into the reverberation characteristic of a studio, including the shape of the room and features of the surfaces, that size alone was not enough to determine the reverberation time. Moreover, because styles of popular music changed over time, flexibility was now considered important in studio design. Recognition that the recording studio was now a critical factor in a record's success meant that acoustical engineers became increasingly important to their design, but even they recognized the bottom line as the buying public:

> The excellence of a recording studio has but one criterion—the public's acceptance of its product. The studio is just a part of that path from artist to audience and audience back to artist. Along this line or loop are the musicians, the studio, the recording apparatus from mike to disc, the platinum ears, and finally the purchasing public's tastes. His tastes determine the popularity of the performer, pro-

mote the struggle to provide better studios and equipment, and plague the boys who pit their ears against his pocketbook.[52]

With its emphasis on the public's listening and *purchasing*, this statement could have been written by a record label executive, but its authors were two acoustical engineers with the Johns-Manville Company, manufacturers of architectural materials, addressing the annual meeting of the Acoustical Society of America in 1946 on "Recording Studio Design." Published in the society's journal the following spring, their essay revealed how sophisticated studio design had grown and how significant the studio was to the sound of records—as judged by consumers, not producers or engineers.

Fourteen years after NBC's Radio City studios first stood as the pinnacle of broadcast studio design, engineers remodeled Studio 3A to accommodate the recording of broadcast transcriptions and records for home use.[53] Since transcriptions were meant to recreate the feel of a live broadcast, they required the acoustical conditions employed in broadcast studios. For spoken word (comedy, drama, announcements), it was best to minimize reverberation to not indicate room size. However, for music, the room needed to be reverberant. Consequently, the acoustical treatment incorporated reflective surfaces as well as absorbent areas, movable panels, drapes, and polycylindrical surfaces, all of which could be adjusted to obtain "optimal acoustical results."[54] It was as if the studio, like the musicians' instruments, could now be tuned to meet the needs of a given recording session. Indeed, one thing had become clear to acoustical consultants and recording professionals, that the studio had become "the *final instrument* that is recorded."[55]

In the late 1940s, Columbia Records was a thriving label, but only one division of its parent corporation, the Columbia Broadcasting System, and television was a rising star in the postwar era. When CBS president William S. Paley made the fateful and apparently highly unpopular decision to transform Liederkranz Hall into television studios, the legendary sound was destroyed. According to musician Manny Albam, a coat of paint destroyed the hall's marvelous acoustics, and "that was the end of the studio. . . . Just killed the whole thing, just one coat of paint."[56] Columbia had the Seventh Avenue studios, but these were not sizable enough to accommodate the big recording dates, so Columbia engineers and recording directors canvassed Manhattan for a suitable replacement. They eventually found an ideal venue in an abandoned Greek Orthodox Church on East Thirtieth Street. Built on solid rock, with three layers of inch-thick maple and pine flooring providing a solid wood sounding board, Columbia's 30th Street Studio would eventually earn a reputation as "the 'Stradivarius' of recording studios," thereby bestowing on it the ultimate designation as a musical instrument.[57]

That reputation came only after a good bit of acoustical tweaking by Columbia engineers, who faced the challenge of transforming a vast empty space with impressive but unruly reverberation into a functional recording studio in which that reverberation could be controlled through technology. The biggest challenge engineers faced at 30th Street, according to William Savory, was that the acoustics of the room, measuring ninety-seven feet long, fifty-five feet wide, and fifty feet from floor to ceiling, had to be "brought into focus."[58] Savory described the reverberation as very good for some things, but too long. "As a result, if you were playing something staccato or rather rapid, it would tend to merge with everything else. A string of very distinct sixteenth notes would come back as a smear [because] you were *immersed* in reverberation."[59] Minimizing reverberation required engineers to place microphones as close to the source as possible to get more direct rather than reflected sound. This, however, made it difficult for musicians to judge how loud they should play. Members of the New York Philharmonic, for example, were accustomed to playing with their own dynamics, but as Savory recalled, "you sit them on this thing where they have microphones closer to them, they can't use those dynamics. They have to restrain themselves." To counteract that, at first the engineers tried flat baffles—upright partitions positioned in different areas of the room to break up or to redirect sound waves. Eventually, as he worked with recording directors to achieve the kind of sound they desired, Savory came up with an unusual design of studio baffle: eight-foot-tall parabolic reflectors placed on wheeled tripods so they could be easily repositioned.[60] Savory often put the reflectors behind the musicians so they were unaware of their presence. This gave the recording engineers more control over the sound, and the musicians a better listening environment but not all were pleased with what they heard. Some musicians, "especially the brass men," Savory recalled, "thought it was wonderful . . . it's like having your music under a magnifying glass." Placing the reflectors close to the musicians produced a more direct rather than reverberant sound; moving them back reduced the intimate presence. Some of the musicians thought it was strange, and one violinist told Savory, " 'This is going to make me go home and practice a hell of a lot more. I can hear all my mistakes!' "[61] Just as musicians began to adjust playing style after hearing themselves on record for the first time, technical fixes for acoustical adjustments in the studio caused musicians to change technique, another example of what musicologist and historian Mark Katz described as the "feedback loop" between recording and musical performance.[62]

The decades-old challenge to engineers of positioning musicians in a recording studio, once a matter of crowding and jockeying for position around a horn, had taken on new proportions with the increased sensitivity of microphones and the desirability of "live" rooms for recording. Thus, the problems of controlling sound did not vanish; they simply changed, forcing engineers to contrive new solutions. Because of this, many producers were initially reluctant to use the 30th Street Studio until Columbia Masterworks vice president Goddard Lieberson recorded the original Broadway cast album of Rodgers and Hammerstein's *South Pacific* there in 1949, just as the musical opened to rave reviews and record-setting advance sales.[63] Savory remembered Lieberson telling the engineers he wanted to make "a studio recording that doesn't sound like a studio recording, that doesn't sound like a radio broadcast, but it does sound like a Broadway stage" and for that, the 30th Street Studio proved ideal.[64] From the opening bars of "Bali Ha'i" to the booming male chorus and brass solos of "Bloody Mary," the sound of 30th Street came through loud and clear. It was as good if not better than Liederkranz Hall, and the results impressed other artists and producers who soon began using the studio. When Lieberson hired Mitch Miller to head Columbia's popular recordings department, Miller immediately appreciated the room's ambience. He produced his first Columbia recording session there in February 1950 with Rosemary Clooney and continued to favor the 30th Street Studio, which along with Liederkranz Hall, he considered "the best."[65] Miller apparently loved using the parabolic reflectors, especially with singer Johnnie Ray, who stood between two of the reflectors as he recorded, enhancing the natural reverberation and thus his vocal presence on a number of his recordings (see fig. 9), including his first big hit, "Cry."[66] Ray's exaggerated articulation, a "hyphenated style of singing" suggestive of doo-wop vocalists and early rock 'n' roll performance styles, coupled with the use of studio techniques to enhance his voice, served to create his unique sound.[67]

In the burgeoning popular music field of postwar America, a unique sound that differentiated one artist from another was becoming as important as the song. The studio's natural acoustics and its manipulation by human intervention became ever more important to the record's success just as the "sound"— its overall character and production value—took on greater significance. Certain studios were considered better than others for some styles of music, and producers chose studios on that basis. As one engineer noted, "when you're trying to record in popular music, you're not only trying to record certain instruments, but a certain feel, 'cause that's what sells records."[68] Although difficult to define, this "feel" was something engineers became sensitive to; some could even "tell who made what record, and where it was made" just by listening to it.[69] In recalling his work at Capitol Records' New York studio in the 1950s, Irv Joel said that any engineer who knew what he was doing "could take a record and listen to it and say, 'That was made at Forty-Sixth Street'. . . because of the microphone technique, because of the echo chamber [which was] a physical echo chamber at that time, and it was distinctive."[70] Studios, like artists, had unique characteristics that made them identifiable, and to some extent, those characteristics became imprinted on the work of the artists who recorded in them. In the 1940s and 1950s, the casual listener would be oblivious to the sound of the studio, but engineers and producers, and no doubt the growing contingent of audiophiles, could tell the difference between Columbia's 30th Street and Capitol's Forty-Sixth Street studios and between natural and artificial reverberation. Soon, record buyers would begin to discern more subtleties in recording quality, aided by higher-quality records and more sophisticated home reproduction equipment, and in the late 1950s, stereophonic sound.

Mimicking the Hall: Spring Reverb, "Echo" Chambers, EMT Plates

Not every studio possessed naturally good acoustics, so engineers devised means of compensating for this, often borrowing from the ideas employed in radio and motion pictures. Artificial echo had been used in broadcasting and film sound since the 1930s to create specific dramatic effects.[71] In attempting to achieve certain sound effects for radio programs, engineers cobbled together unlikely combinations of devices. When guitarist Les Paul performed on radio programs in the 1930s, he witnessed two Chicago radio engineers serendipitously invent a makeshift echo effect while attempting to create the sound of thunder for a soap opera. By removing the needle from a phonograph pickup, inserting a long spring, and then hitting the spring with a mallet, an engineer at WBBM created the crashing sound of thunder. As Paul recalled, the experiment didn't end there: "Now his assistant comes along and says, 'Well what would happen if you put a pickup on the other end of it, of the spring? You got a 'send' on one end, a 'receive' on the other—in other words you're driving the spring with a signal. They did it and to their amazement, they had echo."[72]

This type of echo effect, usually called a spring reverb, composed an electromechanical system that was later incorporated in the Hammond organ, invented by Laurens Hammond in 1935.[73] The terms echo and reverb have been used interchangeably and technically are based on the same acoustical phenomena: the repetition of a sound as it bounces off a hard surface. An echo is an audibly distinct repetition that rapidly diminishes, whereas reverberation is many repetitions of an audio signal so close together as to be indistinguishable, thus sounding like a gradual decay. Each evokes a different sense of place and emotional response in the listener, but both conjure up the sense of physical space.[74] Conventional acoustic means of achieving echo and reverberation involved redirecting a signal through a chamber. The signal was then picked up by a microphone and fed back to the control room. The delayed signal could then be combined with the natural voice or used alone. NBC engineers experimented with various forms of this, including pipes of various lengths to achieve different periods of delay.[75] By the late 1930s, engineers experimented with electronic means of creating reverberation using steel magnetic tape.[76]

As producers and artists sought to achieve the reverberant quality of natural acoustics on their records, engineers devised ways to mimic the Big Hall sound, and vast physical space ceased to be an absolute prerequisite. As early as 1937, eccentric bandleader Raymond Scott had achieved what one listener called "a big auditorium sound" on his records, simply by placing microphones in the hallway and in the men's room outside his record company's office, which also served as the label's recording studio.[77] The hard surface of ceramic tile provided the ideal level of reflectivity to produce the desired reverberation. Chicago engineer Bill Putnam used the same idea a decade later when recording The Harmonicats' version of "Peg o' My Heart" at his Universal Recording studio in Chicago but with a more exaggerated effect, making the harmonicas sound drenched in reverberation.[78] In New York, Columbia engineers at the 799 Seventh Avenue studio used the building's stairwell, a seven-story steel and concrete chamber that worked so well they never changed it. According to Bill Savory, Columbia's famous "stairwell echo chamber" was the brainchild of chief engineer William Bachman: "It started on the seventh floor with a microphone, and with a dual cone fifteen-inch loudspeaker on a landing one-half floor down the stairs. Below that, six and one-half floors of concrete and steel— *sensational!*"[79] The only drawback was that building maintenance workers were inclined to walk between floors rather than use the elevator. Engineers had to caution them to not let the doors slam, to walk quietly and to not kick anything, and to stop whistling while climbing the steps. Such makeshift echo chambers, sensational or not, did have their drawbacks.

Many studios custom-built their own echo chambers, as had NBC in the

1930s; consequently, the chamber became one of the more identifiable qualities of a studio.[80] Capitol Records studios, both in Los Angeles and in New York, were known for their distinctive echo chambers. John Palladino was an engineer for Radio Recorders, where Capitol made most of its Hollywood recordings until it moved to its own studios on Melrose Avenue. Radio Recorders built an echo chamber on the roof of its building shortly after Palladino started working there in 1943, at a time when, as he explained, "nobody knew too much about how it should be shaped or anything. But you just worked with it a little bit and put the hardest plaster you could get on there and put a good mike and a good speaker— and that was it."[81] Irv Joel recalled that the New York studio had "a wonderful sounding echo chamber" when he joined Capitol in 1955. The chamber had apparently been built by an Italian contractor, who, Joel recalled, upon learning about the desired project said, "'Oh! I know what you want; you want one of them cockeyed rooms.'"[82] The contractor, who apparently had experience in building echo chambers, brought in broken pieces of cinder blocks and other materials from his truck and situated them to form a ramp on the floor, which he covered with concrete. He built a new wall at an angle at the end of the space, covered all surfaces of the entire room with a skin coat of very fine cement, and after everything had dried shellacked the walls to make it even brighter sounding.[83] Joel reported that the chamber performed beautifully and that some clients even came to use the studio specifically because they liked the sound of Capitol's echo chamber.[84] Sometime later they hired a "famous acoustical consultant" to design and construct a second chamber, but "after a lot of tuning and such we turned that space into a tape vault and bought [an] EMT plate to replace it."[85] Apparently, the professional acoustics expert could not engineer an echo chamber that compared to the Italian contractor's cobbled creation.

By the late 1950s, numerous papers describing techniques and devices to achieve artificial reverberation appeared in technical journals.[86] Soon companies began to market affordable units that offered controlled reverberation, touting them as space-saving, good investments, and capable of achieving classic studio sounds. The Audio Instrument Company of New York advertised an electronic echo chamber that occupied "2 instead of 10,000 cubic feet," and at $1,485 promised to "pay for itself in 3 to 5 months."[87] In 1965, Fairchild Recording Equipment Corporation proudly sold its "Reverbertron" as a compact means of achieving the classic Liederkranz Hall sound: 'Famous for REVERBERA- TION . . .' declared the advertisement's bold heading, along with an eye-catching photograph of the stately facade of the nineteenth-century building.

For years Liederkranz Hall was world renowned for its remarkable acoustic effects and consequently it was in constant demand for recording. But even Liederkranz Hall had its limitations! Engineers could not always control the reverberation quality and time. However if you wanted to record in Liederkranz Hall today it would be impossible because, as with most old landmarks, it's destined for destruction. But . . . don't fret, don't worry! There's a much more practical, effective, and less expensive method to add controlled reverberation to your sound. Now reverberation comes in a compact, portable attractive and rack mountable package 24 1/2″ high by 19″ wide in . . . THE FAIRCHILD REBERBERTRON.[88]

Bridging the gap between the past and the future of sound manipulation in recording, these devices promised affordability, portability, and controllability, key features that appealed to the many small studios popping up in the 1950s and 1960s. However, these ads also represented a growing trend in selling studio equipment, a rapidly expanding market during the 1960s. By drawing comparisons to existing studios, they were both conferring legitimacy on new, untried technology and, in effect, treating the sound of a studio as something that could be packaged and sold.

One device in particular became a popular electronic means of achieving reverberation. The German-made EMT 140 Reverberation Set, developed by Dr. Walther Kühl in 1953, consisted of a large steel plate, one meter high by two meters long, which was hinged in an acoustically dead wooden frame. When it was excited by a loudspeaker-like device mechanically coupled to it, it generated bending sound waves.[89] The sound, with a reverberation time adjustable between one and five seconds, was then picked up by a microphone and routed to the control room. The "EMT plate," after the company's name, *Elektromesstechnik*, proved to be an efficient source of reverberation and echo because of its portability, tunability, and size. Although it was not easy to move, it was more portable than a chamber, and for studios with limited space, it offered the only option. Many engineers and musical clients did not immediately embrace this new form of electronic reverberation. Co-owner of A&R Recording Studio, Phil Ramone, acquired an EMT plate from Harvey Radio, New York's major distributor of professional and consumer audio in the early 1960s. Ramone spent days tuning the plate: "Tuned it, and tuned it, and one day I hit on it and we did a date," Ramone recalled. "And the brass and the strings sounded incredible!" Despite having converted the ladies' room at A&R into a live chamber, Ramone found the EMT "head over heels better." Still, live echo chambers continued to have cachet with recording customers, so a friend built a large cabinet around the EMT, which sat in the fourth floor studio, and when people asked where the chamber was, he told them it was in the basement.[90] Although the EMT plate afforded more control over reverberation than busy stairwells, bathrooms, and hallways, the initial resistance to its use points to the reason Fairchild evoked the name of Liederkranz in its ad campaign for the Reverbertron. Engineers, producers, and artists wanted to know they would get the desired sound from a trusted source. How could these new devices compare to the live chambers that helped make so many hit records? Would this electronic device sound natural? Whether it was based on resistance to new technology or fear of losing authenticity, Ramone managed to dispel the doubts. By camouflaging his EMTs, he conducted what amounted to a latter-day "tone test," and apparently, his clients were none the wiser.

Given the popularity of the echo chamber, it is not surprising that experimentation went on in studios everywhere. Another popular means of achieving echo—true echo as opposed to reverberation—used two tape recorders and came into use after the introduction of the first professional magnetic tape recorders. Referred to as "tape slap," or "slap-back," this echo operated on the same principle as the other methods, but in this case, the time it took for the tape to travel between the record and playback heads introduced the split-second delay.[91] Les Paul used this in the late 1940s on his own multiple recordings. In Memphis, Tennessee, Sam Phillips used it to great effect, recording the first Elvis Presley and Jerry Lee Lewis records, creating a distinctive sound that became known as the "Sun Sound" after the label, Sun Records. Tape slap did not create the effect considered desirable on major label pop records; it produced a distinct echo rather than the reverberant quality of a concert hall, and it worked best on small bands or singers, not orchestras or big bands. It was also a more affordable special effect for small studios that may not have the money to buy an EMT plate or the space to construct a chamber. When in January 1956 Elvis Presley made his first RCA Victor recordings at the company's brand-new Nashville studio, according to Elvis biographer Peter Guralnick, the engineers there "professed ignorance as to just how Sam Phillips had achieved" the slap-back echo that was so integral to Elvis's appeal.[92]

Rock 'n' roll producers such as Sam Phillips and Lee Hazlewood, who created guitarist Duane Eddy's "million dollar twang" by using a cast-iron grain storage tank as an echo chamber, were not trying to create the big hall sound. Their recording techniques became part of the "sound" of rockabilly and rock 'n' roll, styles that appealed to a new generation of listeners asserting their own identity and musical tastes. Born of necessity and limited technology, these techniques were as different from the production quality on major label albums of the postwar era as the music itself. These musical mavericks were not the first to play around with sound in the studio; decades earlier, radio engineers, a classical conductor, and a pioneer of easy listening were experimenting with generating otherworldly sonic environments in the studio.

"Music was not what was played . . . but rather what people heard": Adventures in Microphoning

At the same time that room acoustics became important in the recording studio, the microphone became an instrument to be exploited in its own way. Conductors Leopold Stokowski and Andre Kostelanetz both experimented with sound in the 1930s. Stokowski collaborated with Bell Lab engineers to improve the fidelity of sound transmission and reproduction, and he was the first musician to recognize publicly (and to vehemently protest) that the control engineer was "the conductor" in the broadcast studio.[93] The experiments Kostelanetz conducted with microphone technique explored new sonic realms that had up to then been considered undesirable distortions of the original performance, at least in serious music. While conducting the CBS Orchestra for the Chesterfield radio program, Kostelanetz realized that what radio listeners *heard*--"that is, what went through the studio mike into the control room and out over the air"—was markedly different from what the orchestra *played* in the studio.[94] Kostelanetz made this "proud and surprising discovery" when he decided during a rehearsal to listen from the control room rather than from the studio so that he could hear for himself what happened to the music after it passed through the microphone:

> I gave the baton to Berezowsky and asked him to go through a few phrases we had just done. It was immediately clear that a lot happened. I realized that equally important with the seating of musicians and with carefully marking scores for bowings and other subtleties of interpretation was the placement of the microphone. Some instruments were virtually lost; others came through several sizes too big, so to speak. One of the first corrections was to hang the mike above the whole group of musicians, not favoring any section.[95]

Facing the age-old problem of placement of musicians, which before 1925 involved jockeying for space around the recording horn but now posed problems with the more sensitive microphone, Kostelanetz and the engineers soon realized that each piece of music needed its own treatment. They began using more than one microphone, placing at least one near the string section. Because strings are "aurally transparent," that is, they produce a distinct sound without dominating other instruments, miking them directly allowed other sounds to pass through and be picked up. Experimenting further with microphone and instrument placement, Kostelanetz found that he could create completely new sounds, giving the music a unique identity:

> Colors began to come through that were not there before, because certain instruments combined to make whole new sounds—the blending of flute and horn, for instance. We placed one mike near a trombone, to which a special mute had been attached, another close to an alto flute, and another next to a viola. I would have the trombone played very high and soft, and when the three sounds were mixed in the control room they became a new sonority. No one who wasn't in the studio could say what instrument or instruments they were listening to. Not even Arturo Toscanini.[96]

Referring to the renowned conductor who was famous for his extremely astute hearing, Kostelanetz was making the point that this blending of instruments through the studio microphone rendered them collectively like a completely new instrument. On another occasion, Kostelanetz asked his orchestra to play the beginning of "Claire de Lune" more softly than usual while in the control room the engineer gradually increased the volume, and to Kostelanetz it sounded "like listening to the breathing of angels!" During another rehearsal, the operatic soprano Rosa Ponselle sang so powerfully she managed to overwhelm the sound of the orchestra. Reluctant to ask the diva to restrain her performance, Kostelanetz instead watched as the engineer brought the volume down, controlling her voice "in such a way that the louder she sang, the softer it came through, and vice versa." A similar situation occurred during the broadcast rehearsal of Ravel's *Bolero*. Not knowing the structure of the piece, which begins almost inaudibly quiet and builds in a long, slow crescendo, the engineer felt it was too soft so he increased the volume at the beginning, only to have to bring it way down as the orchestra grew in volume, thereby reversing completely the slow-working crescendo effect of the music. In both of these instances, Kostelanetz concluded, as had Stokowski, that it was not clear who was really in control.[97]

These examples from the broadcast studio presaged later microphone experiments in recording, but the same principle applied. When sound is converted into an electrical signal via the microphone, then back to sound through the loudspeaker, not only does an inherent coloration, or distortion, take place, depending on the characteristics of the microphone as well as the other controls and the speaker, but that coloration as Kostelanetz found, could also be manipulated at will by methods ranging from microphone choice and placement to signal processing with equalizers and compressors. Moreover, the approach each of these conductors adopted toward the use of technology in sound recording and reproduction inspired others who followed, for whom recording, rather than live performance, became the dominant medium. Mitch Miller, who worked with both conductors as an oboist and went on to become the first popular record producer credited with "inventing" singers by the use of recording methods to give them an identifiable sound, believed that both Stokowski and Kostelanetz "knew more about sound and microphones than anyone. I learned plenty from them."[98]

Producers and Engineers: Collaboration and Skillful Manipulation

Miller also learned from paying careful attention to how the recording director and engineer worked when he played oboe in recording sessions during the 1930s and early '40s. At the time their job, like the recording studio itself, was largely functional: to capture the performance by judicious microphone placement and careful setting of levels, and to make sure no technical mishap occurred that would result in the need to re-record. Miller recalled a recording session on which he was oboist:

> I was in the room playing, but I can tell you what they did! They set a level, and they looked for peaks, while you were rehearsing. And then the producer would be listening for balance and all that . . . but producers then worked only to try to make it comfortable. There was no producer I know who came along and said, "Why don't you try [it] this way?"—like to improve what they played. They were just [there to] get it down.[99]

Although rock producers of the 1960s have been cited as the first to create "studio sounds," notably Phil Spector and the "wall of sound," Mitch Miller, working years earlier, was arguably the first to change the producer's role in popular recording from the A&R man, who matched the "artist" with the "repertoire," to someone who actively shaped the sound of the record. Historically, records sold on the strength of a song's popularity, not the performer's reputation, a fact that dated back to late-nineteenth-century origins of the business of popular song and music publishing.[100] Records were "little more than sheet music on shellac," and the performers who sold records based on their name recognition had achieved that through radio airplay, live performances, or film appearances.[101] Departing from standard practice, Miller believed that the record not only should document an artist's live stage show but also should offer the listener something unique, something they could only experience by listening to the record. A classically trained musician with a pop sensibility, Miller has been derided for some of his later musical choices, but by perceiving recording as a new medium to be exploited, even resorting to gimmicks to make his records stand out or to convey a particular emotion, he was ahead of his time. He was the first to use harpsichord, French horns, and English horns on pop records, and he tried to build excitement into everything he produced. He did so, according to engineers who worked with him, "by driving the arranger, the musicians who played it, and usually the singer."[102] Apparently, he also could be extremely demanding of engineers. In a 1953 *New Yorker* profile, Robert Rice witnessed Miller in action during a Frankie Laine recording session, shouting out directions on how to balance instruments as he hovered over engineer Tony Janak in the control room: "Where's the organ? That's too much. A little less echo. Now the chorus comes up. Now Frank. Help him out on those low notes, for God's sake. More guitar. More guitar. Damn it, I said more guitar!"[103] By the time Miller screamed, "Money!" when he heard the right mix, Janak was exhausted.

Miller's career as producer began when in 1947 John Hammond hired him to head the classical department of Keynote Records, a new independent company that specialized in jazz and folk music. When Keynote merged with Mercury Records, Hammond put Miller in charge of Mercury's popular recordings, relocating the company's recording activity from Chicago to New York.[104] Like nearly all the new small labels that emerged after the end of the war, Mercury did not have its own studios, so they rented the Majestic Records studios, and the services of chief engineer Bob Fine. In Fine, Miller found a willing and extremely capable collaborator in his quest for new and unique sounds. One of the first records they made together was "Again," sung by Vic Damone. At that point, Miller knew little about recording technique, but he knew he did not want to make a record like those made in the 1930s where every vocalist sounded, in his words, "like they're singing through a hunk of wool." So when they recorded Damone, Miller recalled, he asked Bob Fine how he could "put a halo around the voice." Describing an imagined sound in the 1940s, before the sound being described had become commonplace, and long before any shared language had developed, must have proved challenging.[105] Fine apparently knew immediately what Miller was after and used the bathroom echo chamber technique, which satisfied Miller's attempt to add ambience to Vic Damone's vocal track, and the result drew the curiosity of other record companies. "We would just confuse the competition—they didn't know what was going on. They were trying to figure out what mikes we used," when, in fact, Miller said, the real trick had to do with knowing how to exploit "aural phenomena," the use of electronics in getting the sound to, as he put it, "reach the consciousness of the listener." A perfect example of this, he said, was the Damone session for Mercury. During the recording, musicians in the studio included eight violins, three violas, two cellos, a bass, and woodwinds. One of the violinists thought things just didn't sound right in the studio. Miller recalled, "Sascha's playing . . . he's looking around, he's shaking his head, he comes in and he listens to a playback. He says, 'Out there is nothing!' He says, 'I come in here it's the New York Philharmonic with Caruso!'" What sounded weak and lifeless in the studio, because the musicians were separated by gobos (baffles) to isolate their microphones, came through the control room monitors as something completely transformed. Just as Kostelanetz had discovered during his Coca-Cola radio show days, Miller saw the importance of "doing everything for the listener's ear. On that part of the aural phenomena," he noted, "there is more phenomena than anything."[106] Although the music was created in the studio, the transformation took place in the control room.

With improved sounding records and increased emphasis on live rooms or other means of achieving reverberation, by the 1950s, the era of the dead studio had ended and the importance of the control engineer in recording was on the rise. Kostelanetz knew it and included the engineers in his memoirs by referring always to "our experiments" and "our discovery." Stokowski was so convinced of the engineer's importance in the sound of a musical performance that he demanded his own controls for his NBC broadcasts.[107] Mitch Miller considered the engineer "the strongest link in the chain," and was among the first to suggest they receive credit on albums, although company executives balked at this suggestion. Engineering credits on record albums did not become common until the 1960s.[108]

By 1947, it had become evident, at least to those in the recording industry, that the recording engineer played an important part not only in getting a technically good recording but also in enhancing the sound of the record and in some cases the natural ability of the performer. One of the first official definitions of the job of "Recording Engineer," published in the American Industries Series in 1947, emphasized the value placed on skill due to the "ever-increasing quality standards within the industry." In addition to technical expertise, the best in the profession also possessed considerable knowledge of music, "since the finest of performances can be hopelessly distorted if the recording engineers are less than entirely successful." All this was not much different from what had been expected of recording engineers since the 1890s, but something new had crept into the engineer's job description: "On the other hand, performers whose popularity exceeds their abilities can often be materially assisted by skillful recording engineers, who, by exercising good judgment in microphone placement, recording characteristics and level, and similar factors, can make the resulting records sound far better than the singer ever did in real life."[109]

According to this definition, skilled engineers could work magic in the studio, but similarly, they could ruin a good take. Recalling his days recording Charlie Parker and others in the 1940s, producer Teddy Reig claimed that, if the engineer was not appropriately "pampered," an engineer "could destroy . . . in ten seconds"[110] the sound an artist spent a lifetime developing. Although few engineers expected to be pampered, Reig's claim speaks to the increasing power musicians and producers attributed to them. In 1947, recording engineers had limited technology to assist them. They could choose the type of microphone and where to place it to balance the musicians, use equalization to compensate for differing microphone quality or compression to make broad dynamic ranges "fit" into the grooves of a record, and of course, try various means of achieving reverberation, which was most important in aiding the singers "whose popularity exceeded their ability."[111] Recording engineering was still as much an art as it was a technical skill. Like the film industry's monitor man, the control man of the recording industry needed to possess "the brain of an engineer, but the heart of an artist."[112]

Conclusion

In 1952, *Newsweek* profiled "the men behind the microphones," emphasizing the role played by recording engineers in the studio. The improved sound of records, the author pointed out, was due to "engineering cleverness" and the technological revolution of the previous five years, which had come about from the LP, Vinylite plastic, and tape recording. Now it was possible not only for recording to more faithfully "represent music as it is played—or, better yet, as it should be played," but for records to go beyond the live performance, to become a "new medium" in which not only the recording engineer but also the listener had more control over what was heard.[113] Between the new ambient rooms and improved recording and reproduction technologies, records in the era of "high fidelity" sounded more spectacular, more vivid, closer than ever to the long-strived-for concert hall realism that Bell Labs engineer Joseph P. Maxfield first described with the introduction of electrical recording. Or did they? As the article pointed out, high fidelity proved to be anything but faithful to a live performance. Despite company advertising rhetoric, high fidelity was a sonic illusion and would only become more so over the next two decades. This achievement rested not only on new technology and live rooms but also on the engineer's skillful manipulation of the controls. By 1952, recording engineers acquired new tools and techniques, lending them even more control over the recording and the potential to make more creative use of those tools.

Control Men in Technological Transition

Engineering the Performance in the Age of High Fidelity

*A*fter World War II, a revolution took place in the recording studio. In the spring of 1948, the first high-quality magnetic tape recorders entered broadcast and recording studios, and Columbia introduced the 33^1/$_3$ rpm long-playing microgroove record (LP). In 1952, binaural
recording and reproduction systems were demonstrated at the Audio Fair, and
RCA Victor began experimenting with stereo recording sessions in 1954.[1] Before
the decade was out, stereo recording and reproduction had become a reality.
These three innovations—tape, the LP, and stereo—have long been recognized
as the technological foundation of "high fidelity" records for consumers, but
their effect on studio recording was nothing short of revolutionary. With tape,
recording engineers now had more control during recording, and the ability to
manipulate the recording *after* it was made. Through editing and re-recording,
engineers could now alter any part of a recorded performance without sacrificing the entire original because the recording medium was both editable and
reusable. No longer were musicians required to perform flawlessly from start to
finish; now mistakes could be erased, edited out, and new parts overdubbed with
relative ease. The length of tape on a reel allowed for many minutes of recording
time, and since the LP could accommodate up to twenty minutes of music per
side, songs, at least on albums, were no longer limited to four minutes. Radio
airplay still required standard length tunes, and when RCA Victor introduced
the 45 rpm, seven-inch-single format in 1949, three minutes became the ideal
song length.[2] Recording in 2-channel, then 3-channel stereo, engineers were

forced to re-conceptualize the spatial arrangement of instruments and voices in the studio and the balance in the recording. Increasingly artists and recording directors looked to the man in the control booth as the authority on what was or was not possible in the studio. The emphasis shifted from the studio where the musicians played to the control room, now considered the "nerve center of the recording operation."[3] As the distinctive "sound" of individual records assumed greater importance, the role of the recording engineer expanded beyond "technical expert" to in some cases artistic collaborator. More than ever, recording required teamwork among musicians, producers, and engineers.

From Disc to Tape

Until 1948, professional recording was made directly to wax or lacquer-coated discs. The quality of the lacquer-coated aluminum disc, commonly but erroneously referred to as an "acetate," improved during the 1930s, and by World War II lacquer discs replaced the wax cake as the original recording medium.[4] Wartime restrictions necessitated substituting glass for aluminum, and a shellac shortage led to severe cutbacks on the output of phonograph records and radio transcriptions, with output reduced to zero by December 1942.[5] The Armed Forces V-Disc program continued to produce records on Vinylite, a flexible and nearly unbreakable material superior to shellac, which had already been used for radio transcriptions and would later replace shellac in commercial records.[6] Discs were capable of reproducing a wider range of frequencies than ever before but still limited in terms of playing time and offered no practical means to correct flaws without re-recording and thus sacrificing what otherwise may have been a fine performance. Tape recorders and the long-playing record promised to overcome both of these problems.

Both magnetic tape and wire recorders were used during World War II in military as well as civilian applications. In the United States, Bell Labs in New York, the Armour Research Foundation in Chicago, and Brush Development Company in Cleveland all carried out research and development on magnetic recording.[7] By 1945, Brush produced the first commercial tape recorder in the United States, the Model BK-401 Soundmirror. This tape recorder was not without problems, however. When editing "The British Crisis" for CBS radio in September 1947, engineer Joel Tall used two Brush BK-401 Soundmirrors, but found them so noisy he had to remove the power pack from one of the units and place it eight feet away to reduce the noise to barely audible levels.[8] By 1947, it was clear to audio engineers that magnetic tape recording and reproducing systems held great promise because of the ability to edit, erase and reuse tape. None of these early recorders, however, could approach the sound quality of existing disc recording equipment, and it was believed that considerable design and development work needed to be done on both recorder and tape to achieve superior sound quality in magnetic recording.[9]

In 1947, a group of engineers in California were conducting such work, and a prototype recorder, based on the German Magnetophon discovered by US servicemen after World War II, had reached the final testing stage.[10] Ampex Electric Corporation succeeded in producing the first studio quality tape recorders in the United States and were quickly embraced by the entertainment industry. The first large installation of the Ampex Model 200A recorders took place at ABC studios in Chicago, New York, and Hollywood in April 1948. The Ampex recorders operated continuously from April through September, approximately seventeen hours per day recording programs for delayed broadcast. The Bing Crosby Show became the first radio program taped for later broadcast. Crosby had long sought a means of recording and editing his programs that surpassed the disc-to-disc method. Dubbing, as it was called, was a laborious process that required re-recording a previously recorded program while editing out unwanted portions. After he and his production staff heard a preview demonstration of the Ampex machines, Crosby paid in advance for the first twenty production Model 200 tape recorders and Crosby Enterprises became the Western representatives for Ampex. The Ampex Model 200 was marketed as "the machine that put Bing Crosby on tape," but its appeal was based on a record of rugged reliability, sound quality, and ease of editing, all of which appealed to recording engineers.[11]

Quickly, word about the Ampex machines' performance spread throughout the recording industry.[12] Capitol Records was the first label to order Ampex equipment, soon followed by RCA Victor, Columbia, Decca, and MGM.[13] Tape recording seemed a godsend, for it solved many of the problems associated with disc recording. It was capable of higher fidelity than disc recording because frequency response was not limited by the inertia of mechanical parts, dynamic range was not limited by the dimensions of the groove, and surface noise and needle scratch were eliminated from the original recording.[14] Initially, record companies continued to record original masters on disc, using tape as a backup, as they were unwilling to rely on an unproven technology as the primary recording medium.[15] This conservatism was well placed. Despite claims of durability and reliability, as with any new product, many technical problems needed to be worked out in the design of these Ampex machines and in the composition of the tape. One problem was tape "print-through," in which the signal from one layer of tape transfers to the adjacent layer of tape on the reel.[16] Another problem involved the tape transport speed, which was initially the same as the German Magnetophon, thirty inches per second. This resulted in higher sound quality but only about fifteen minutes of recording time per reel of tape. Moreover, since editing involved considerable movement of the tape back and forth across the recording heads to locate precisely an edit point, a process called "scrubbing," over time this created excessive wear on recording heads, which resulted in loss of fidelity, requiring Ampex to make continual improvements.

Ampex soon had plenty of competition. Fairchild Camera and Instrument Corporation, one of several companies founded by millionaire inventor and entrepreneur Sherman M. Fairchild, first introduced a quality machine that operated at fifteen inches per second, and Columbia Records engineers used this rather than the Ampex recorder to make early LP masters.[17] Other companies in the professional magnetic recording field included Rangertone, Magnecord, Presto, and RCA, and in 1949, Ampex produced the improved Model 300, a smaller, less costly version of the Model 200 with improved head design and tape quality, and operating speed of fifteen inches per second. This recorder, along with the subsequent Model 350, became the preferred studio recorders for years to come.[18] By 1950, every major recording studio had converted to tape as its primary recording medium.

The introduction of magnetic tape ushered in exciting new possibilities for recording and thus directly affected those who made records. At the same time, the $33^1/_3$ rpm microgroove long-playing record offered the classical music lover the possibility of uninterrupted symphonic listening, the jazz lover could hear the kind of extended solo improvisation work until then enjoyed only in live performances, and the popular record buyer could hear more songs from a given artist than would have been released in the single-record format. The LP offered consumers more musical choices; however, the long-playing record also had a profound and much-anticipated effect on performers, composers, and on the work of engineers and producers.

Record Length, Time Limitations, and the LP

Since Edison, recordists had attempted to lengthen the playing time on records, but until 1948, the standard 78 rpm record measured ten inches in diameter and held approximately three minutes of music.[19] Symphonic works had to be recorded on multiple discs and released as multi-disc albums, with annoying breaks between and often within a movement. Even popular artists deplored the time restrictions. Although most musicians accepted time limits as a necessary part of making records, these restrictions could cause even the most experienced artist frustration. Producer Teddy Reig recalled that time was the most important consideration in making 78s, "the worst thing was watching that clock run down to about 2:30 or 2:45; you never knew if they were going to finish on time. I used to have cardiac arrest during that last fifteen seconds." Reig found this especially true with the real giants, such as Benny Goodman, Count Basie, and Benny Carter, whose sense of tempo was so consistent they seemed to Reig to possess "built in metronomes. No matter how many takes you did, they would never vary more than four or five seconds."[20] Some performers refused to change tempo for artistic reasons. When Frank Sinatra recorded "Body and Soul" in 1947, his producer wanted him to speed up the tempo to remain within the desired three minutes. Sinatra refused, insisting that any change of tempo would "kill the feeling."[21] Rosa Ponselle declared she would have taken more liberties to express emotion in her performance without the cuts necessary to stay within the time restriction.[22]

Artists mostly accepted the time limitations out of necessity, not by preference. Many deemed it an impediment to creative expression. In his memoir of the Jazz Age, Ralph Berton recalled that musicians such as Bix Beiderbecke hated the three-minute time restrictions of early ten-inch 78 rpm records because there was no room for improvisation—the essence of jazz. "The soloist who, like Bix, liked to 'stretch out' in eight or more successive choruses," Berton declared, "was severely hampered by this limitation, which might be likened to having to make love on an escalator, finishing by the time you reach the top."[23] In April 1938, Commodore Music Shop owner and record producer Milt Gabler assembled a group of musicians in the Brunswick Records studio in New York to record a jam session on twelve-inch discs, rather than the usual ten-inch format for jukebox play, offering the musicians an additional two minutes of playing time. Months before, Gabler had started Commodore Records, the first independent jazz label in America, to record some of the lesser-known jazz players that the major labels overlooked or songs that they rejected. It was Gabler who recorded Billie Holiday's "Strange Fruit" when her label, Columbia, refused because of the song's controversial subject, lynching.[24] Gabler's recording experiment, documented in a 1938 *Life* magazine photographic feature devoted to the growing popularity of swing music, gave the players room to improvise, to record unwritten songs that evolved from musical ideas sketched in on the spot. Revealing how novel improvisation must have been to *Life*'s readers, the article

Figure 1 In the Music Room, Building 5, Thomas Edison's West Orange, New Jersey, Laboratory, 1905. *Left to right*, Albert Kipfer, A. T. E. Wangemann, and George Boehme make experimental recordings, testing different shapes and sizes of recording horns. *Courtesy, Thomas Edison National Historical Park.*

Figure 2 Recording at the Edison Studio, 79 Fifth Avenue, New York City, March 30, 1916. Conductor Cesare Sodero (*left*, standing on stool) was Edison's music director from 1914 to 1925. Metropolitan Opera tenor Jacques Urlus sings into the horn protruding from the curtain as he reads the music hanging from one of several music holders while the musicians play in very cramped quarters, some with eyes on Sodero, others on their music. The unseen recordist was somewhere behind the curtain. *Courtesy, Thomas Edison National Historical Park.*

Figure 3 A Western Electric operator sits at the forerunner of the modern mixing console. Here he adjusts the amplifier system "to obtain the correct loudness" during a Columbia Phonograph Company recording session in 1925. *Source: The Scientific Monthly* 21 (Jan. 1926): 76.

Figure 4 "Jam Session at Commodore." On March 23, 1940, photographer Charles Peterson documented a recording session for Milt Gabler's Commodore label at the Decca studio on West Fifty-Seventh Street. The session resulted in a seventeen-minute version of "A Good Man Is Hard to Find" spread over four sides. Pictured here: Artie Shapiro on bass; Joe Marsala, Eddie Condon on guitar, Jess Stacy on piano; and George Wettling on drums. *Courtesy, The Charles Peterson Jazz Photo Collection.*

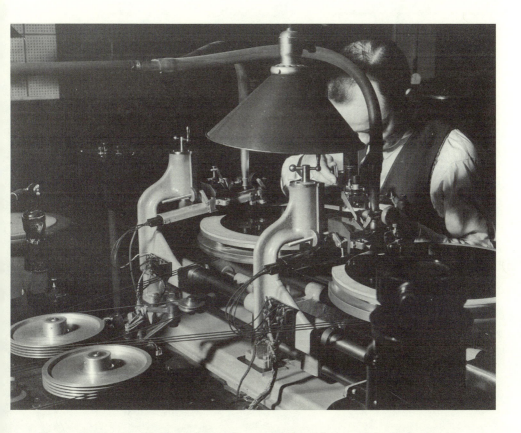

Figure 5 Engineer Cecil Bridges operates the cutting lathes at Brunswick Studio for the Commodore recording session on April 30, 1938. The large tube running across the top of the photo and bifurcating down to each lathe suctioned the lacquer chip away from the disc as it was being cut. The gravity-fed gears governed speed more precisely than inconsistent electric motors. *Courtesy, The Charles Peterson Jazz Photo Collection.*

Figure 6 These pages from *The Phonograph and How to Use It* (New York: National Phonograph Company, 1900) show the Edison Spring Motor Phonograph (*left*), Index of Parts (*below*), and Instructions for Operating the Edison Spring Motor Phonograph (*opposite*). As the index of parts and instructions for making a record show, home recording in the early twentieth century required some technical dexterity and considerable patience. *Courtesy, Kelvin Smith Library of Case Western Reserve University.*

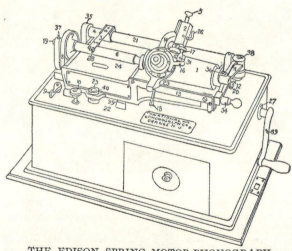

THE EDISON SPRING MOTOR PHONOGRAPH
INDEX OF PARTS.

1. Brass Mandrel to hold wax cylinder. (Always assembled with Main Shaft.)
2. Turning Rest, to shave cylinder.
3. Feed Spring.
4. Back Rod.
5. Sapphire Knife Spring Knob.
6. Main Shaft. (Always assembled with Brass Mandrel.)
7. Main Shaft Pulley.
8. Drive Belt.
9. Start-and-Stop Switch.
10. Speed Adjusting Screw.
11. Swing Arm.
12. Swing-arm Center.
13. Straight Edge.
14. Speaker Arm.
15. Speaker.
16. Speaker Lever.
17. Speaker Adjusting Screw.
18. Speaker Arm Lift Lever.
19. Main Shaft Center.
20. Swing-arm Center Adjusting Screw.
21. Back Rod Sleeve.
22. Top Plate.
23. Speaker Clamps.
24. Phonograph Body.
25. Body-holding Screws.
26. Shaving Knife Lever.
27. Winding Key Sleeve.
28. Feed Nut.
31. Speaker Adjusting Screw Lug.
33. Winding Key.
34. Lock Bolt.
35. Back Rod Set Screw.
36. Swing-arm Center Set Screw.
37. Main Shaft Center Set Screw.
38. Swing-arm Spring Washer.
39. Top Plate Lug.
40. Body-holding Screw Washers.

To Record Machine at rest.   Open speaker clamps (23) and
insert recorder with the speaker lever (16) pressed
up against lug (31).

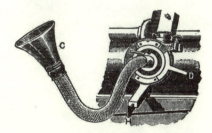

Press up speaker lever D, place the speaking tube C on the
Phonograph, and lower the lift lever H.

Press up lift lever (18).   The *numbers* refer to cut on
page 70.   Throw down the lock bolt (34) and open swing arm
(11) wide.

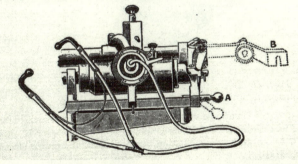

Figure 7 Presto's introduction of the instantaneous disc drew more amateurs into the field, giving rise to small studios during the Depression. The J5, or "Junior," model was one of Presto's early home recorders, introduced around 1935. The pin protruding from the recording head was for a removable aluminum weight used when embossing on aluminum blanks (*top*). Cutting lacquer blanks (*bottom*) did not require the weight but did produce a fine thread of lacquer during the cutting process. *Source:* John F. Rider, *Automatic Record Changers and Recorders* (New York: John F. Rider Publisher, 1941).

Figure 8 A passion pays dividends. *Left to right,* George Avakian, John Hammond, Benny Goodman, and Bill Savory inspect the lacquer discs of Savory's air check recordings of the Goodman Band's broadcasts. When Goodman heard them, he was so impressed with the sound quality he urged Columbia to release them, which they did in 1952 under the title *Benny Goodman 1937–38 Jazz Concert No. 2.* This picture was taken at the release party held at Columbia's 30th Street Studio, October 1952. *Courtesy, The William A. Savory Collection, The National Jazz Museum in Harlem.*

Figure 9 Singer Johnnie Ray and producer Mitch Miller listening intently to a playback. Miller leans against one of the parabolic reflectors designed by Bill Savory, Columbia 30th Street Studio, 1950s. Photo © Herman Leonard Photography LLC. *Courtesy, Sony Music Entertainment.*

Figure 10 Quincy Jones contemplates a score during a session in Studio "A" of Fine Recording, the ballroom of the former Great Northern Hotel on West Fifty-Seventh Street, New York, 1959. Recording sessions in Fine's Ballroom began in August 1958 and ran almost continuously through 1971. *Photo by Milt Hinton, © The Milton J. Hinton Photographic Collection.*

Figure 11 Eddie Condon leads a Decca Records session, 1950, produced by Milt Gabler. This is the same studio pictured in figure 4 from a decade earlier, now with no brocade-covered panels, more microphones, new paint, and a doorway into the control room where microphone cables were previously plugged into the wall. Musicians (*left to right*) are Cutty Cutshall on trombone, Condon on guitar, Wild Bill Davison on cornet, Peanuts Hucko on clarinet, Gene Schroeder on piano, Fred Sklow posing at the bass (bassist Jack Lesberg arrived too late for the photo shoot), and Buzzy Drootin on drums. Gabler stands at the doorway into the control room. *Courtesy, The Charles Peterson Jazz Photo Collection.*

Figure 12 Eddie Smith inspecting the grooves as he cuts a 78 rpm master on the Scully lathe at King Records, Cincinnati, mid-1950s. *Courtesy, The Edward J. Smith Photographic Collection.*

Figure 13 Thomas Boddie, first control room in the basement of his Pierpont Avenue home in East Cleveland, 1950s. He purchased the console, stripped of everything but the knobs and VU meter, rebuilt the electronics, and did the same with other components. *Courtesy, Boddie Recording Company Records, The Western Reserve Historical Society.*

Figure 14 Cream recording session, Atlantic Records. *Left to right*, Atlantic Records executive Ahmet Ertegun, producer Felix Pappalardi, and engineer Tom Dowd look out from the control room as the British rock group Cream record "Strange Brew" at Atlantic's New York studio, April 5, 1967. Dowd has his hand on the faders of the mixing console, small enough for him to "play it like you'd play a typewriter or a player piano." On the right stands the Ampex 8-track recorder. *Photo: Michael Ochs Archives / Getty Images.*

Figure 15 Control room of Columbia's Studio "B," the big room at Fifty-Second Street, 1966. Leonard Bernstein seated at the elaborate in-house designed and built console with round control knobs at a time when all new consoles featured linear faders. Paul Simon stands to the right of Bernstein as they listen to a playback of a Simon and Garfunkel track. Bernstein had an appreciation for all kinds of music, including rock, and his attention helped rock to be taken more seriously as an art form. *Photo: Douglas R. Gilbert / Getty Images.*

reported that, "like most good Swing, the music was literally composed by the player as he played."[25] The Commodore recordings were not exactly "long-playing" since only about five minutes of music could fit on a twelve-inch, 78 rpm side, but in March 1940, another jam session, this time at the Decca studio on West Fifty-Seventh Street, resulted in a seventeen-minute version of "A Good Man Is Hard to Find" spread over four sides (see fig. 4). Although some of the performance may have been lost as the engineer changed discs, it at least represented the improvisational freedom that jazz musicians had longed to be able to record.[26]

RCA Victor had put a long-playing record on the market in 1932, but its poor quality and the subsequent drop in record sales the following year led to a quick demise.[27] However, Edward Wallerstein, the same executive who pulled the RCA version off the market in 1933, catalyzed the successful development of the $33^1/_3$ microgroove long-playing record at Columbia Records fifteen years later.[28] The development of the Columbia LP began in the CBS labs under Peter Goldmark but was completed by a team of engineers led by chief engineer William Bachman in the recording division at 799 Seventh Avenue. The key technical components that made the LP a reality were microgroove cutting and a heated stylus, but the commercial success came about from a combination of factors, including wide distribution, the availability of an affordable player by Philco, and the second American Federation of Musicians' recording ban effective January 1, 1948, that made dealers and distributors hungry for new product.[29] Unlike the previous recording ban of 1942–44, this time the record companies had more advance warning, and they made good use of it, recording virtually around the clock in the months leading up to January 1, 1948, when the ban went into effect. "They were recording twenty-four hours a day in any place that had a turntable and a cutter head," recalled Tom Dowd, and since the ban affected transcription companies as well, they sought the assistance of the best independent recording studios. One of these was run by one of the few women recordists in the industry Mary Howard, who went from secretary to engineer in the NBC recording department when wartime labor shortages gave her the opportunity to become a disc cutter. After the war, she established a studio in her apartment, and from there put in a 126-hour week recording air checks for transcription companies in the days leading up to New Year's, a period referred to as the "acetate Christmas."[30]

Even before the strike was announced, Columbia teams were at work transferring its existing 78s and sixteen-inch $33^1/_3$ rpm standard groove masters to microgroove LPs to prepare for the LP's public debut. The process proved trickier than anticipated because the CBS labs' belt-driven turntables did not synchronize properly. Record producer Howard Scott of Columbia Masterworks and technician Paul Gordon devised a system using a handmade clock face on the platters, a stopwatch, good timing, strong finger snaps, and teamwork to synchronize the transfer process. As a classically trained musician, Scott was able to read the musical score and mark precise splice points where one side of the 78 ended and the other began. Even then, inconsistent pace of some conductors meant that Scott had to rely on his musical timing, snapping his fingers to cue Gordon at just the right time.[31] When Columbia and CBS introduced the LP in a press conference at the Waldorf Astoria in June 1948, the impressive accomplishment masked the amount of both technical and tacit skill that went into the new record's creation.

The microgroove LP, running at 33⅓ rpm, the same speed as electrical transcriptions, offered playing time on a single twelve-inch disc of up to nearly forty-five minutes using both sides, enabling an entire symphony to be available on one record. Pressed on Vinylite, a lighter and more flexible material than shellac, they were nonbreakable and virtually free of needle scratch, an unfortunate result of the filler in shellac which one recording engineer likened to "softened asphalt scraped up off the roads."[32] However, Mercury Living Presence record producer Wilma Cozart Fine, another pioneering woman in the male-dominated recording industry, recalled that some early vinyl was also inferior, and it took some time before high-quality vinyl was developed, which was particularly important for classical music.[33] The increased fidelity of the record set a new standard in listening, and the longer duration eliminated the need to speed up or cut songs and enabled jazz musicians to improvise. According to producer George Avakian, the success of the LP in popular music rested in large part on adopting a concept, a theme that linked the songs on the album instead of merely assembling a quantity of songs. Initially inspired by Arthur Godfrey's TV Calendar Show, in which he sang a song for each of the twelve months of the year, this thematic concept, along with affordable prices, an available player, and wide distribution became key factors in the public acceptance of popular LPs.[34]

The use of magnetic tape for recording, the Vinylite disc for playback, and better home reproduction equipment improved the overall sound of recorded music for consumers, giving birth to the era of high-fidelity records. Whereas disc recording had achieved a frequency range of 50–8,000 cycles per second, tape could record flat from 30 to 15,000 cycles per second, thereby capturing not only the fundamentals but also the overtones of musical instruments and voices, resulting in greater clarity and depth. As Andre Kostelanetz had discovered that, in broadcasting, what was played in the studio was not necessarily what went out over the radio to listeners, the wide spectrum of sound captured on recording tape did not necessarily mean the record buyer would hear it if along the subsequent path to final commercial disc fidelity was lost. For this wider frequency response to be enjoyed by the record buyer, many other improvements in recording and playback apparatus were necessary. Columbia's engineers were mindful of this in developing the LP, and by the time the long-playing disc was ready for commercial release, much had been done in the manufacturing process—from improvements in the electroplating to chemical compounds, tooling, and materials—to gear up for the higher quality required.[35] Before the record could be processed, the studio engineer had to cut a lacquer master from the original tape recording. Here, too, improvements in the cutting apparatus ensured minimal noise or loss of signal in the transfer from tape to master disc. The use of a heated stylus, which cut the lacquer like a hot knife cutting through butter, provided a smooth surface to the groove wall, thus eliminating most surface noise.[36]

Microphoning

In addition to the natural acoustics of the studio and the musical instruments and voices, a vast array of other instruments came into play in recording. These were the control man's instruments, which had grown dramatically since the days of the recording horn, sound box, and disc cutter. Now engineers deployed different types of microphones, mixing consoles, amplifiers and pre-amplifiers, equalizers, compressors, limiters, and of course tape recorders, cutting styli, lathe, blank discs, cables, vacuum tubes and other electronic components, and loudspeakers, referred to in the studio and control room as monitors. These and other related technical components of recording were selected according to engineers' taste and judgment. Among the many choices facing recording engineers was the selection and placement of microphones. Partly because the sounds they picked up were now more audible on record, and partly because so many new and sensitive microphones were introduced after the war, the choice and placement of microphones became an art form, assuming greater importance in the 1950s than before.

Condenser microphones that contained a small amplifier and power supply were the first microphones to accompany the electrical recording system in the 1920s. In the late 1920s, Western Electric introduced the first high-quality dynamic microphone, the 618-A moving-coil microphone.[37] The simplicity of the dynamic microphone over the condenser microphone and amplifier rendered it more practical for many applications of disc recording that required omnidirectional microphones.[38] The most popular microphones for recording in the 1930s and 1940s were the RCA velocity, or "ribbon," microphones: the diamond-shaped RCA 44-BX bidirectional and the bullet-shaped RCA 77-A unidirectional microphone. In the 1950s, as magnetic tape became the primary recording medium, new and better-sounding microphones became available and like the high-quality tape recorder, some of these came from Germany. In 1947, Georg Neumann, a German company, developed a new type of condenser microphone capable of picking up a wider dynamic and frequency range than any other previous microphone. The Neumann U47, marketed by Telefunken in the 1950s, became the vocal microphone of choice for Frank Sinatra, among others, and the first microphone to make single-mike recording of symphonic music possible because of its sensitivity and extraordinarily even directional pattern.[39]

Clair Krepps, the engineer who set up Capitol Records' first New York studio, was among the first recording engineers in the United States to use the Telefunken mike. In 1950, Krepps recalled, an elderly German man came to the Capitol mastering room in New York with a box, offering a microphone for Krepps to try. Because it was made for German electrical systems, Krepps had to modify it from twenty-five-cycle to sixty-cycle current. Then he decided to use it in a recording session with Peggy Lee and the Stan Kenton band at WMGM, the studio Capitol used at the time for its live recording sessions. Krepps believed the mike would provide support for Lee's relatively soft voice. "The reason I used it on Peggy, was I knew that she was a band singer that was accustomed to standing in front of the orchestra, you know. And I worried with Pete Rugolo with his ten brass section, six saxophones, and Eddie Safranski on bass, and Shelley Manne beatin' the drums, and Peggy was singing with a very soft voice, that you'd never hear Peggy unless you had a directional microphone. And it worked!"[40]

Knowing how instruments could overpower a vocalist and what he must do to compensate for this, Krepps demonstrated the tacit knowledge all good engineers needed in this period of rapid change. Because the directional pattern of the U47 could be adjusted, it was ideal for orchestral as well as close-miking of either voices or instruments. In 1951, Mercury released the first record in its new classical catalog, the Living Presence Series. Engineer Bob Fine had worked with Mitch Miller on Mercury's popular recordings in the late 1940s, putting the "halo" around Vic Damone's voice by using a bathroom echo chamber. In Chicago, Fine did not have to mimic the hall's ambience; he had the real thing, for he was recording the Chicago Symphony performing Mussorgsky's *Pictures at an Exhibition* in Orchestra Hall, their first recording in three years.[41] Fine recognized that the best way to capture both the performance and the natural ambience of the room was by using a single U47 positioned in just the right spot in the hall. His use of the single microphone was a drastic departure from the multi-miking that had become common practice, and, in the process, Fine made recording history. The number of microphones used in symphonic recording had gradually increased, necessitating the close-miking of instruments, a technique that eliminated the ambience of the hall. To make up for this loss of ambient sound, another microphone might be used to pick up the room tone, but this, as one audio engineer described it, was "a hot-house variety of sound, a sterilized electronic version of the music of the concert hall."[42] Fine's single-microphone approach was revolutionary and became a new standard for classical recording.

Popular recording, however, employed more microphones and effects. As more microphones came on the market, recording engineers confronted the problem of variable quality and response of microphones. Every microphone sounded different, and this required recording engineers to learn which mikes best suited an instrument's range, much as the acoustical recordist chose specific horns and diaphragms for recording certain ensembles. Microphones could be used with more specificity, but not every engineer had a wide choice of microphones, nor were all microphones of equal quality. Consequently, in the mastering process—the cutting of the master lacquer from the original tape—the mastering engineer sometimes used an equalizer to make the sound as close to natural as possible, or at least what the mastering engineer assumed to be natural. Clair Krepps, the engineer for Capitol, specialized in mastering records. He used equalizers in the 1940s and '50s to compensate for various inconsistencies in the original recording. Equalizers, which are signal-processing devices used to change the frequency response of the signal passing through it, were used regularly to compensate for the loss of high-frequency signal on the inner grooves of a record, called diameter equalization, but they were also used to correct the sound of certain instruments. Engineers could use equalizers to boost a weak sounding bass or to add life to a dull-sounding horn section. Essentially, Krepps explained, "an equalizer is a tool to correct the deficiencies of microphones, room acoustics, or for special effects."[43] Rarely were equalizers used during recording, but when they were, engineer and studio owner Walter Sear noted, they were not called equalizers, "they were called 'corrective devices' [because] they were used to correct mistakes you made when you recorded."[44] Recording directly to full-track (monophonic) or 2-track (stereo) tape, they would be used only if the engineer had trouble with the microphone or encountered some other problem during recording, but they were not used routinely and they were not normally part of the control console. To use them, the engineer had to patch them in by plugging into the patch bay, a rack-mounted panel with a jack field, similar in appearance to a telephone switchboard, enabling the engineer to connect, or "patch in," different devices, depending on the needs of the particular recording session.

The early equalizers, made by Cinema Engineering for the movie industry, posed problems when used in mastering phonograph records. They were noisy, made audible clicks when adjusted, and caused a full twenty-four decibel loss, which required the use of another amplifier to correct. Krepps, who hated using them, supported the development of the first quality equalizer for recording, made by Oliver Summerlin and Gene Shenk, two New Jersey audio engineers. Summerlin had known Krepps since the two served together in the Navy, and subsequently, the two built the Capitol Records mastering studio in New York in 1946. In 1949, Summerlin left to work for Audio and Video Products Corporation on Fifth Avenue, distributors of Ampex recorders and four years later formed a business partnership with Shenk in West Englewood, New Jersey, building production line oscillators. Krepps had just moved from Capitol to MGM Records and was in the process of building another mastering room when he decided that rather than commute to Manhattan every day, he could rent bench space in Summerlin and Shenk's shop to do his initial wiring. One day, Krepps mentioned the need within the recording industry for a good equalizer. Summerlin and Shenk agreed to build one, and Krepps provided $250 seed money with an MGM purchase order. Their company, Pulse Techniques, developed the first quality outboard equalizer, the Pultec, which even today is a highly valued piece of vintage gear.[45] Explaining their lasting appeal, Walter Sear described the care with which Summerlin and Shenk built these early equalizers: "They were two guys who set up a production line, but every unit that came off, they'd sit there and diddle around—you know, change a resistor here," and make various minute adjustments "until they got *sound*. Well, they had ears; they knew what it should sound like when it was right."[46] By "sound" Sear simply meant that they heard the Pultec perform as they intended. Summerlin and Shenk had designed an equalizer that was undetectable when used.

Many engineers, instead of relying on equalizers, cultivated a microphone technique whereby they used specific microphones for specific instruments based on the kind of sound they wanted, or what they had learned through experience worked best for a particular instrument or voice. Al Schmitt first learned microphone technique early in his career, out of necessity. Apex Recording, the studio where he worked, had only one equalizer and using it affected the entire recording. Because there was no way of patching it to affect the sound of individual instruments, Schmitt learned to get the sounds he wanted by microphone choice: "If I wanted a brighter vocal sound, I used a brighter mike. If I wanted a warmer bass sound, warmer piano sound, I used maybe a ribbon mike."[47] Because ribbon microphones did not have the full frequency sound that tube or transistor mikes had, Schmitt said they were more forgiving of imperfections in a vocal performance. To rely on microphone choice and to avoid equalization altogether, Schmitt admitted, an engineer needed to know his microphones, what they sounded like and where they should be placed.[48] Several engineers likened this ability to "get sounds" by careful choice of microphones to a painter mixing colors on a palette, and the music critic and audio enthusiast Edward Tatnall Canby considered the art of microphoning the equivalent of any interpretive art, arguing persuasively that engineers should receive credit along with the composer or performer.[49]

Engineers knew that microphone choice and placement were critical factors in recording, but the singers' microphone technique also helped. Performers regulated their relative volume by leaning in to the mike on low or soft notes and backing off a bit on the high or loud notes, like a more subtle version of jockeying for position that was necessary around the acoustical recording horn. Frank Sinatra is widely regarded as the first to master microphone technique. Mitch Miller, who worked with Sinatra in the studio, said "Sinatra couldn't be heard from here to ten feet away. . . . None of them [could]. In fact, it's part of the *art*," Miller emphasized, "part of the art of recording electronically. That's why [famed operatic bass Ezio] Pinza sounds so ridiculous, 'Sohm enchaahnted eve-e-ning,' you know?"[50] Forty years earlier, Pinza's classically trained voice would have ideally suited the recording horn, but through the microphone, it seemed out of place. However, Sinatra grasped that singing through a good sound system required a different technique from an un-amplified stage performance, and he became the first performer to know how to use a microphone for dramatic effect.[51]

While the wider range of good microphones gave engineers greater flexibility, it also posed difficulties for the recording engineer faced with balancing an array of instruments and voices. Recording brass and strings, for instance, became quite a challenge without the use of baffles, movable upright panels also referred to as *gobos*, to isolate the instruments from one another. Capitol Records engineer John Palladino recalled that recording Frank Sinatra at Capitol's Melrose Avenue studio, a converted broadcasting studio, always posed a challenge because Sinatra did not want to be isolated in a vocal booth and insisted on recording live with the band, just as he would in a live performance, but this left engineers little control over the relative volume of his voice. The challenge for engineers in this scenario was to capture and balance all the instruments, including Sinatra's voice, so that the overall mix worked. They positioned him on the floor in front of the stage where the musicians performed to elevate the instruments enough to prevent their direct sound from leaking into the vocal microphone and thus overwhelming Sinatra's voice. This "very simple splay" only partially worked because it did not keep the sound of the orchestra completely out of the vocal mike, and because arranger Nelson Riddle used a lot of strings and a lot of brass, Palladino recalled, "it was a constant fight, you know, to make all those things come out."[52] RCA engineer Ray Hall agreed that the most difficult challenge was recording brass and violins at the same time, because the brass would overpower the violin mikes. Because of the time constraints on the sessions, they did the best they could by using baffles, but eliciting the cooperation of the musicians in how they played and where they stood was important.[53]

Some engineers came up with quite inventive solutions to the problems of instruments leaking into vocal microphones. Atlantic Records engineer Tom Dowd recalled that, in the 1940s, "no one said 'This is the way you record, that is what you use.' With his Navy experience, Clair [Krepps] knew how to go about figuring it out."[54] Krepps's experience jury-rigging what he needed to get a job done served him well when faced with problems in the studio that presented no clear technical solution. During one recording date at MGM studios, he had difficulty miking the singer so that she could be heard over the band. Realizing that the problem was due to the RCA 44BX bidirectional mike, which was picking up the band as well as the vocalist, Krepps knew he could not place her any closer than about eighteen inches from the mike because that boosted low frequencies. If he turned her mike 180 degrees, it would be out of phase with the band's microphone.[55] Krepps decided the only solution would be to find another way to prevent the leakage. "So the band took a break," he recalled, "and I got on the elevator at the fifteenth floor, went down the street and ran up and down Fifth Avenue until I found a drug store, and I went in and bought a box of Kotex and took them back to the studio, and taped one on one side of the microphone and it worked! . . . It cut down the sound of the band. And you can imagine, a bunch of musicians. It took five minutes to quiet them down. . . . She laughed along with them."[56]

One might speculate about how Krepps, a married man no doubt familiar with the accoutrements of a woman's menstrual cycle, came up with this solution. Had he made a subconscious word association with the idea of "leakage" or "signal bleed"? Or had he simply recalled that the shape, size, and density of this cotton batting would ideally fix his problem? Krepps's solution certainly was novel and practical but not likely to become standard operating procedure. To cure this recurring problem, studio engineers built string shells resembling a movable enclosure cut in half, like a half-shell, and vocal booths. Ray Hall recalled that RCA built isolation booths for vocalists similar to those used on the television quiz show *The $64,000 Question*, but these were cramped and confining, and singers felt uncomfortable and preferred to hear themselves with the band.[57]

Recording Consoles and Black Boxes: The Control Man's Instruments

Of all the components in the recording chain, the control console was the last to be produced on a large scale and the most frequently customized. The first control panel was the Western Electric amplifier system, a vertical rack about six feet high with volume controls and meters. By the 1930s, recording studios used broadcast consoles that combined a sloping horizontal desk with more controls and switches and an upright back panel where the VU meters and other controls were mounted. Western Electric, Raytheon, Gates, General Electric, RCA, and a few other companies manufactured these for radio, and as with other radio technologies, recording studios adopted them. Unlike the early Western Electric racks, designed for standing use, the broadcast consoles were desk high, enabling the engineer to be seated.

The engineering departments of Columbia and RCA Victor custom-built the consoles for their studios into the 1950s, around the time independent manufacturers began producing various components for the recording industry. The Langevin Manufacturing Corporation, a New York company that produced a range of broadcast audio facilities, began building consoles specifically for recording studios in the 1950s, incorporating an improved volume control mechanism. The change in this particular component of the console demonstrates how one small improvement, the design of the volume control, could have a much larger effect on the engineer's work. The most commonly used volume control was the German-made Daven fader. According to Bill Stoddard, a recording and design engineer, "They were stiff, and they had a terrible feel, and they were just awful. They were for submarine commanders, not mixers . . . they probably will last forever [but] they just weren't very sensitive or very arty." Both Langevin and Cinema Engineering introduced a different kind of volume control, a slide-wire rotary fader that Stoddard said felt like "they floated ... they had a beautiful feel and they were very conducive to getting very arty on the console." Characteristically, Stoddard relied on the musical instrument metaphor when describing this control. When using the Daven faders, which required a fair amount of force to move them, he explained, "You just didn't feel like you were playing on a Steinway, it was like you were playing in mud," whereas the slide-wires "you could work with your little finger."[58]

Ease of console operations enabled engineers to manipulate multiple controls with one hand, which became increasingly important as simultaneous operations were expected of engineers. In a paper presented at the 1955 annual meeting of the Audio Engineering Society, Philip C. Erhorn, an independent audio design engineer who began custom-building consoles in the mid-1940s, described how new technology had changed recording methods, necessitating increasingly complex mixing facilities. In seeking different and more commercial sounds, clients had begun to demand "all kinds of exaggerated sounds [involving] the use of echo chambers, program equalizers, sound effects and sound effects filters, a variety of microphone types, vocals isolated and treated independently of the music pickup, tape-live multiple dubbing, and of course, virtually instantaneous playback facilities to both control room and studio personnel." With so many possibilities, and the studio clock ticking, Erhorn said, "the engineer must be prepared to conjure up various effects at the flick of his wrist."[59] Within five years of the universal conversion to tape, much had changed in the control room; the engineer had more options from which to choose and more operations to perform to make a record. The client, artist, producer—whether for phonograph records or commercials—expected to be able to use the latest in recording techniques and to be able to hear the results on the spot. Not surprisingly, Erhorn's guiding design principles of operating ease and convenience were aimed at making things easier on the engineer, but streamlining these procedures also benefited the artist, the commercial client, and the record company. Studio time, as well as musicians' fees, cost money, and technological delays could be frustrating and costly.

In 1956, Tom Dowd built his own console in one of the most creative examples of recycling available materials. From the time Atlantic Records began in the late 1940s until 1956, the company used their offices at 234 West Fifty-Sixth Street for some recording dates, pushing their desks against the wall to make room for the sessions, and used the Capitol Records' studio for larger recording sessions that took place in New York. Beginning in 1952, Dowd recorded everything in both monaural and stereo, using an Ampex full-track model 400 recorder, and a portable staggered-head 2-track Magnecorder. Stereo records were several years in the future, but stereo tapes for home users were available sooner, and Dowd reasoned that Atlantic would be prepared with a stereo catalog to meet future demand once stereo records and home stereos became a reality. Recording in both formats, however, required two sets of controls, one feeding the mono machine and the other feeding the stereo. The impossibility of monitoring both machines simultaneously through headphones meant he mixed by sight: "I was more or less mixing by what I saw on the meters for the stereo machine as opposed to what I heard on the mono machine. I mean it was hairy! But this was how in the early days it went." When Atlantic owners Ahmet Ertegun and Jerry Wexler relocated their offices and gave Dowd the entire floor, he redesigned the studio and control room. He did not like the design of the available consoles like the Raytheon he had been using because the heat from the tubes and the amplifiers gave him, as he put it, "sunburned knees." Dowd reconfigured the control room to suit his needs. Recalling something he had learned in the physics lab at Columbia University, he devised a system where he had nothing but faders in front of him, and all the electronics were positioned behind him. During the move and reconstruction of the office, Dowd wired racks of amplifiers, preamplifiers, and line amplifiers. "When the physical demolition and reconstruction took place—this is the god's honest truth—they were taking out a door to put in a double-weight heavy soundproof door," he recalled, "and I looked at it and thought, I don't have a god-blessed console to put these stinking knobs on." When the worker took the door down, Dowd had it cut to fit the space in the control room, drilled holes for the knobs, and that "plain old wooden door" served as his console.[60]

Designing one's own controls and salvaging an old studio door for a console typified the technological ingenuity of independent sound engineering at the time. Before the 1960s, when independent studios and record labels began to mushroom across the country, few companies designed equipment strictly aimed at the recording studio market because it barely existed. The major record companies had technical departments to build what their recording engineers needed. At Atlantic Records, and other small studios, the engineer *was* the technical department, and innovation was born of necessity. In addition to moving the amplifiers and tubes away from his knees for more comfort, Dowd also changed the configuration of the console to make it easier to manipulate the controls:

I got tired of reaching eight three-inch knobs with one and a half inches between them, it was like three and a half feet wide, like my hands are flying left, right, and I started aligning things—put this instrument here, it's one you never ride, put that one on the eighth fader, put that one, put this one . . . eventually, the same people that I bought the amplifiers from sent me a catalogue and they had slide wires in it. . . . And all of a sudden I had a console that was twelve inches wide. And my fingers sat on top of each one of the knobs—I could play it like you'd play a typewriter or a player piano.[61]

Dowd's use of the musical metaphor, like Stoddard's reference to "playing a Steinway" and other engineers' reference to painting with microphones, indicates how recording engineers considered their work to be as much an art as it was a technical skill.

When engineer Frank Laico joined Columbia Records in 1946, the recording consoles, built by the company's own technical department, had six microphone inputs. As he pointed out, this "was not a whole lot, and if you had a band out there, or an orchestra, you had to be damn sure where you were going to put those microphones so that you'd get the result that was required. And oftentimes you had to do a lot of make-shifting. In those days it was a really experimental time."[62] Laico said that the consoles did not have built-in compression or equalization, but the technical department would develop whatever the control engineers needed, which they in turn could patch into the system. "We used to call them the 'black boxes,' and we would request whatever we needed, and they would try to satisfy us. Most of the time they did."[63] As recording engineers employed greater numbers of microphones, they used compression to prevent the orchestra from overpowering solo vocalists, but as Laico discovered, proper use of compression, which essentially reduces the dynamic range of the musical performance, required a certain skill:

If you didn't use it properly, you'd squash the voice so that when the voice is supposed to be at its loudest, and the vocalist or the musician was playing at their loudest, if you didn't use the compressor properly you'd squeeze it down so that it sounded like nothing. And that wasn't the purpose for that. So you had to know how to use your equipment . . . and you have to realize the boundaries and stay within them. And that's where I, myself, was very successful [because] it never sounded like I was using it. And that's the secret.[64]

One of the problems with these early compressors was that you could hear them working, so that many engineers avoided using them altogether. Al Schmitt used what he called "hand compression" or "hand limiting," which involved learning the song so that he knew the dynamics and could anticipate what was coming during the recording session. "When the singer sings loud you back her off a little bit, or when they're soft you bring 'em up," Schmitt said, "But you had to learn the song one run-down through."[65] Engineers who had musical training read the score and knew when certain instruments came in so that they could make the necessary adjustments during the song.

Capitol Records' John Palladino recalled that more microphones and more instruments introduced the need for separation and careful control during the recording, either by recording a track and adding lighter instruments or vocals later or by using isolation. In many recordings, he said, "you can hear what's happening there and what the poor mixer had to put up with—drum sounds started to get smeary, bass sounds bad, strings were never loud enough in relation to something else, the vocals were fighting different things." Because the engineers at that time could not preset anything, they had to "actually be turning stuff on and off when the need arose. You couldn't just let it go and hope it was going to work out all right."[66] Until the introduction of multi-track recording, engineers had to actively mix during the recording because post-mixing was not yet possible. Achieving the right balance of instruments and vocals constituted the engineer's most important skill, for it was virtually the only means of controlling the relative sounds being recorded. Although the introduction of tape made it possible to edit out mistakes and splice in corrections, editing did not permit engineers to adjust levels of different instruments other than their relative balance in relation to the other instruments *at the time of recording*. Later, multi-track recording made it possible to record each instrument on its own track, but before that, according to Palladino, they managed by a combination of control of the arrangements and cooperation of the orchestra, but the engineers never had complete control.

Stereo, the End of the 78, and the Coming of Multi-track

In 1954, RCA Victor became the first American label to begin experiments with stereo recording sessions to determine its technical and commercial potential.[67] Studio owner, recording engineer, and inventor Bob Fine began 3-track stereo recording for Mercury's Living Presence Series in 1955 using custom-built Ampex 3-track recorders.[68] As early as 1952, Tom Dowd had recorded 2-track stereo tapes of Atlantic's mono sessions using a staggered-head Magnecorder.[69] Several other companies released stereo reel-to-reel tapes long before there were stereo records because no one had yet devised a suitable stereo-cutting system.[70] Knowing that consumers were not likely to embrace a new format without some assurance of compatibility with their existing record players, the record companies came to a remarkably quick consensus on a standard system through their new trade association, the Recording Industry Association of America (RIAA). Founded in 1951, the RIAA was the first formal trade association of record companies in the United States. Its initial mission was to set standards of recording practice, and one of its first accomplishments was to establish in 1954 the standard recording curve, or equalization curve, for the recording and playback of vinyl records.[71] In 1958, after six months of demonstrations and discussions surrounding two competing systems of stereophonic disc recording, record company representatives came to a quick agreement on the standard.[72]

By the fall of 1958, Westrex made the first stereo cutters available to professional studios. Bill Stoddard recalled that Universal Recording in Chicago immediately acquired one and retired their Scully 78 rpm lathe, which he disassembled, refurbished, and set up for stereo cutting. To Stoddard, this signaled the end of the 78 rpm era, and indeed, it was not far off. In 1959, a three-year survey of juke box operator record-buying trends revealed that the number of 78s purchased for juke box play had declined from 19 percent of the total in 1956 to 9 percent in 1957, and 1 percent in 1958. By 1959, with operators buying more stereo discs, both extended plays (EP's) and 45 rpm singles in equal number, it was expected that the 78 rpm would virtually disappear.[73]

However, while many recording engineers embraced the possibilities of stereo, there was a certain amount of resistance to change and not all recording engineers welcomed it. Bill Savory, who left Columbia to become chief engineer for Angel Records found that even major labels resisted: "At first, the sort of national anthem around Capitol was that 'stereo doesn't mean anything—it's just a way of selling two speakers instead of one.' The English wouldn't stand for that, though."[74] The English company EMI owned Angel and acquired Capitol in 1956, the year the construction of the new Capitol headquarters in Hollywood was completed. The first recording facility of its kind built literally from the ground up on the corner of Hollywood and Vine, the Capitol Tower resembled a giant stack of records with a spindle protruding from the top. The three specially designed studios and four shock-mounted sublevel reverberation chambers provided, according to Capitol's vice president of manufacturing and engineering, "acoustically controllable and electromechanically flexible" studios that fulfilled "the esthetic considerations important to the artist and the practical engineering considerations of concern to the modern producer."[75] However, they were designed to record only monaurally, a surprising decision given the pervasive recognition within the industry of the inevitability of stereo. Around 1957, Savory was sent to California as senior audio engineer for Angel Records, which used the Capitol studios, and in that capacity built a separate stereo control room below the mono control room. "The only two-story control room in the industry—a two-story control room so they could do monaural recording in the top story, they had their own microphones, their own mixing console, their own tables, their own everything—they could do mono, and nobody could interfere with them," because it seemed that part of the management continued to believe "we're gonna have to live with mono forever."[76] Public acceptance of home stereo playback systems was a few years off, and during the early 1960s, albums were released in both mono and stereo, sometimes only "simulated stereo" from a mono master, but the days of mono recording ended in the 1960s.

For the control engineer, stereo was not a simple matter of adding another speaker in the control room and dividing the recorded signal between left and right speakers. It entailed a new way of listening and envisioning, not only how the instruments should be miked but how the overall sound should be planned. As the number of tracks increased rapidly from 2- to 3-track stereo, then to 4-track, 8-track, 16-track, and more, engineers had to constantly revise their outlook. As Palladino recalled, more channels and more control meant that engineers had to decide what stereo should be:

> Is stereo something on the left? Is it a mono sound on the left, and then a mono sound on the right, and then a combined sound in the middle? I mean, what do you do with the thing? Because, as soon as you put . . . a mono sound on the left, . . . you haven't really improved the sound very much of that particular section. The only way you improve the sound is [if] other leakage then happens in the room, that one mike may accent mostly that sound but then there's also the sound of that section leaking over into the other side into the other mike. So, there was always a battle of whether you double mic something . . . put two mikes on something so you get a stereo sound on that particular thing or if you try for mixing various mono sounds and placing them anywhere you want but [making] just really, an electrical placement. I mean, you'd have so many more dB of this. In the center it was equal; if it was a little bit mid-right, it would be a little bit more to the right, so you were trying to artificially place a lot of these things.[77]

Palladino describes what I would call the mental architecture of the sound engineer, one that was undergoing continuous change in this period. The ability to control the sound meant that decisions had to be made as to how to control it and to what end. Increasingly, the producer and artist became involved in these technical decisions, becoming more and more conscious of the technological options they had in creative decision making, just as the engineer began to offer, by virtue of his familiarity with and control over those options, a certain level of creative input.

With the advent of stereophonic sound and the completely new approach to recording it, record companies immediately recognized the need to modernize their existing recording equipment and to build additional facilities. In New York, Columbia Records modernized the 30th Street studio, added new editing and mastering cubicles at 799 Seventh Avenue for stereo recording, and raised the penthouse roof to heighten the ceiling of the new studio they made from combining two small studios on the seventh floor. Columbia engineers built all the recording equipment, including three special studio control consoles containing modern and versatile facilities for recording, one for Seventh Avenue, one for 30th Street, and one for their new West Coast facility. This last studio, Columbia's first in California where they previously only rented studios to accommodate artists in the Los Angeles area, was the former CBS Radio Studio A in the KNX building on Sunset Boulevard. This studio, 115 by 65 feet with forty-foot ceilings, was larger than 30th Street and represented a move away from the broadcast studio design of some thirty years before. With the audience seats and stage removed, walls stripped of acoustical treatment, only the balcony remained, giving the new studio what Columbia engineers considered to be revolutionary reverberation characteristics. The control room, which measured eighteen by twenty-two feet, was considered huge for the time and allowed both recording engineer and producer to monitor stereo recording.[78]

When CBS bought the American Record Company in 1938, the company inherited three recording engineers: Vincent Liebler, Adjutor Theroux, and Cecil Bridge. Twenty-six years later, the Columbia Records recording department had grown to a staff of eighty-four recording engineers and thirty-four backup and administrative positions.[79] This reveals not only the prosperous growth of the company but also the division of labor and specialization that accompanied it. Their redesign of the studio to accommodate stereo recording also indicated the trend toward larger control rooms, a development that suited the increasingly collaborative work of engineers and producers.

Engineering the Performance: Tacit Knowledge and the Art of Controlling Sound

With tape, the LP, and stereo, the work of recording engineers, and the working environment of recording studios, changed forever. Like a Rube Goldberg contraption, one change in recording technology triggered improvements and innovations in other parts of the recording chain, adding flexibility as well as complexity, simplifying some steps while adding more steps to the process. The higher fidelity of Columbia's LP and of the seven-inch 45 rpm record developed by RCA Victor introduced the public to a new listening experience, but what made that listening experience even better was the way recording engineers and record producers made the most of the recording studio in exploiting the wider spectrum of sound that was now recordable.

Capturing that sound, even with the improved technological tools at their fingertips, was not a simple matter, and recording engineers continued to learn their craft through a combination of experience, intuition, and experimentation. Caught up in a wave of technological enthusiasm and the desire to improve the sound of records, the cohort of recording professionals that built the recording industry after the war worked in a rapidly changing technological culture. Because so many shared experience in the armed forces, they had a shared discourse, a way of communicating efficiently, and a sense of camaraderie that helped to forge a cooperative, collaborative working environment, even in the competitive postwar economic climate. As would later be true of the early computer hacker culture, recording engineers readily ignored corporate rules about proprietary secrecy to help a fellow engineer work through problems. For a time, at least, they shared knowledge to improve the technology and to move it forward.[80]

Technological change in the workplace is not always welcome, but engineers had long anticipated, even eagerly awaited, the benefits of both tape and the LP. Because tape technology was so new, and the recorders were not designed by recording engineers but by mechanical and electrical engineers, most of whom did not have a complete understanding of its use, both studio and design engineers were forced to adapt. Everyone was working without a road map. Consequently, recording engineers frequently modified, fixed, invented, and devised novel solutions to the problems that high-fidelity recording introduced.

In 1944, *Business Week* reported that War Production Board figures showed the electronics industry had experienced a war boom unlike any other industry, including aviation. The report stated that there was little concern that expanded capacity would exceed postwar demand since unlimited job opportunities awaited in the postwar economy. Topping a list of some sixty jobs expected to need personnel was Radio Broadcasting, followed by Worldwide Telephony and Telegraphy, Radar ("the most glamorous electronic gadget"), and Sound Recording. A wide range of trained workers would be available to fill these jobs, and those with knowledge of the latest electronic techniques and access to secret military equipment had a distinct advantage.[81] Subsequent demands exceeded predictions. By 1953, *Audio Engineering* reported a shortage of engineers in practically all fields and that by 1954 there would be twenty-five thousand new jobs in civilian industry for engineers and only about half that number of new graduates to fill those positions. The proposed solution was for industry "to use its engineers more efficiently as engineers, and to put non-engineering routine work in the hands of those who are not qualified for the more rigorous engineering work."[82]

By the 1950s, the recording studio involved a multitude of jobs, not all of which required trained engineers, and many who entered the recording profession had either no training or had started out as amateur radio operators or technicians. The involvement of these self-taught audio enthusiasts factored in the growth of the recording field, which expanded so rapidly after World War II that no one could be expert in every aspect. The field of audio encompassed a broad array of engineering jobs, but few of those entering the recording studio had degrees in engineering. More often, they had experience and an affinity for the job. According to Tom Dowd, who had a musical background, some knowledge of physics, and a healthy curiosity, there really was no training for recording engineers, nor was there any recognized professional standing for some time. Those who lacked a radio license were considered second-rate engineers, or apprentices, whereas those with radio licenses were considered full-fledged engineers who consequently were given more responsibility. Dowd argued that was no assurance of quality: "The fact that he had a radio license didn't make him a better engineer when it came to having ears or hands. . . . There was a pecking order that was established by the union that said that this man can't do this, and this man isn't qualified to [do that], etcetera."[83]

Dowd's observation gets to the heart of what makes a good recording engineer; that blend of technical expertise and creativity embodied in the individual with "the brain of an engineer, but the heart of an artist."[84] The term that best describes the cognitive map of recording engineers is *tacit knowledge*—the unarticulated, implicit knowledge gained from experience.[85] If one could "disassemble the intuition" of recording engineers, as did sociologist Douglas Harper of the garage mechanic in *Working Knowledge*, certain similarities would be evident, especially a deep understanding of materials and an ability to fix things by unconventional means.[86] Just as mechanics use a working knowledge of materials and machines to do their work, recording engineers need a working knowledge of the behavior of sound and the machinery of its propagation. Eugene Ferguson persuasively argued in *Engineering and the Mind's Eye* that nonverbal cognition and visual thinking are central to design engineering praxis.[87] Recording requires a similar tacit knowledge, an *aural* thinking that takes place in the mind's *ear*, as it were, of recording engineers. A broad range of hearing is important, but more important is the ability to detect sounds embedded within a dense matrix of others. This ability may be inborn, but surely it is also acquired by doing the job.

RCA's Ray Hall was the only African American engineer anyone remembers at a record company studio in the immediate postwar period and probably one of the few anywhere until the 1960s. Certainly, he was among the few—black or white—who were actually trained engineers. Hall graduated from Manhattan's prestigious Stuyvesant High School in 1942 and then attended City College until he was drafted the following year. While serving with the Marine Corps in the Marshall Islands, Hall was accepted into the V-12 Officer Training Program and returned to the United States to attend Purdue University. After his discharge from the Marine Corps, Hall completed his degree at Purdue in electrical engineering in 1949, after which he returned to New York. One day Hall read an article in the *Amsterdam News* about a pilot program sponsored by the Urban League seeking blacks with scientific and technical backgrounds to apply for jobs at various companies. Hall applied, got an interview with RCA, and was accepted into its engineering training program. The training required Hall to work in a variety of locations: at the Camden office, which focused on government projects, Hall was asked "to design some kind of rocket." He passed the trial and eventually became one of the few to successfully complete the training program. He was then interviewed by Al Pulley, the head of RCA Victor recording studios in New York, who needed someone to work on phonographs belonging to the Victor artists. "So for the first three or four months," Hall recalled laughingly, "I went around repairing—I had my college degree and everything—repairing phonographs . . . it was an honor, because I went to so many great artists' homes—Stokowski, Horowitz—I met so many artists. Before I started I didn't know a hell of a lot about phonographs, but you learn as you go. There's nothing to it, really."[88]

Hall's skills did not remain underused for long; by 1950, he was working on recording sessions as second engineer, maintaining the equipment and assisting the primary engineer on conductor Arturo Toscanini's recording sessions at Carnegie Hall and then at Manhattan Center with engineer Fred Elsasser recording vocalists Mario Lanza, Kay Starr, and Harry Belafonte. After working on many sessions as a "second man," Hall recalled the first chance he got to be head engineer: "It's almost like being a co-pilot; then the pilot says, 'Take over,' you know?"[89] Recording engineers frequently cite the flight metaphor to describe feelings of trepidation at the responsibility suddenly thrust on them. This is not surprising. Except for the greater expanse of physical space, the control room environment resembles a cockpit, not only in the instrument panel but also in the window engineers and pilots face in their work.

Engineer John Woram began at RCA in 1959 after a stint in the Navy and electronics training at the RCA Institutes. Like Hall, Woram began at the bottom of the corporate ladder, as an inspector in the quality control department. Woram's job involved the following: after the recording was made, the mastering engineer cut a test lacquer, which would then be sent to quality control where Woram would check the grooves, by sight as well as by sound, for any excessive levels of distortion. If all was right, the disc was sent to the producer, and if he approved it, a second disc was cut and sent to the factory for processing and pressing. From this job, Woram moved up to disc cutter and then in 1963 became recording technician. At this point, Woram recalled, his job got interesting, and sometimes nerve-racking.

During one date with the Boston Symphony recording Tchaikovsky's Violin Concerto, the first movement ran nineteen minutes, but the tape reels Woram was in charge of operating ran only fifteen minutes. To capture the entire piece, Woram had to perform an intricate dance with the two tape machines, which were always in operation on any session, one master tape and one for backup. Both machines would operate simultaneously, but when the tape was about thirty seconds from the end, he would cut the tape on one machine and quickly load a new reel on that machine. Once the first machine was recording again (which had to happen before the second machine ran out of tape), he then put a new reel on the second machine. Hence, there were two periods of time when only one machine was operating while the other was being loaded. As the assistant recording technician, he was also responsible for cueing up the playback, which the conductor would normally want to hear after the take was complete. To ensure this ran smoothly, Woram had to cue up the second portion of the tape while the first was playing. The recording engineer then made sure the playback levels would be satisfactory to the conductor. While that was happening, Woram listened through headphones to the second portion, which he had to cue up so that it would provide a smooth transition when the first tape ran out. While Woram did all of this, the recording engineer was, as he put it, "trying to figure out how to keep from falling asleep." It wasn't that he was lazy, or unwilling to help, nor was it due to job description restrictions. It was simply a one-person operation; two people trying to operate the tape machines would be "like two people trying to drive a car."[90] Eventually, Woram became a recording engineer, the highest studio position at RCA, at which point he found there was not much to do, at least on classical music sessions. The engineer "couldn't play around much" since the conductor and orchestra director made the decisions about what the recording balance should be. Once the microphone placement was determined, consistency dictated that it stay in that position, sometimes for years.

As with many large corporations, the major record labels became increasingly hierarchical as they grew larger and more successful, both in terms of creating different divisions and subdivisions within each of those divisions. Some divisions of labor in the recording studio were based on need. The hierarchy of tasks at Columbia Records and RCA Victor was based on experience, training, and skill but also on union regulations—National Association of Broadcast Engineers and Technicians (NABET) for RCA engineers and International Brotherhood of Electrical Workers (IBEW) for Columbia, Capitol, and Decca. Engineers were required to join the union once they were hired by those labels. The division of labor gave each engineer a particular "jurisdiction" over which he presided (recording engineer, tape operator, mastering engineer, etc.), and in the control room, only engineers were permitted to operate the equipment. This arrangement grew increasingly awkward beginning in the 1950s with producers such as Mitch Miller, who had definite ideas of what he wanted a record to sound like and itched to get his hands on the controls. If an artist chose to record at an independent studio, as did many rock groups beginning in the 1960s, a union engineer from the artist's label was required to be present, even though the engineer often just sat in the control room.[91] Not every major record label followed union rules. At Capitol Records studio in New York, which was IBEW because Capitol Records in California was IBEW, engineer Irv Joel admitted that they were "the most non-union union people there were" and that everyone would pitch in to do whatever job needed to be done if the occasion demanded.[92]

The unions had always advocated training, but few formal education programs existed until the 1960s. Numerous efforts to promote the German *Ton-meister* degree program, which involved training advanced music students in the skills of sound recording, failed in the United States.[93] However, there had been a few "sound schools" on the West Coast. In the late 1930s, John Palladino enrolled in the Music Department at Los Angeles City College, but ended up spending most of his time in a recording studio housed in the Physics Department.[94] During World War II, two engineers from Electrical Research Products Division of Western Electric taught a group of courses at the University of California at Los Angeles, as part of a government war-training program, later publishing a technical book based on their courses.[95] The University of Hollywood, founded in 1946 by Howard M. Tremaine, a film sound specialist, offered an eighteen-month program in sound and audio engineering, culminating in a bachelor of science in Audio Engineering.[96] Most American recording engineers of the immediate postwar period, however, learned the trade from scratch with their first civilian jobs. Even in the late 1950s prejudice against musical training for recording engineers persisted in some companies. While he was still in quality control at RCA, John Woram tried to convince the management in the recording department to allow him to take coursework, paid for by RCA, in music theory at Columbia University. He knew he wanted to be a recording engineer, and he reasoned that since he was going to work with musicians, he ought to know something about music. His boss, however, vehemently opposed the idea, saying, "What would you want to take that for? It has nothing at all to do with the job!" This was the prevailing sentiment at that time, and it took a bit of convincing before Woram finally received his employer-paid musical training.[97]

The Studio Working Environment: Teamwork and Collaboration in the Trading Zone

The 1950s were an exciting, challenging, and rewarding time for recording engineers. Although they worked behind the scenes, getting no album credit as did producers and artists, engineers were highly respected within some professional circles according to Clair Krepps; and because the artists and most producers did not understand what the engineer did or how he did it, just that he could make things sound better, "he was a genius." Like the staff producers and company executives, recording engineers in that era wore white shirts, neckties, and jackets; Krepps said he "wouldn't think of sitting down at a console without my jacket on. Not a sport jacket, a suit jacket. My poor wife had to iron five white shirts every week."[98] This formal attire reflected not only pride in their work but also the overall corporate culture of American society in the 1950s, the era of the organization man, class consciousness, and social conformity. In *The Status Seekers*, an analysis and critique of 1950s class hierarchies in all areas of American society, Vance Packard observed that even labor unions shed their long-held ideal of the "leather-jacketed, open-collared" worker and developed their own complex hierarchies "with many staff people in neckties."[99] This scenario was about to change radically in the 1960s, but in the studios of American record labels big and small in the postwar era, formality was the order of the day.

Krepps also insisted that engineers were the only ones determined to improve the sound of records, whereas many artists and producers believed the song was the most important factor in a record's success. The sound of a record comprised many elements, including the performers' style, the musical arrangement, and instrumentation, but the technological variables over which recording engineers or mixers had control—the studio acoustics, types of microphones used, and any effects such as echo or double-tracking—became increasingly important to a record's success and often its most identifying feature. As recording engineers devised the means of achieving a particular sound, recording artists grew ever more dependent on the engineer's skill and judgment, even to the point where one observer speculated that in the future the engineer "could take over the recording industry and dispense entirely with the musician."[100] Similar observations were made in the early twentieth century, although the replacement for the musician then was not the recordist but the phonograph. Edison's Tone Tests engendered "visions of future voiceless and instrumentless operas and concerts," and thus "no future need of paying gigantic salaries to mere human beings."[101] As it turned out, the phonograph did not replace live music, nor did artists' salaries decrease, but audio engineering became an essential part in all three: recorded sound, live performances, and the artists' successes.

Another cultural shift during the 1950s involved fundamental changes in musical organizations, and this affected the working environment of recording studios. During World War II, big bands had to cut back on touring because gas and rubber rationing imposed limits on travel and because band members had enlisted. Bands split up as members sought steady employment. Musicians who had been on the road for more than a decade simply tired of the rigors of touring. Bassist Milt Hinton found steady employment in recording sessions for popular vocalists and other artists who needed top recording musicians. Hinton, an African American who played with Cab Calloway's big band from 1935 to 1950, said that recording and radio work provided unparalleled artistic opportunity for blacks because "[a] guy hears you, he don't see you."[102] Black artists, no matter how famous, still suffered the sting of segregation and racism well into the 1950s, and although bandleaders had long ago crossed the color barrier, mixed bands were not universally accepted in some areas of the country.

White musicians also found studio work more appealing. Saxophonist Manny Albam, who in 1938 at age sixteen began touring with bandleader Muggsy Spanier and went on to play with numerous other bands, including Bob Chester, Georgie Auld, and Charlie Barnet, quit the road in 1951 and devoted full time to writing and arranging music for recording sessions.[103] In the 1950s and early 1960s, studio work steadily increased in New York City, Los Angeles, and elsewhere, as singers usurped the prominent place big bands once held. Because they needed musicians and arrangements for their recordings, skilled session players and musical arrangers were in high demand. Session players were often not the same as those who performed with the artist on the road because studio work required particular talents. The ability to perform efficiently and effectively in the recording studio required excellent sight-reading skills, a certain kind of personality and creative ability within the bounds of the session's particular requirements, and punctuality that not all musicians possessed. During his career as an active session musician, Hinton said that there were at least three hundred regular session players in New York who seemed to come from all different musical backgrounds. "No matter what their background, all the studio guys were impeccable musically and shared the ability to discipline themselves in all aspects of life. They knew studio work wasn't art, but they were willing to use their skills to perform a service."[104] As with any creative endeavor, steady income and a home life meant sacrificing some degree of artistic freedom, and session work proliferated with the rising recording industry and the boom in advertising.

Tape and the LP had far-reaching effects, not only on music listening and recording but also on patterns of employment for musicians and studio personnel. Milt Hinton credited LPs and high-fidelity sound with opening up opportunities for black musicians in studios. Although he encountered bigoted people in recording as in other businesses, most of those in positions of authority "seemed more concerned about sound than skin color. . . . It was only ability that mattered." Hinton believed that because high-fidelity recording made the sound clearer, listeners heard more, so although musicianship could be overlooked during live performances because the visual aspect was most important, "when the same person went into the studio to make a record, it was a different story. The visual was gone so the sound had to be impeccable. That's why they'd hire the best—black or white—it didn't matter. As a result, throughout the fifties and into the sixties I made records with many people—singers and bandleaders—who I never could have played with publicly."[105]

Hinton was not the only African American who benefited from the increased demand for skilled studio musicians, nor was the source of that demand limited to singers and bandleaders. Many rock groups employed session musicians because not all members of these bands were strong enough players for recording. Earl Palmer became first-call session drummer in Los Angeles from the late 1950s through the 1970s, playing on a wide range of rock and pop records with artists from Little Richard to Frank Sinatra, as well as television themes such as *77 Sunset Strip* and *The Brady Bunch*.[106]

Another consequence of the shift from name bands to celebrity vocalists was the rise in the producer's importance, because artists rarely had either the concept or the say in how they should sound. When Brooks Arthur, who started as a vocalist and then became an engineer and producer, first worked for Decca Records' founder Jack Kapp in the 1950s, the term *producer* was not yet commonly used. Rather they were A&R men, which stood for *artist* and *repertoire* because they developed the artist and selected their material, and they made a profound impression on Arthur when he first entered the business. "The A&R guys, they were the kings, they were running the show," said Arthur. "And the artist was actually even more beholden to the A&R man than everybody was to the artist, unless the artist was of tremendous magnitude. But in general, the artist was just humbled by the luxury of being able to work with this A&R man, 'cause the A&R man was *the star*."[107]

Arthur was a keen observer. As he watched people like producer Milt Gabler in the studio, or those he encountered while making his own record, he became aware of a tacitly observed protocol. If he wanted to change anything the violin section played, he had to talk to the concertmaster. If he wanted to change anything about the way the rest of the musicians played, he had to go to the general music contractor who had booked the band. All this Arthur learned by listening, observing, and "pick[ing] up the nuance of how it was all going. It was like there was an order."[108]

A similar pecking order prevailed in classical recording, where the recording director, or producer, was considered the boss.[109] Recording director Howard Scott of Columbia Masterworks thought that the most important figure in classical recording was the producer because of the many responsibilities that position entailed. "Musically," Scott noted, "the producer was the key . . . he was the catalytic agent between the man who did the recording, the engineer if you will, and the artist who performed." Scott worked with some of the greatest conductors and most outstanding musicians of the twentieth century, including George Szell, Leopold Stokowski, Leonard Bernstein, Glenn Gould, and Isaac Stern, and thus learned to deal with different temperaments and demands. He considered his job "part psychiatrist, part doctor, part genie, part everything. And [the producer's] engineer was an extension of the artist, and the producer and the engineer—they were integral parts, they all made the record." As recording director, or what came to be known as producer, of Columbia's remote recording sessions, Scott was responsible for every aspect of the date:

> I was responsible musically for the sound, the balance, the reverberation, the artistic performance depending on the ability of the artist, of course, and splicing the material together—taking out wrong notes, wrong passages, and putting in right notes and right passages. Technically, I supervised the placing of microphones with and by my engineers. It was a joint effort. I decided how close or far the placement should be, or how many microphones should be used on a section, or even an instrument. I was the one who made the choices of microphones to be used, with the engineer.[110]

Frank Bruno, Scott's engineer on these remote recordings, agreed with Scott that they worked together on various technical decisions, but it was Bruno and his engineering colleagues at Columbia, Adjutor "Pappy" Theroux and Ed "Buddy" Graham, or Irv Joel at Capitol, who saved many a remote date by arranging the transportation and proper installation of the equipment, which sometimes arrived in pieces. It was up to the session engineer to make everything operable, which sometimes required reassembling equipment that had broken during shipping. Bruno recalled that he would open the trunks and "the machines used to look like Heathkits, there were parts laying around. I'd find a condenser and say, 'Where did they use this value condenser?' [I had to look] through the schematic and see where it fell off. Those were the fun times."[111] Despite, or perhaps because of, such challenges, Bruno loved his job. As two people who worked together, Scott and Bruno had different perspectives on their relative importance to the recording session, but they worked as a team with clear notions of their individual responsibilities, and no apparent sense of rivalry or competition.

Arrangers for big bands had long been valued and were some of the highest-paid members of their organization. In 1928, Paul Whiteman's weekly payroll listed $375 to arranger Ferde Grofé, but only $150 to his vocalists, among them a young Bing Crosby.[112] The arrangers assumed even greater importance with popular singers. The record company A&R man not only looked for potential hit songs but also sought top arrangers, those who not only knew their craft but also were sensitive to what worked best for different artists and were hip to the latest styles. On observing how a recording session worked, one reviewer concluded that choosing the right arranger was one of the most important decisions an A&R director could make, "since the success of most popular tunes depends largely on arrangements."[113]

Arranger Alan Lorber emphasized how important the musical arranger became to the overall sound of a popular record in the pre-rock era, when vocalists dominated early rock 'n' roll and doo-wop. In the early 1960s, "you couldn't make a recording without an arranger," who wrote every note that was heard on the record other than the melody. Lorber described how he created "a framework for the sound," for the instrumentation and the orchestration that was chosen for a particular song, which, he believed, "dictated the needs of the creative process and where and what kind of studio would be chosen to make that record."[114] Successful arrangements for recording sessions required more than musical skill. The size of the room and the composition of the band were critical factors that arrangers had to accommodate. John Palladino spoke about the limitations on what he as an engineer could do in the studio and what the arrangers had to bear in mind when writing musical arrangements. Small studio recording posed limitations, especially when using strings with a band, and arrangers quickly learned they could only do certain things in the studio. This was why, Palladino emphasized, "a lot of the arrangers—[Frank] DeVol, Paul Weston, Billy May, or Nelson Riddle—I mean, all those guys *knew how to write for the studio*, you know? They learned it by actually doing it and finding out what they could and couldn't do."[115] Knowing how to tailor an arrangement to fit the acoustical parameters of a given studio could make the difference between a good recording and one that challenged the engineer to do his job. Phil Ramone, a violin prodigy turned engineer and later producer, recalled that some engineers might even be fearful of certain arrangers because of their particular style, "because of the big broad strokes that [they] would put in the arrangement [which] wouldn't work in a tinier room." It would then be up to the engineer to suggest that a larger studio was needed, but "as pop music became tougher and louder and crazier, you were totally dependent. And the arranger became like king for many, many years."[116] The most sought-after popular music arrangers wielded tremendous power over a session. During the 1950s and into the early 1960s, Ramone said, New York had four or five pop arrangers on call for everybody. These arrangers, he said, were "the kings of the sessions, and they dictated who the engineer should be." A little known fact of recording in those days, Ramone revealed, was that engineers "worked to please the arranger. The producer was always happy if he got what he wanted out of the date, and you [the engineer] were kind of anonymous . . . the arranger was the important guy out there."[117]

While some arrangers wrote for the room, they did not always consider the demands their orchestration might place on the recording, how the instrumentation they chose might make the engineer's job more difficult. "Quincy [Jones] or any of the arrangers at that time, they wrote for the room, but they didn't worry about how the recording was," Ramone recalled, and if the engineer could not make it work, "the producers would think you were just inadequate." Consequently, the engineer had to establish enough authority and rapport with the musicians to enlist their help by adjusting their playing as needed, something Ramone described as "a co-op affair between you and the guitar player, or you and [the drummer]—you say, 'You know, you really gotta slam that low tom for me because it's not cutting through. . . .' Instead of sitting back like the king and only sitting in your chair and adjusting equalizers and limiters. That's not the way you got people to play." [118] This cooperative effort between recording engineer and session musician was an arrangement that required trust, respect, and the shared assumption of the engineer's authority in how to get the desired sound. However, Ramone pointed out, he could not gain this by remaining ensconced in the control room. Arranger Manny Albam agreed: "For me, any engineer that just sits in the control room and doesn't go out and listen to what something sounds like, I would never go back and use the guy."[119] Songwriter Jerry Leiber told Tom Dowd, "'Man I can't stand going in the studio when the engineer won't get up to see what's going on! . . . just twirls knobs.'"[120] The engineer was not a control man simply because only he could operate the controls; he needed, above all, to use his ears and his ability to understand what the artist and the producer hoped to achieve in the recording. His job was like that of a translator, as Albin Zak noted, "of musical ideas, human presence, artistic personalities, the sounds of instruments, voices, and rooms . . . from their original state into the medium of the recording."[121]

The frontispiece of Peter Galison's cultural history of microphysics, *Image and Logic*, features a cartoon that depicts the various ways in which different members of a research and development team viewed the magnetic particle detector, the result of their combined efforts.[122] Each perceived and defined the technology in terms of its relevance to his or her own field of expertise. It should be apparent from the foregoing recollections that the different perspectives of each member of a recording team could be depicted much the same way. Producers, engineers, session musicians, arrangers—each perceives his contribution to the success of a recording session as critical, and in a sense, they are all correct. It should also be clear from these recollections that they all considered themselves part of a team. Like the physics lab described by Galison, the work of the recording studio required coordination between its separate spheres— the success of a recording rested on the joint efforts of engineers and mixers, musicians, arrangers, conductors, producers, all coming together with a common goal of making the best record they could. During the 1950s in which so much was changing both technically and artistically, the recording studio constituted even more of a laboratory than in the experimental acoustic days, when each new innovation was designed to achieve greater control over the sound, introducing not only more choices but also increased challenges to adapt old methods to new technologies. The recording studio was like a "trading zone," as Galison described the physics laboratory, where different technical and artistic traditions worked cooperatively. The recording business constituted especially at this time a "multitude of worlds" in which individuals from remarkably diverse cultural, social, educational, and economic backgrounds collaborated.[123]

The concept of a trading zone might also apply to the informal trading network that organizations such as the Audio Engineering Society and the Institute of Radio Engineers' Professional Group on Audio fostered. In the early years, recordists and the companies for which they worked closely guarded their proprietary designs—of the sound box, cutting apparatus, recording techniques, and microphone placement. During and after World War II, and particularly with the formation of these professional societies in the postwar period, there appears to have been more sharing between recording engineers in competing firms. This kind of informal know-how trading between rivals became an important source of innovation.[124] As Tom Dowd put it, "We were friends, swapping stories . . . there was a new breed of people . . . there'd be this free exchange of information that was not political, or handcuffed to a company."[125] Two major catalysts for this openness included the wider availability of technology as more manufacturers entered the recording equipment market and the overwhelming desire of recording engineers to know more about what they were doing and how they could improve on it. When Clair Krepps first set up the Capitol Records mastering studio in New York, with nothing but his Navy signal corps experience, he was able to call the chief engineers of RCA Victor and Columbia, Al Pulley and William Bachman, who "would stay on the phone for a half hour answering the questions that I would throw at [them] on how to do this and how to do that."[126]

Engineers were motivated by the desire to make better-sounding records and by a certain kind of technological enthusiasm, and why not? Since they received no credit for their work, nor any financial incentive for their contribution to a recording, what would prompt them to compete with one another? Competition and secrecy certainly surrounded hit songs, and engineers were instructed not to cut artists personal demos of the songs they recorded until after their official release. Clair Krepps once succumbed to the "feminine wiles" of Joni James, giving her a copy of her recording although he had been specifically instructed not to do this until it was released. She promised she would play it only for her boyfriend, but proceeded to play it at a party that night and, as Krepps recalled, "within twenty hours our competition was putting out the same song. I almost lost my job over it. When people talk about secrecy in the record business, that was the secret. You never wanted your competition to know what songs you were recording."[127] Between engineers, however, competition usually took the form of friendly one-upmanship.

Conclusion

If the late 1930s and 1940s brought improved record quality and more lively acoustics in recording, the immediate postwar era brought revolutionary change in recording technology even further improving the fidelity of records and changing the nature of studio work, magnifying the role of the engineer and producer and the importance placed on teamwork. The ability to manipulate sound and to alter the recording after the fact made it possible to achieve identifiable sounds in popular music and increased precision in classical recording. These trends resulted from technological developments, but they also fueled further innovation and increased the complexity of the recording process, putting greater responsibility on the engineer. More than any other time, the 1950s laid the technological groundwork for the creative freedom that flourished in the following decade. The song remained important, the music still at the heart of recording, but "the Sound"—in terms of quality as well as individuality—assumed greater importance than ever before in making records and in forging careers. Technological innovations and engineers' inventive adaptations to their use set the stage for the exploitation of the recording studio as a new creative medium in the 1960s.

The big band era was essentially over by 1955, destroyed by the rising expense of keeping large bands on the road, by the emergence of small combo jazz, solo pop vocalists, and a growing number of very active recording studios where top musicians found steady work as session players on records, on advertising jingles, and for film and television. A listening public eager for something new became increasingly drawn to consumer audio, "high fidelity" sound, stereo record players, and radio-phonograph consoles. Advances in audio engineering had changed forever the nature of sound recording, of musical performance, and of listener perception of music, and the recording studio was at the center of that transformation. The major label studios based in New York and Los Angeles maintained high standards consistent with their technological capabilities, training young engineers for months and even years to maintain consistency in the company's particular sound. The labels also maintained an active location recording schedule; some transported their equipment while others used local studios. Nashville became a hotbed of recording activity as country music's popularity escalated; every label sent A&R men to record at Nashville's Castle Studios from the late 1940s through 1953, and by 1956, RCA had built its own studios there.[128] But as blues, rhythm and blues, rockabilly, and rock 'n' roll, became increasingly popular with younger audiences, the fidelity in studio recording that engineers had long worked to attain became inconsequential to some listeners. The rise of small labels and independent recording studios ready to serve new, untried talent during the 1950s and early 1960s brought a whole new meaning to unique sound.

The Search for the Sound

Rhythm and Blues, Rock 'n' Roll, and the Rise of the Independents

*D*uring the 1940s and 1950s, control men and staff producers at Columbia, RCA Victor, Decca, Mercury, and Capitol refined their recording techniques and established slick production values for the classical, popular, and swing artists signed to those labels. This same period also witnessed the rise of many small record companies established to serve a growing consumer market for boogie-woogie, rhythm and blues, country, and rock 'n' roll—working-class music considered outside the mainstream musical culture and largely misunderstood by established record labels.[1] Like many other small-business enterprises in postwar America, independent labels found success by exploiting niche markets for specialty products, which, in the world of recording during the 1940s and 1950s, meant gospel, blues, rhythm and blues (still classified as "race" music until 1949), jazz, hillbilly/country, and rock 'n' roll.[2] With the black migration to northern and western cities both during and after World War II for work in the auto, steel, rubber, and defense industries, African Americans brought their taste in music as well as their musical talent, and the supply as well as the demand for black musical genres spread throughout America, not only among black listeners but among white record buyers as well. The same was true of hillbilly music, which by the late 1950s had developed into a Nashville-based country music industry with mass cultural appeal, a development that began over a decade earlier with a vibrant local recording scene.[3]

In addition to population migrations, other forces at work contributed to America's changing musical landscape. In the early 1940s, two industry boycotts, one in radio broadcasting, the other in recording, opened the door for the music of these small labels to reach a wider audience. The first of these, the 1941 radio boycott of all ASCAP-licensed songs, gave the music that had been rejected by that performing rights organization, but signed by its rival BMI, wide distribution on national radio. Overnight, the music of America's most popular composers such as George Gershwin and Cole Porter disappeared from the airwaves, replaced by the music of Hank Williams and Wynonie Harris.[4] The second boycott, the previously mentioned musicians' strike of 1942–44 during which no union musicians were permitted to record for commercial records or transcriptions, caused a cessation of any new recordings by the majors and opened the market for small labels willing to sign agreements with the musicians' union while the major labels refused.[5] While seemingly tangential to the technological development of recording, these events influenced musical culture in America by giving a boost to the styles of music that had previously only been heard in regional markets, recorded almost entirely by independent recording studios, and released on small labels.

Some of these independent record companies, such as King Records in Cincinnati, grew into little recording empires, vertically integrated operations that controlled under one roof every aspect of making records from recording to pressing to shipping.[6] Motown Records in Detroit discovered, recorded, and developed artists and maintained a stable of session musicians that became integral to the Motown Sound.[7] Most small labels, however, did not assume such ambitious proportions. Some, such as Chess Records in Chicago and Atlantic Records in New York, started with nothing but an office that occasionally served as the recording studio until the company's success warranted building studios of their own.[8] Some independent record labels in major entertainment centers rented studio space, hired engineers, and used pressing facilities owned by the major labels, who thereby cut their overhead, but in doing so, they nurtured the competition that would eventually pose a major challenge to their market supremacy.[9] Outside the major entertainment centers, operations such as the Memphis Recording Service grew from owner-operator Sam Phillips's part-time recording hobby into a commercial studio and only later evolved into the Sun Records label.[10] Most of these independent labels, such as Atlantic, Dot, Modern, Gone, Elektra, Roulette, and Veejay—literally hundreds by 1959—used a growing number of independent recording studios both within and outside of the major entertainment centers.[11] Unlike the rash of small labels that sparked the blues craze in the 1920s, most of which folded or consolidated during the Depression, this new crop of independent labels and recording studios spearheaded the transformation of the American recording industry and musical culture.

Independent recording studios in the 1950s ranged from highly professional facilities such as Universal Recording in Chicago and Fine Recording in New York, to radio stations, basements, storefronts, and backrooms of radio repair shops. The postwar availability of affordable, quality tape recorders and the emergence of new musical genres that captivated young listeners, inspiring many to have musical aspirations of their own, created the ideal conditions for the proliferation of small studios. These differed from the many transcription and production studios of the 1930s and 1940s, some of which continued after the war, but most did not survive the end of the transcription business, which faded with the rise of the disc jockey in the 1950s.[12] The independent studio business that mushroomed after World War II grew to meet different needs, recording everything from advertising jingles and orchestra performances, to polka and rock 'n' roll. Many of these offered remote recording services, and by the 1960s, several companies specialized in on-the-spot recording for high school bands, churches, weddings, and even funerals. The so-called vanity records market flourished during the 1960s until the later introduction of consumer cassette recorders gave these customers the ability to make their own recordings. Within two decades of World War II, recording technology had reached the masses.

Some independent studios produced high-quality recordings that rivaled the major flagship studios such as Columbia and RCA, whose technical departments created the necessary "black boxes" and equipped their studios with the most sophisticated technology. A few independent studios were highly organized, efficient, and technically advanced with a staff of talented mixers and skilled maintenance engineers. In the major recording centers of America, these independents started to compete with major studios. Other independent studios, located outside the industry hubs, were owned and operated by a single proprietor, who hired perhaps one or two assistants. These studios symbolized opportunity for aspiring singers and songwriters who might otherwise never have the chance to make a record. The most celebrated example was Elvis Presley, who first came into Sam Phillips's studio to make a vanity record for his mother. By providing the springboard for nascent talent, independent studio owners injected new blood into the record industry. At the same time, the increasing number of musical hopefuls encouraged by the meteoric rise of rock 'n' roll and the growth of the record industry gave these studios ample work, creating the demand for more. Offering creative freedom, an experimental, can-do approach to making records, sometimes from equipment cobbled together from secondhand or home-built components, studio entrepreneurs were often risk takers who appreciated the kind of music that was not being recorded by major record labels. What they lacked in technical sophistication they made up for with enthusiasm and inventive problem solving.

Independent studios increased dramatically with the postwar availability of affordable tape recorders that made professional-quality recording attainable by less technically inclined amateurs, who were more interested in recording unique musical expression than in understanding how electrons flowed through vacuum tubes or the intricacies of disc cutting. Before World War II, recording was a serious amateur pursuit that required considerable technical facility because most of the equipment had to be home-built, assembled, and maintained. Driven by their interest in radio and electronics, a fascination with how machines worked, and a desire to harness sound from the air and engrave it onto shiny lacquer discs, these early amateurs embodied a technological enthusiasm for building and tinkering with both equipment and sound. By the 1960s, recordists had become "recording engineers," yet ironically, many of those who entered this profession often knew less about electronics and the basic technical foundations of recording than their self-taught predecessors. For them, the music, more than the technology, had provided the lure. Yet, for many in the 1950s, opening a recording studio involved equal parts technological enthusiasm and artistic inclination. Sam Phillips described his driving force this way: "I am a sound freak. I could play around with sound forever."[13] Playing with sound became an important factor in the distinctive sound of his Sun Records, just as "chasing sound" enabled Les Paul to create his own unique musical identity through recording techniques and sound effects.

This chapter examines the evolution of four different independent recording studios—three in Cleveland and one in New York City—each with a different history, each serving a slightly different clientele but all sharing certain attributes. Collectively, they offer a window into how the musical culture of the postwar period came to be defined by the recorded performance. Independent recording studios met the growing demand of aspiring musicians and songwriters to disseminate their music beyond the local market. These studios symbolized opportunity for studio owners and operators as much as for their musical customers, and these stories reveal the ambitions of small-business owners at a time when perceived opportunities frequently outweighed actual financial success. Whether the studio was in the heart of the recording industry or in a midwestern steel town, the musical clientele only provided one portion of revenue, for advertising became the bread-and-butter business for most of these independent operations, as it did for some major record company studios who served advertising clients when the facilities were not booked for the artists.[14] Yet it was the music customers, not advertisers, who drove technological change in these independent studios, and with no corporate or union restrictions on how they recorded or who touched the controls, independent engineers collaborated with musicians, experimented with sound, and willingly tried new approaches to recording.

From Polka to Punk: The Cleveland Recording Company

In the mid-1930s, Frederick C. Wolf, a Czechoslovakian émigré and radio announcer, started a modest recording studio in downtown Cleveland, Ohio, with a Presto transcription disc recorder. Wolf was neither an engineer nor a musician; he was a businessman, a political mover in Cleveland's ethnic communities, and an enthusiastic promoter of the music of his homeland.[15] During the 1930s, Cleveland was a hotbed of ethnic music in America, earning the title of "Polka Town."[16] Every Sunday from eight o'clock until noon, Wolf and his fellow nationality announcers, Joe Pales, "Heinie" Martin Antoncic, John Lewandowski, and Jimmy Rose alternated using the mike to broadcast their respective half-hour shows of Czech, Slovak, Slovenian, Polish, and Hungarian music. In 1936, the group formed a business partnership, Nationalities Broadcasting Association, and offered an award to ethnic broadcasters to encourage quality programming.[17] Wolf and his associates bought the Sunday morning "dead air" time at a special rate, acquired sponsors, hired local musicians they knew, and broadcast the music live over radio station WGAR. One of those musicians, Sheldon Henderschott, remembered Wolf as "a very sharp businessman," who during the Depression sold advertising time by walking up and down Broadway, going into each store, and accepting payment in scrip, a commonly accepted substitute for currency during the Depression, as well as in cash, in turn paying the musicians in half scrip and half cash.[18] Finding it increasingly difficult to get up in time to make the Sunday morning shift, Wolf decided to record his Czech music programs and then capitalized on his investment in the disc recorder by charging his fellow nationality program announcers to record their own programs, thus was born the Cleveland Recording Company.[19]

In 1938, Wolf recorded the polka musician and bandleader Frankie Yankovic when the future "Polka King" and Grammy Award winner was still working in factories by day and playing in local taverns by night.[20] After Yankovic was turned down by both Columbia and RCA Victor when he sought recording contracts, he recorded his first two 78s—three Slovenian polkas and a waltz medley—at Cleveland Recording. The records sold well locally as did four more sides he recorded the following year. In 1943, Yankovic joined the US Army, but he continued to cut records at Cleveland Recording during furlough. With no time to rehearse and limited funds, Yankovic made the most of his Cleveland Recording date, recording thirty-two sides in one afternoon session. "We had no time to fool around," Yankovic recalled, "if somebody hit a wrong note, we just kept going." When one of the musicians suggested recording a song over because the musicians had made too many mistakes, Yankovic insisted, "Leave the clinkers in. People like it better that way."[21] Such a casual attitude toward the performance did not reveal a lack of perfectionism so much as Yankovic's conviction that the mood of the recordings should be as spontaneous as their live performances.[22] This would not have been acceptable practice at Columbia or RCA, but since Yankovic—not a record producer or company representative—was the leader, the decision was his. Even in the best circumstances, mistakes were virtually unavoidable in the days before magnetic tape recording and the editing and re-recording it enabled. Yet imperfect as they may have been, those early Cleveland recordings got Yankovic signed to Columbia Records in 1946 and established Cleveland Recording's reputation as *the* place for Cleveland's polka musicians to record.[23]

In 1950, Wolf fulfilled a long-held ambition to open his own radio station devoted primarily to nationality broadcasting, and he relocated Cleveland Recording to the new studios of radio station WDOK on the fourth floor of the Loew's State Theater Building, an ideal site for his recording studio.[24] Soon after he opened the station, Wolf hired Kenneth Richard Hamann, a young navy veteran, as an engineer for both WDOK and Cleveland Recording. He could not have made a better choice. Hamann had dabbled in electronics since childhood, built a dual turntable setup for the "teenage canteens" sponsored by his high school, and studied at Case Institute of Technology before enlisting in the Navy where he received training in aviation electronics. He also held a FCC (Federal Communications Commission) First Class broadcast engineer's license, was a member of the Institute of Radio Engineers, and had a strong musical sensibility.[25] Initially, Hamann was one of several staff engineers who split his time between the transmitter site, which had to be monitored every moment they were on the air, and running recording sessions as they were booked. Cleveland Recording still operated part time, and most of the work served advertising agencies and corporations such as Ohio Bell, Kirby, and Westinghouse, companies that required synchronized recordings for the 35-mm strip films they used for employee training purposes.[26] Hamann became increasingly drawn to the recording side of his job, steeping himself in audio technology and experimenting with various ways to extend the studio's capabilities. He built equipment in his basement, including monitoring and switching devices to add to Cleveland Recording's facilities in an effort to overcome the limitations of disc recording. Eventually, Hamann made the equipment so complex that only he knew how to use it, thus inadvertently building in a kind of job security, but he insisted "it *had* to be in order to do some of the things that we were getting into recording-wise."[27] Keenly aware of the achievements of contemporaries through his involvement in the Professional Group on Audio special-interest group of the Institute of Radio Engineers, and the Audio Engineering Society, Hamann sought to achieve state-of-the-art technology by whatever means necessary, thus giving Cleveland Recording a competitive edge.

Hamann soon got the chance to expand his engineering skills when in 1952 Wolf purchased an Ampex 300 monophonic tape recorder, which as he put it "opened up all kinds of vistas" for recording, editing, and sub-mixing.[28] Throughout the 1950s, Hamann worked on developing stereo methods of broadcasting and recording and was awarded a patent for a binaural system of disc recording.[29] Like other hi-fi hobbyists at the time, he experimented with "ping-pong stereo," recording the roller coaster at Euclid Beach Park and other environmental sounds on the Ampex 2-channel recorder the studio acquired in the mid-1950s. At Cleveland's Second Annual High Fidelity Fair, held at the Hotel Statler in November 1957, Hamann presented "A Visit to Stereo-Land," a demonstration of his recordings for local high-fidelity fans.[30] As early as 1954, he had begun stereo broadcasts over WDOK using both the AM and FM channels and, in October 1958, had devised and engineered the world's first 4-channel stereo broadcast, a live performance by the Dukes of Dixieland from the Modern Jazz Room over radio stations WERE-AM and -FM and WDOK-AM and -FM in Cleveland.[31] Scholars have analyzed the high-fidelity craze from the perspective of home consumers, mostly male, whose penchant for creating the ideal listening environment for high-fidelity sound led them to build unique stereo equipment for the home.[32] Hamann represents another kind of hi-fi fan, one who created unique listening experiences for the public as well as for his own home-listening environment.

In 1959, impressed with the quality of the German 78s on Deutsche Grammophon they had received from Decca Records A&R man Lenny Joy, Wolf and Hamann traveled to Europe where they met with the German audio manufacturer Georg Neumann and visited recording sessions at Teldec and Deutsche Grammophon. They returned to Cleveland convinced that the Germans, as Hamann put it, "were not doing anything we were not capable of doing, except that they were paying attention to detail—every little thing—and the results showed."[33] Cleveland Recording had been using Neumann microphones since the early 1950s, but they now acquired a Neumann cutting lathe, and Hamann designed a recording console based on the flat desk design he'd seen for the first time in a Berlin beer hall that during the week doubled as a recording studio. So he purchased the necessary components and built a flat desk, 10-channel input, 3-channel output console with linear faders that enabled him to adjust several channels with one hand during mixing and incorporated equalization and limiting for each individual channel rather than the entire board. Since the major studios of Columbia and RCA were at this point still using consoles with round volume-control knobs, and even independent equipment manufacturers such as Langevin continued to produce this type of console as late as 1968, Hamann's design was technologically somewhat advanced for 1963.[34] He also assembled a remote recording unit, a six-foot-high, twenty-two-inch-square rack of portable recording gear he transported in the back of a station wagon, using it to record performances at the Oberlin College Contemporary Music Festivals conducted by Aaron Copland and Igor Stravinsky. By the early 1960s, Cleveland Recording had both studio and remote facilities that met high standards and a skilled engineer with unbounded enthusiasm and ingenuity.

Hamann's first rock 'n' roll recording session occurred shortly after he and Fred Wolf returned from Germany. Decca Records had commissioned Cleveland Recording to tape Bill Haley and the Comets' performance at Cleveland's Masonic Auditorium. This experience proved to be a logistical nightmare for Hamann, who had to navigate the heavy portable recording gear up a narrow flight of stairs into the auditorium's projection booth. Well into the 1960s, rock concerts continued to be held in auditoriums built for acoustical performances, not electronically amplified rock concerts, so Hamann learned some important lessons. First, he was not prepared for the volume or intensity of rock 'n' roll. He had already learned that some of the louder passages of classical pieces distorted on tape, and he determined that the microphone inputs of his mixer were too sensitive for the Neumann U47 mikes, which featured pre-amplifiers. His first remedy was to introduce plug-in attenuators between sections of the mike cable, but he eventually designed his own mixer so the high microphone output would not overload it. The other lesson he learned recording the Bill Haley concert was the wide variation in output volumes of the group's instruments and the ensuing difficulty of capturing drums and electric guitars as well as acoustic bass on tape. With only four microphones at his disposal, and one reserved for the show's announcer, disc jockey Bill Randle, Hamann learned that recording rock 'n' roll required techniques far different from those he'd used recording a symphony orchestra or a polka band.[35]

By the early 1960s, the studio's clientele began to change, reflecting trends in musical culture sweeping America. The commercial clients continued to be the primary source of Cleveland Recording's revenue, and polka groups kept coming in, but now there were more bookings for popular singers such as the Tracy Twins, rhythm-and-blues groups, and rock bands, ranging from high school kids to semiprofessional bands and singers wanting to record demonstration records to send to record companies. Like so many baby boomers, Cleveland's teens had been "jolted by the original rock revolution in the mid-'50's," a local photojournalist observed, and "were starting to get the idea that they could make this music themselves [and] were putting down the accordion and picking up the guitar or sax."[36] Groups such as Joey and the Continentals, Rocco and the Flames, the Montclairs, and Tom King and the Starfires made records at Cleveland Recording and released them locally. The Montclairs' instrumental "Happy Feet Time" was the first of the studio's records to hit the national charts in 1965, and The Outsiders's "Time Won't Let Me" rose to number five on the *Billboard* pop charts in 1966.[37]

That recording was significant for Hamann because it represented hours of effort and experimentation to achieve the proper mix and years of accumulated technique recording polka, classical, and now rock music. To reach the point of making a properly balanced 3-channel recording of a song that featured drums, bass, guitars, organ, trumpet, saxophone, tambourine, and vocals—all engineered so skillfully as to maintain crisp presence on the final record—Hamann had gone through what he called "an evolutionary process." Some of the early recordings, as he put it, "were, generally speaking, pretty awful," but each time he made a record, he learned something and tried to build on the experience. Not only did he have to devise ways to accommodate the louder drum and guitar sounds of rock 'n' roll, as he had learned recording the Bill Haley concert, but the limitations of three channels forced him to anticipate the final product and to develop sub-mixing techniques to achieve the correct proportion of drums and bass on the first track, so that subsequent tracks would not overpower them.[38] Like the early acoustical recordists who were forced to arrange instruments around the horn to capture a well-balanced performance, Hamann and other engineers of the early 1960s made similar accommodations with limited tracks and skillful, well-planned overdubbing.

By 1966, three years after Hamann had completed building his 3-channel mixing console, Cleveland Recording acquired a Studer 4-track recorder, which Hamann modified to 8-tracks two years later. With multi-track recording capabilities, instruments now could be recorded more discretely and thus controlled more fully in the final mix. In July and October 1967, Hamann recorded two more songs by Ohio groups, which both reached the *Billboard* "Top Ten" during the week of February 3, 1968: The Lemon Pipers' "Green Tambourine" and the Human Beinz's "Nobody but Me." A proliferation of local rock groups recorded at Cleveland Recording in 1968, including the James Gang with Joe Walsh and the following year Detroit musician Terry Knight with members of his former rock group, The Pack, who now called themselves Grand Funk Railroad. The four demos they made at Cleveland Recording marked the beginning of a long and lucrative association between the studio and one of its most successful clients. After Knight, who acted as producer and manager, secured Grand Funk a record deal with Capitol on the strength of their Cleveland Recording demos, they returned to Cleveland to record their first album. By then, Hamann had managed to talk Fred Wolf into buying a 16-track Ampex recorder with some of the proceeds from the sale of WDOK—a somewhat extravagant move that paid off.[39] Between 1969 and 1973, Grand Funk recorded seven albums at Cleveland Recording, each of which sold in excess of a million copies. With record sales like that, the group and its producer could have had their choice of studios, but Knight preferred to work at Cleveland Recording with Ken Hamann. When Don White, an engineer Knight had previously worked with at Cleveland's Audio Recording Studio, tried to lure his former client back, Knight declined by explaining: "When you're winning at poker, never get up from the table."[40] Knight believed he had found a well-kept secret, telling a *Billboard* magazine columnist in 1971 that he continued to work at Cleveland Recording because it was "a highly developed—as yet undiscovered—studio. . . . I find it technically to be one of the five top studios in the country."[41] Considering that Knight could have had his pick of studios, this was high praise for Cleveland Recording, but it had as much to do with Ken Hamann's engineering skill and his studio's laid-back atmosphere, as it did with the technology at their disposal. During this period, increased competition between independent studios for clients manifested in the race for more studio gear and in the aesthetic "vibe" of the studio and skills of the engineer.

For his part, Hamann credited Fred Wolf for having the risk-taking spirit that led him to finance studio improvements, despite being, as Hamann put it, "all thumbs" in the studio. "We shuddered every time he put his hands on certain pieces of equipment."[42] But it was Hamann who continually adapted and built equipment and accommodated the increasingly experimental recording ambitions of his music clients. One early example of Hamann's willingness to experiment, and of the extravagance that characterized much later rock studio work of the 1970s, was a session during which the rock group The Human Beinz chopped up a piano in the basement of the Euclid Avenue building while Hamann miked the process, feeding the lines up to the fourth-floor recording studio. In other sessions with the James Gang, Hamann recalled that Joe Walsh was "forever experimenting" with getting new and different sounds for his guitar and his voice. "We tried out different ways to modulate his voice. We ran his voice and/or guitar through a Leslie speaker, tried all kinds of different things, unusual things; tried to get new sounds or interesting sounds that worked."[43]

This quest for sonic singularity had not begun with rock guitarists. In the late 1940s, Les Paul created an identifiable guitar sound based on recording technique, which he later had to find ways to recreate on stage.[44] Novelty tunes exploiting studio tricks and gimmicky stereo-exploitation records such as Enoch Light's *Persuasive Percussion* albums flourished in the 1950s. But the concerted effort to find new sounds in the 1960s, in part a consequence of the availability of new signal-processing devices, and in part the need to distinguish one's style, was best described by James Gang drummer, Jimmy Fox, as "the search for the sound."[45] After the release of the Beach Boys' *Pet Sounds* and the Beatles' *Sgt. Pepper* albums, many rock groups embarked on a quest for unusual sounds that were often un-reproducible in live performances. The "quest for fidelity," at best always an elusive goal, had never been the holy grail of rock music, in which the record had become a creation, no longer simply intended to document a well-rehearsed or even an inspired performance. Musicians sought to create new sounds, to layer and shape a song using the tools of recording technology. In the process, a new symbiotic working relationship had developed between musicians, engineers, and producers in the recording studios of the 1960s, which had a profound effect on the music they made. As Hamann aptly summed it up in 1995:

We were working together during those years, during the late '60s mostly, and it became much more of an art. I would say, in retrospect, a true art form, where the engineer and the record producer and the musicians were working together as a

team to create music, sounds that portrayed some kind of image. Just like a painter
... with colors on a palette. We were doing the same thing. And we were able to. We
were told more than once that our *freedom*, the freedom that the producers and
musicians enjoyed in our studio, was far beyond what they could have in either
New York, Nashville . . . [or] Los Angeles, in particular, because the unions out
there would prevent anybody from getting their hands on the controls.[46]

The symbiotic relationship between engineer, producer, and musician that Hamann describes, and the freedom from union restrictions that were imposed
at major labels, sums up the appeal of the independent studio in the 1960s.
Ken Hamann was not a musician, he was an engineer, but he approached audio engineering as an art.[47] Perhaps because of that he did not cultivate the kind
of shrewd business sense that may have precluded giving away studio time to
struggling musicians, some of whom left Cleveland Recording for the big studios once they had achieved national acclaim. His own technological enthusiasm fed off creative musicians, without whose prompting Hamann would undoubtedly never have tried certain things. Cleveland Recording was significant
for Fred Wolf's foresight that recording was the coming field, and for Ken Hamann's ability to keep the studio technologically competitive with the best in
the country by adapting, innovating, and building his own equipment. Although,
at times, Hamann ran the studio on a shoestring budget, Cleveland Recording
outlived many of its contemporaries.[48]

Wolf remained deeply involved in the studio until the late 1960s, when declining health forced him to retire from the studio business entirely. In 1970,
Hamann and another engineer, John Hansen, bought Cleveland Recording and
moved the studio to a new location, spending six months renovating an old
Chevrolet dealership on Cleveland's Near East Side. John Hansen handled commercial clients, and Ken Hamann continued to record rock groups such as Wild
Cherry, the James Gang, Tiny Alice, and Pere Ubu, but the string of hit records
he had engineered in the 1960s was not to be repeated.[49] Musicians were notoriously poor-paying customers, and it became increasingly evident that the
commercial clients were keeping Cleveland Recording afloat. Tension mounted
between the "rather raucous" music clients and the "rather straight-laced"
commercial clients, as well as between Hamann and Hansen over money matters, finally leading to what Hamann characterized as "a divorce," when, in 1977,
the two owners decided to go their separate ways.[50] They divided the studio
equipment, Hansen kept the name Cleveland Recording, and Hamann opened
a new studio, Suma Recording, in an old country home east of Cleveland. Cleve-

land Recording continued in operation until Hansen's death in the early 1990s, and Suma Recording continues to record clients ranging from the experimental rock group Pere Ubu to commercial jingles.[51]

"Hum Your Tune. Hits Start at Schneider Recording"

The forty-year history of another Cleveland studio, Schneider Recording, offers a view into a different kind of independent studio that flourished in the postwar period. Although Hank Schneider began recording polka musicians and advertising clients, just as Fred Wolf had in the 1930s, Schneider Recording remained a small, family-owned and operated business, offering a range of services in addition to recording. Hank Schneider was an engineer as well as a musician, but his studio never became as technologically advanced as Cleveland Recording. From the time he opened the studio in 1945 until his death in 1984, Hank Schneider and his wife, Kay, served a wide range of clients, from advertising agencies and corporations to professional musicians and amateur songwriters. They maintained a conservative approach to running the studio, never investing in more equipment than the business would support. By keeping it small and personalized, they weathered business fluctuations that put many other small studios out of business. Schneider Recording Studio Laboratory, as it was called in the early years, could indeed boast that it recorded the original audition records for several hit songs of the 1950s, as well as album masters for companies such as Decca and RCA Victor.[52] Precisely because its doors were open to even those who had done no more than sing in the shower, Schneider provided an invaluable service to songwriting hopefuls who otherwise would never have entered a recording studio. Kay Schneider recounted that just about every songwriter who came through Schneider's door would say something like, "Now I wouldn't even be here if my tune wasn't better than 'Stardust.'"[53] Whether it was or not did not matter to the Schneiders. Their interest was in providing a service, not in profiting from the songs they recorded, or in promising to place songs with artists, as did the "song sharks" of Tin Pan Alley.[54]

Hank Schneider possessed a unique background that combined technical training and professional musicianship that ideally suited him for the newly emerging audio engineering field. As a young boy, he built crystal radio sets and later opened a radio shop in Oak Park, Illinois. He started his own band at age fifteen, played trombone with Benny Goodman and Gene Krupa in Chicago, and later performed with orchestras on Mississippi riverboats. From the 1920s through the early 1940s, he arranged music for live radio broadcasts, and during

World War II, he served with the Army Signal Corps. After years on the road as a touring musician, Schneider and his wife settled in Cleveland at the St. Regis Hotel. Schneider rented a downtown office space, initially as a place to do his musical arranging while he played with local radio orchestras and club bands. Soon he purchased a Presto disc recorder and a Brush crystal microphone, listed as Henry C. Schneider Recording Studio in the telephone directory, and offered his services as lead sheet arranger, musician, and recording engineer. Between that modest advertising and Schneider's reputation among local musicians, by the early 1950s Schneider Recording became so busy that he was forced to give up performing entirely except for studio work. The rise of television, the recording industry, and the radio disc jockey hastened the end of live radio broadcasts and thus the end of radio work for musicians and arrangers, but disc jockeys, television, and record promotion also contributed to the popularity of country, rock'n' roll, and rhythm and blues, genres that inspired musical hopefuls who came to Schneider and other small studios to make records.[55] Moreover, Hank Schneider is an example of a working musician displaced by technological change, who saw in that technology the opportunity for an alternative career.[56]

In 1947, Hank Schneider bought a Brush BK-401 "Soundmirror," the first magnetic tape recorder sold in the United States and one of the first commercial tape recorders on the market after World War II.[57] Although this was not a sufficiently high-quality machine for recording music, Schneider used it for spoken-word recording, beginning with a twenty-six-week series, "In Baseball Today," narrated by Cleveland Indians' pitcher Bob Feller, that was carried by forty radio stations.[58] Schneider recorded the weekly series on tape, but because radio stations had yet to convert to tape and still used transcription equipment, the program had to be transferred to discs.[59] To supply the forty transcription copies per week, the Schneiders bought four Fairchild disc recorders, but even then, the job of cutting forty discs still took two people five hours to complete. Determined to maintain a conservative business model and to keep overhead low, the Schneiders did not hire assistants.

In 1950, having outgrown the original location at 1514 Prospect Avenue, the Schneiders moved to a larger space down the street, a former hotel with hardwood floors, acoustically far superior to the cement floors of its former location. Schneider soon acquired his first music quality tape recorder, a Presto RC-7, and began to serve more music customers, recording polka musicians Ernie Benedict and Johnny Vadnal for RCA Victor, as well as making master recordings for Decca, Capitol, Mercury, and Continental. Schneider recorded many types of nationality and folk music and Cleveland's many ethnic communities—Serbian, Syrian, Greek, Hungarian, German, and Italian—provided a ready clientele.[60] Very quickly, the Schneiders' reputation for personalized service also drew aspiring songwriters who could not play or read a note of music, but who eagerly responded to the sign by the studio entrance that declared: "Hum Your Tune. Hits Start at Schneider Recording."[61]

In 1952, a young high school student from Cleveland Heights, Don Howard, came into the studio with a song he wanted to record for his father's birthday. The record he made at Schneider's studio, "Oh, Happy Day," was a somewhat off-key, out-of-tempo, sentimental yet sincere ballad that caught listeners' ears and became an instant local radio hit, selling so many copies that the local pressing company, Triple-A Records, could not manage its distribution, and the song was subsequently picked up by a national label, Essex Records, and became a national hit.[62] Howard had performed the song himself, but many of the clients Schneider attracted did not play an instrument, and one songwriter even came to the studio with lyrics written on toilet paper. Kay Schneider remembered that "so many of these people could not write music at all. They would hum the tune to my husband, he would write it down in lead sheet form, which is the first thing you'd do—you can publish it with that . . . and I would type the lyrics underneath."[63] Typical of these clients were George and Pearl Lendhurst, a local husband and wife songwriting team who brought several song ideas to Schneider's studio in 1957. By this time, Schneider had upgraded his recording equipment to include an Ampex Model 350 tape recorder and Telefunken U47 microphone, providing considerably higher sound quality than either the Soundmirror or Presto recorders. He also started working with Dick Glasser, a musician from Canton, who with Schneider would provide the musicianship and arranging for songs their clients could only hum. Glasser sang and played guitar, Schneider provided other instrumentation, and the two made numerous audition records by multiple recording their parts, becoming what Mrs. Schneider called "a little band."[64] The audition tape of one of Mrs. Lendhurst's songs performed by Dick Glasser so impressed Phil Chess of Chess Records in Chicago that he called the Schneiders to secure the rights to release it. Because Glasser was under contract to another company, Chess Records could not release the Schneider demo, but another company snatched up the song and within months of the time she and her husband entered Schneider's studio in 1957, Mrs. Lendhurst's "Be-Bop Baby" became television teen idol Ricky Nelson's first hit on Imperial Records, climbing to number three on the *Billboard* pop charts. Glasser, who wrote several songs covered by other artists, eventually moved to California and became a producer for Liberty Records, the parent label of Imperial Records, and later worked for Warner Brothers Records.[65]

The Schneiders gained no fame or fortune from the success of these recordings, nor did they seem to care. On the contrary, they steadfastly avoided doing promotional work for the artists they recorded although they were happy to recommend songs they felt had potential to the record company executives with whom they dealt. Although they advertised that "Hits Start at Schneider Recording" the studio, like Cleveland Recording, paid its bills by doing radio and television commercials for General Electric, Goodyear Tire & Rubber Company, and Ford Tractor, and through jingle writing and arranging. Schneider was also a capable engineer, adept enough to modify his Presto cutter to run at 45 rpm and to devise a chassis punch to cut a large whole in the seven-inch acetates he recorded not long after RCA first introduced the seven-inch-single format in 1949 and before commercial blanks of that size were available. He also built his own recording console using a modular design approach. Audio engineer Brian Telzrow, who apprenticed with him in the 1970s, recalled that Schneider "would buy a whole shelf full of these little preamplifiers which would then go into these little individual faders. So he would just be able to make a console that was as large or as small as he wanted, just by simply adding more volume controls and connecting, ganging them together."[66] But Schneider made few major capital improvements after his 1953 purchase of the Ampex recorder and Telefunken mike. The equipment served his purposes, and until his death in 1984, the Schneiders operated the studio as a family operation, with Mrs. Schneider running the front office and offering her vocal and keyboard talents whenever needed. At a time when most small studios were either expanding to ride the booming record industry wave, or closing their doors because they could not compete with the larger studios, the Schneiders maintained a personalized operation that gave aspiring musical and engineering talent a place to start.

The Rube Goldberg Approach to Building a Studio: Boddie Recording

One such aspiring engineer was Thomas Boddie, a radio and television repairman who built his own studio and record pressing plant with his wife, Louise. As African American entrepreneurs, the Boddies faced formidable obstacles in attempting to build their business in racially troubled Cleveland, yet they maintained an ambitious and optimistic approach to their enterprise. In the early

1960s, Boddie Recording became the destination of soul, gospel, pop, country, and bluegrass musicians from as far away as West Virginia. With neither the publicity of hit records that might attract more business, nor the lucrative business of commercial clients, the Boddies could not afford to compete technologically with the larger studios. Tom Boddie built, modified, or purchased used equipment exclusively. Like Schneider Recording, Boddie Recording represented the quintessential "mom and pop" studio of a former era, serving clients who typically could not afford the downtown studios. They combined technological enthusiasm with the optimistic "grand expectations" associated with postwar America's burgeoning economic growth and the American dream of private enterprise.[67]

Tom Boddie discovered his knack for tinkering and a fascination with electronics at age fifteen when he succeeded in repairing a broken crystal radio set. Inspired to learn more, he spent hours in the library reading *Popular Mechanics* and then returned to his grandparents' Quebec Avenue home to build various gadgets in the attic. Like most electronics enthusiasts, Boddie loved to create things as well as take them apart to figure out how they worked. As he put it, "I guess I was struck by Rube Goldberg, you know how he makes one thing 'work' another thing."[68] Boddie was less interested in the unnecessarily complex nature of the inventions depicted in the popular cartoon series than he was with the cause and effect of mechanical and electronic devices. With no formal training, experimentation became a major component of his self-education regimen. While at Fairmount Junior High School, Boddie helped install and run the school's sound system. He rigged up a home-built electric guitar for a neighbor with a set of earphones and a radio speaker, and then became the soundman for the guitarist's band, a job that entailed holding the band's only microphone up to the sax player when he played a solo, thereby "makin' a job for myself," as Boddie characterized it.[69] After junior high school, Boddie entered East Technical High School and majored in industrial electricity because it included a course in radio, the closest he could get to studying electronics. After graduation, Boddie applied for a civil service job and worked briefly at Wright-Patterson Air Force Base in Dayton, Ohio, designing test equipment for a Link trainer, a small, single-person, stationary mock aircraft designed to teach pilots instrumental flying. Six months later Boddie was drafted. He asked to be assigned to the signal corps, but after basic training, the army assigned him to the quartermaster corps and eventually shipped him to Japan.

After his discharge, Boddie returned to East Tech for a course in radio and television. He hoped to find employment as an electrician, but like most other

African Americans in 1950s America, faced the covert discrimination of a white-dominated industry. As he recalled, "At that time black folks couldn't walk into a place and get a job. So I had to *make* my job . . . the newspaper was just loaded with ads for electricians and things like that. And the excuse was, 'We want somebody with five years experience.' . . . That was what they told black folks."[70]

Despite a booming postwar economy, blacks still faced discrimination in the workplace, and certain unions, including plumbing, construction, and electrical work, prevented blacks from becoming members, effectively barring them from a job because those unions maintained closed-shop status.[71] So Tom Boddie set up his own repair business in the basement of his Pierpont Avenue home and began to service clients through Lou Cole's furniture store on East 105th Street. While fixing radios for a nearby pawn shop, Boddie spied his first recording machine, a used Wilcox-Gay paper disc recorder. This was, as Boddie described it, "the primitive of primitive disc recorders . . . an 'el cheapo,' so to speak," that he had to play around with to get working again, just as he had tinkered with the crystal set.[72] Recording had quickly become a serious hobby for Boddie. As soon as he got his mustering out pay from the service, he bought a Rek-O-Kut transcription disc recorder and began to pursue what became his lifetime passion.

Boddie had heard of both Cleveland Recording and Schneider Recording, and as he put it, "the recording studio bug got to me." Through his children's godfather, the custodian for Schneider Recording, Boddie lined up an interview with Hank Schneider, who was so impressed with Boddie's ambition he quickly became his mentor. Whenever he could, Boddie went to Schneider's studio and talked to him about everything related to recording and electronics. Whenever Schneider bought a new piece of equipment, he showed it to Boddie, who then borrowed the manual and went home to build his own version from the diagrams. Boddie also kept his ears and eyes open for liquidation sales, acquiring the control board from radio station WXEL after it had been stripped of the electronics, leaving Boddie only the panel, knobs, and VU meter. Boddie installed his own amplifiers, and with this rebuilt radio console, the Rek-O-Kut disc cutter, a used Magnecord tape recorder, an Ampex 350 full-track recorder, Western Electric and RCA44BX microphones, and a home-built mixer, Boddie set up his first recording studio in his basement repair shop (see fig. 13). At the time, he recalled, "it was just for me to look at 'cause I didn't have any customers," but soon Boddie got his first paying job recording the operatic tenor Jan Peerce performing with the Jewish Singing Society at Severance Hall in 1959.[73]

Within the year, Boddie bought a home on Union Avenue and installed his studio in a small house in the back lot, a former dairy that had been converted to an auto body shop by the previous owner. Working as an organ repairman by day, Boddie transformed his studio by night, installing windows, wiring, sound-proofing, walls, and a control booth. In the early 1960s, Tom met his future wife, Louise, through Bill Hawkins, a mutual friend and Cleveland radio's first black disc jockey. Louise had long been involved in local politics and had worked her way up to become the first woman president of the Cleveland Business League, the oldest minority business association in Cleveland. After the Boddies married in 1963, Louise took over running the business so that Tom could devote more energy to recording and building his studio. Through his job servicing organs for the Hyde Piano Company, Boddie acquired a used Hammond B2 organ and a set of chimes—the only studio with a full set of chimes in Cleveland, and perhaps anywhere outside of major radio studios—and Boddie Recording soon attracted gospel groups. Because of their low hourly rates, the Boddies also attracted mu-sicians that could not afford the downtown studios, novelty groups such as the Turfits, who Tom described as "a bunch of guys who wore mini-skirts," and Harry and the Apes, who sported bearskins. They recorded hundreds of demo records for soul and rhythm-and-blues artists, which they intended to send to Motown Records in Detroit, hoping that the more famous black-owned record company might pick up one of these acts. They also attracted a number of blue-grass and country and western groups, traveling singing groups, and bands who gave Boddie Recording a reputation as "Little Nashville."[74]

Boddie also did remote recording—everything from his first paying gig re-cording Jan Peerce, to the O'Jays at Leo's Casino and the Who at Public Hall, to funeral services, sermons, and bar mitzvahs. The custom recording business proved to be a steady source of income during the 1960s for Boddie as well as others, such as former high school teacher Jack Renner who later co-founded the audiophile label Telarc Records. For those like Renner, whose main cus-tomer base was high school bands and orchestras, the vanity disc recording business virtually disappeared after the introduction of cassette recorders in the 1970s because, as Renner explained, "kids' parents could sit in the audi-ence of all the concerts they played all year long. They could do their own. Who wants to pay six dollars for an LP that only has forty-five minutes?"[75] Although the sound quality did not equal the recordings made on a portable Ampex, pro-fessionally mastered and pressed on vinyl, audio cassettes could record up to two hours of music. Better recording quality, for many customers who simply wanted a memento of the event, was not worth the added expense. Recogniz-ing that this new recording format could become another source of income, the Boddies purchased a cassette duplicator and did a brisk business in the 1980s providing on-the-spot cassette copies of church services for congregants who wanted to buy recordings of the sermon and choir performance they had just attended.

Local music groups recording demos or wishing to sell their own records at their live performances still wanted limited runs of 45-rpm discs, and in the early 1970s, the Boddies decided to start their own pressing plant. This was at the height of the oil embargo that severely reduced the availability of vinyl for making records, and the Boddies were finding it difficult to fill the demand. Boddie borrowed a thousand dollars to buy stock in a pressing plant owned by a local family, who within two months decided to dissolve the company. Unwilling to lose his investment without a fight, Boddie offered to run the plant for free during the evening, while still repairing organs and running his studio. After the pressing company went bankrupt, the Boddies mortgaged everything they had to secure a $30,000 loan to buy the pressing equipment, set up their own plant, and convert the studio to an 8-track facility. Again, Boddie did the work himself, installing plumbing in the studio, automating the presses, and building the electronics to add to the Ampex 3-track recorder he acquired in the 1960s while Louise hung drywall and eventually took over pressing the records. Customers who had done their own recording flocked to Boddie's pressing plant, where "for less than the price of a dinner roll per unit, anybody could walk in off the street, fill out an order form, drop off their one-fourth-inch tape, and return a week later to pick up boxes of real-life records bearing their name."[76] By providing an invaluable service to aspiring recording artists, the Boddies fulfilled their own aspirations, and other than the exhausting work and grueling loan repayment schedule, the Boddies seemed to have forged the kind of enterprise they had long dreamed of and worked hard to achieve.

But the dream was short-lived. After Cleveland's race riots in the late 1960s, whites grew increasingly reluctant to venture into black neighborhoods, and by 1974 the Boddies had lost nearly all of their white clientele.[77] The vinyl supplier stopped shipping their orders, initially claiming it was due to the shortage created by the oil embargo, since vinyl is a petroleum-based product, but he finally admitted that he feared he would lose his white buyers if they found out he was selling to blacks. Then the Boddies started receiving mysterious letters with no return address, offering to buy their pressing plant, and a troubling call from the loan officer at their bank, informing them that word was spreading to put the Boddies out of business. As Louise remembered, he told them "'there will never be another Motown Recording in this country.' And that's what we were aspiring to do, [to] work our way up. We weren't trying to cut anyone out. We were just trying to survive."[78] Although there is no evidence to suggest that Berry Gordy considered Boddie Recording a threat to Motown Records, these incidents illustrate the racism that black businesses continued to face in the civil rights era.[79] To the Boddies, these events all added up to some kind of conspiracy. Undeterred, they bought a grinder to regrind the vinyl waste from making their records, and when that ran out, bought records from the store to regrind to press their own. The final blow came when Tom Boddie, too long burning the candle at both ends, suffered a stroke, putting an end to both the pressing plant and the recording studio.

If Ken Hamann could be said to have run Cleveland Recording on a shoestring, Boddie ran his studio on a gossamer thread. Although he enjoyed the moral support of his fellow studio owners, Hank Schneider and Ken Hamann, this could neither alleviate the financial burdens nor prevent the discrimination the Boddies faced. The Boddies believe that, with financial backing, they could have succeeded, but with a limited budget, they could not afford new equipment. In the 1980s, they bought a new Sony cassette recorder for their location dates, but as Tom neatly summarized it, "everything else we built or modified or resurrected from somebody's scrap pile."[80] This was no way to build the kind of studio that would attract high-paying customers, particularly the advertising agencies and corporate clients that were the mainstay of Schneider Recording, Cleveland Recording, and Audio Recording, three studios that managed to survive in a city that had no major entertainment industry. Yet Tom Boddie was able to start a studio with nothing but his own enthusiasm and ingenuity and, with the help of his business-minded wife, make it a viable enterprise for more than a decade through hard work and willpower. The Boddies were among the most enterprising and tenacious small-business owners, overcoming both economic challenges and racial discrimination. Their story illustrates the powerful lure of the recording business at a time when small studios and labels saw a chance to develop.

"Hit Factory": Bell Sound Studios, New York City

As the center of the recording industry through the late 1960s, New York City provided more opportunities for small studios to thrive after World War II than did Cleveland. New York's independent studios served a similar range of clients: musicians, songwriters, and advertising agencies, but they also attracted recording artists with record contracts who sought a looser recording atmosphere than the studios of their record label. Bell Sound became one of the first independent studios to give the major labels competition.[81]

In the 1940s, Al Weintraub and Dan Cronin were classmates at Brooklyn Technical High School and aspiring radio engineers. At Tech, as it was commonly known, both were involved in New York's educational radio station WYNE, and after graduation both joined the municipal radio station, WNYC, as full-time radio engineers. In 1950, along with another partner and a six-hundred-dollar investment, Cronin and Weintraub opened Bell Recording Company at 73 Mott Street in New York's Chinatown. Bell Recording began, as did so many other independent studios in the postwar period, by recording air checks, weddings, and bar mitzvahs. They soon outgrew the tiny storefront in Chinatown, and Weintraub moved the company to Brooklyn a year later. After Cronin was drafted into the army, Weintraub quit his radio engineering job and evening courses at Brooklyn Polytechnic Institute to run Bell Recording. The Brooklyn studio was too small for live recording, but small record companies began to call when they needed a master disc cut and the other New York studios were booked.[82] By 1954, Cronin had returned from Korea, and Bell Recording moved back to Manhattan, this time to a West Eighty-Ninth Street location large enough to accommodate live ensemble recording. With a live studio in the city and a reputation they had developed from the disc cutting work, Bell began to serve more music industry clients, recording Faye Adams's "Shake a Hand" and Frankie Lymon and the Teenagers' "Why Do Fools Fall in Love?" By 1955, Bell moved to West Forty-Sixth Street and Eighth Avenue, in the heart of midtown where nearly every other New York studio and the majority of record companies and advertising agencies were located. Within five years, the newly named Bell Sound had grown from two engineers operating out of a Chinatown storefront to nine employees working in what *Billboard Magazine* called "probably the hottest studio in the nation."[83]

Bell's popularity grew with the era of rock 'n' roll. Along with a handful of other independent studios such as Associated and Dick Charles, both demo studios, Bell Sound was where the young songwriters and producers of the Brill Building gravitated, and Bell was by far the biggest operation. In 1958, Weintraub and Cronin contacted another former Tech classmate, Dave Teig, who had been working as a radio announcer in Petersburg, Virginia. Teig knew nothing about the studio business or recording; at Tech, he had been an aspiring actor, not an engineer. Weintraub assured him studio work would be an easy transition from radio, so Teig accepted the job as studio manager and, like so many others in the fast-growing recording industry, learned the business by doing it. He apparently had his work cut out for him for the studio on Forty-Sixth Street was so popular it was often teeming with waiting clients. If a session ran overtime, he recalled, "you'd have fifteen musicians standing out in the lobby" because there was no waiting room. The freight elevator was so small they had to hoist two concert grand pianos through the studio window when the piano duo Ferrante and Teicher recorded at Bell Sound, an act that Teig recalls often ran over their scheduled time. "I used to have to go in [to the studio] and say, 'Listen, there's a whole group and a whole session waiting to come in. You guys gotta move!' They'd say, 'One more take! One more take!' "[84] Studios always built in a fifteen-minute or half-hour "bumper" between sessions, to clean the studio, to empty ashtrays, or to set up microphones for the next session, but in a busy studio like Bell, sticking to the schedule often proved difficult. Teig's description of the musicians waiting in the lobby, of pianos being hoisted through a window, and of anxious pleas for extra time illustrates how Bell Sound, like the recording industry overall, was rapidly burgeoning.[85]

By 1957, the company had seventeen employees, and Bell's owners were forced to expand, moving into 237 West Fifty-Fourth Street between Eighth Avenue and Broadway. They started on the fifth-floor annex and gradually took over two more floors. By the late 1960s, Bell Sound had sixty employees, three studios, four editing rooms, five mastering rooms, a film room, live echo chambers, an office and tape library on the ground floor, receptionists on the second and fifth floors, a full-time maintenance crew, and a sales department. Its operations were as streamlined as those of Columbia Records or RCA Victor, but Bell did not have staff producers, nor did it manufacture, promote, or distribute the records recorded there. Another difference was due to its size and the relationship Bell had with its customers. At Columbia and RCA, an engineer was assigned to a session.[86] At Bell, customers could request favorite mixers; some would only work with the engineer of their choice and were even willing to pay a premium for this. Eddie Smith remembered that mixers were on call, "it was like being a fireman or something," to the point where "your time wasn't your own."[87] Eventually, Bell began charging the client for the engineer's overtime and allowing the engineers to know the booking schedules for the coming week. The mixers were assisted by young trainees, referred to as "button pushers," who worked their way up through an apprenticeship program, often with no previous musical or electronics experience.[88] Bill Stoddard described the assistants' duties:

They would set up the microphones, plug them into the console, check them out, some of the guys would even write with a crayon on the faders what [instruments] they [controlled]. They ran the 'take' sheets, they got the tape in the control room and ran the tape machines, even though the engineer started and stopped the machines. The button pusher would sort of oversee that whole thing, so the engineer could just turn his head aside to the button pusher and he would say, like "Tape 7 is coming up" or something so you would have the slate right.[89]

The button pushers, like most apprentices, did the more mundane yet essential tasks, allowing the mixers' job to run more smoothly. This job specialization was one of the biggest differences between Bell Sound and other independent studios. Whereas most small studios required everyone to be able to handle every aspect of recording, engineers at Bell were hired to do a specific job. Stoddard, who had owned his own studio in Cincinnati, then worked for Universal Recording in Chicago and Fine Recording in New York, recalled this as the biggest change he noticed when he came to Bell: "a mixing engineer was just a mixer, and not expected to make masters and [mixing] was where the 'glamour' was!"[90] By the early 1960s, mixers were indeed becoming, as Stoddard put it, "the glory boys. Theirs was the top job in any studio."[91] They worked closely with the producers and recording artists, whose sound was ultimately—and literally—in their hands as they operated the controls. Yet at Bell Sound, Stoddard insisted, they all worked as a team. "The value of the mixer to the studio was how the clients accepted him. His talents were in his 'bedside' manner," but the mastering engineers were just as important to the overall success as the studio men.[92]

The mixers controlled the sound, so the recording artists relied on their ability. This had always been true, but only grew in importance as recording grew more technically complex. But technical ability, as Stoddard noted, was not enough, and some mixers did not succeed, no matter how technically skilled. Dave Teig described the mixer as "a technician and a diplomat," because mixers had to know their craft as well as human nature and how to work with musicians who could become quite demanding and unreasonable. In the pressure-cooker environment of the recording studio, Teig noted, "everything comes out about their [true] personality."[93] Some mixers could get carried away with the star quality of the job. One Bell Sound mixer, who apparently considered himself above the required nine-to-five, Eddie Smith recalled, "got to be a prima donna," and although he was a good engineer, he began to arrive just in time to do the date and nothing else.[94] This kind of attitude could not last long at Bell, which despite the task specialization, was not a rigid hierarchy nor a union shop. Owner Al Weintraub mixed dates along with his employees, as Stoddard recalled, "he was one of the troops."[95] So they fired the arrogant mixer and hired Eddie Smith, who had been at King Records in Cincinnati for the previous ten years, doing everything from mixing to playing and arranging music to cutting masters (see fig. 12). When he came to Bell in 1961, Smith recalled, "the wonderful thing about it was that's all you did. Your job was to make the client happy and do the dates. . . . I did Dinah Washington, 'cause Henry [Glover] wanted me to mix her. But aside from that, everything was, I'd do the dates and they'd take it downstairs and mix it."[96]

Stoddard remembered Bell Sound as "a happy place to work," aided by the fact that the studio was so popular; at times the studios were booked solid for weeks at a time and even weeks in advance. On many days, Stoddard recalled, mixers did three 3-hour dates back to back and still wound up on overtime.[97] Bell was a well-organized operation, with a positive atmosphere conducive to making records, but customers were not drawn to the studio solely by mixers who were both technicians and diplomats. Equally important was the studio's reputation for cutting-edge technology and, especially, its distinctive sound. In his memoir, *What a Wonderful World*, record producer Bob Thiele described how Bell's sound distinguished it from other studios and, ultimately, inspired him to change the way he produced.[98] In January 1957, Thiele brought Buddy Holly to Bell Sound to record "Rave On" and "That's My Desire." Thiele had been to the Clovis, New Mexico, studio of Norman Petty, the independent studio owner who became producer and manager of Buddy Holly and the Crickets. Petty's own trio had recorded for Columbia, and with royalties from one of their records, he built his home recording studio to exacting standards.[99] Thiele considered Bell "one of the hottest and most distinctive studios of the era," which he recalled was "unlike the traditionally configured studios at Decca, or anywhere else." Thiele produced for Coral Records, a Decca subsidiary, and probably worked in Pythian Temple's large ballroom with the ambient quality that had been so desirable for recording big bands and orchestras. By contrast, what Thiele liked about Bell Sound, and what differentiated it from other studios at that time, Thiele recalled, was its "revolutionary 'dead sound'—sound that wasn't traveling all over the room—which allowed producers to isolate the individual musicians and achieve a literally unheard-of depth of sonic separation, vibrancy, and excitement."[100] The key to this vibrant sound from an acoustically dead room, explained Bell mixer Eddie Smith, was, ironically, the lack of reverberant quality; the dead sound "allowed engineers to use lots of echo" from the chambers, echo that could be controlled. Later, Smith recalled, "the engineers plastered the walls and the room got very live."[101]

At other recording studios, Thiele had become frustrated with the way the sound in the control room differed from what he heard in the studio, which always sounded more exciting to him than the final product. "It seemed nobody knew how to control a bass or an electric guitar or drums and the cymbals," Thiele remembered, "and every time they continually got lost when the record was made."[102] So Thiele chose Bell Sound to record the song "Sugartime" with the McGuire Sisters and a sixteen-piece big band, believing Bell's "innovative isolation capacity" would at last enable him to record the rhythm precisely as he heard it. However, because he was able to hear each instrument so distinctly, Thiele recalled, "I was completely startled, confused, bothered, and bewildered by the unaccustomed clarity of every separate musical element in my ears." He requested so many retakes that not only the artists but also the arranger and normally "implacable" studio musicians grew aggravated. "Then, and very slowly, I start[ed] to get it." Thiele realized it was because he could hear the rhythm, which was "the heartbeat of every pop record," as he had never before heard it. Ultimately, he dismissed the brass and saxophone players, keeping only the rhythm section to achieve the sound he was after. "One take, and it was perfect!"[103]

Thiele's story provides insight into the pop recording process just at the moment when rock 'n' roll's popularity was beginning to influence other musical genres, changing the way producers approached recording.[104] It was also an awkward moment. From his account, Thiele appears to have been as concerned about offending Neal Hefti for jettisoning his band arrangement as he was with getting the right sound. Extremely conscious of how he was taxing the patience of the professional musicians, he seemed to have been almost apologetic about asking them to do take after take. This was not the way these seasoned professionals were accustomed to working in the studio. This was still the 1950s, when the rule was, as singer and vocal arranger Anne Phillips recalled, "three 3-hour sessions, four tunes a session, full orchestra, and that's it. You're done! . . . nobody ever practiced ahead of time, they were all great readers. All the record dates I'd been on, it was literally one run-through and you were ready to record."[105] As Thiele found, the new technology demanded a new approach, and as this technology continually changed, so did the methods of recording.

By the 1960s, multi-track recording made it possible for more instruments to be recorded discretely and mixed later. Many engineers tried to keep the instruments as separate as possible, but others, like Eddie Smith, used mike leakage to their advantage. Smith recalled how the sound he achieved at Bell became something other studios tried to emulate: "I used to be doing a date, and a guy from Capitol would be over and they'd say, 'Look, how did they get this sound? Why can't we get this sound?' Then the next day, guys from Columbia: 'What did they do to get this sound?' They'd bring their boys over . . . 'cause all the people wanted to work at Bell because of the sound."[106] Bell's reputations traveled quickly. An engineer from Philadelphia would come to New York, sneak into the studio, and try to learn the secret to getting the big drum sound, "'cause at that time, everything had to sound like it was in a palace," Smith recalled. What Smith did that made it sound big was the opposite of what Bob Thiele appreciated with the small rhythm section. The secret to Smith's big drum sound was allowing the drums to leak into the string mikes, which along with the live echo chamber gave the drums a really big sound. Bell's echo chambers, he recalled, "were really live!" and the source of friendly rivalry among the engineers. Musical arranger Alan Lorber recalled that they knew which echo chamber was the best, and each one would rush to turn it on for his session. If someone else got to it first, "they used to, you know, yell at each other, 'You're in my echo chamber! Get out of my echo chamber, it's my echo chamber for today, it's not your echo chamber.'"[107] The echo chambers at Bell, like those at Capitol Records in Hollywood, the stairwell at Columbia Records Seventh Avenue studio in New York, or the cockeyed room built by the Italian contractor for Capitol Records in New York, were absolutely integral to a studio's distinctive sound and one of the most identifiable features of any studio in the 1950s.

The final component of Bell Sound's popularity, the other factor in Bell's legendary sound, was the state-of-the-art recording technology that had always been part of the studio's appeal. Al Weintraub's partner, Dan Cronin, had always been the studio's resident technical genius. He designed an early noise reduction system for magnetic tape, the first automatic width and depth control for stereo disc mastering, and in the late 1950s, the first solid-state professional recording consoles when the industry was still based on vacuum tube technology.[108] Shortly thereafter, the studio moved from 2-track stereo recording to 3-track on half-inch tape and quickly to 4-track. Bell manager Dave Teig recalled, "Everyone had to jump on the bandwagon to compete," and although Bell converted to 8-track, it did not take the next step to 16-track recording along with other New York studios in the late 1960s.[109] Instead, Weintraub turned his attention to high-speed duplication of 8-track tape cartridges, a consumer format that was popular until the late 1970s.[110] He hired an engineer from Ampex, Mort

Fujii, to design and build the duplicating equipment. The duplication plant was a moneymaker, but the studio also had to be technologically competitive to continue to be profitable. Fujii and Cronin then built a 12-track recorder unlike any other in that it used two-inch tape. At the time, 8-track recorders using one-inch tape were the accepted standard, and 16-track recorders using two-inch tape had been installed in a few studios. The advantage of using two-inch tape and only 12 tracks was that it produced a much better signal-to-noise ratio than the 16-track recorders because of its wider track width. Thus, thought Weintraub and his associates, Bell Sound would be able to compete favorably with New York's 16-track studios like Regent Sound and the Record Plant because their distinctive technology would draw customers; they would lead rather than follow the pack.

Ironically, the studio's technological leadership ultimately served to undermine its reputation and led to the exodus of both mixers and clients. On December 26, 1967, they christened the 12-track recorder and initially it proved to be quite a success, but only with the advertisers, not the music companies as Bell's owners and engineers had expected.[111] Then, recalled Teig, "we ran into a major problem." The problem with the 12-track, "that bastard twelve," as Anne Phillips remembered it, was its very uniqueness. Because there was only one, and it was not compatible with any other recorder in the city, clients could not take their tapes elsewhere for mixing or overdubbing if Bell was fully booked, as it usually was. Even if they did all of the work at Bell, from recording to post-mixing, they could not be assured of the machine's availability.[112] Several former employees and customers recalled the 12-track as "the beginning of the end" for Bell Sound, but it was not the only factor. In January 1968, Dan Cronin was killed when his private plane crashed as he flew to Upstate New York for the weekend. His death was both a personal tragedy and professional loss from which Bell Sound never recovered.

Al Weintraub and Mort Fujii now began to focus on the duplicating plant at another location in Manhattan and frequently took some of Bell's maintenance crew to make sure the equipment there was operating smoothly, thus neglecting the studio equipment, leading to breakdowns and more unhappy customers.[113] At the same time competition increased, Dave Teig recalled, "studios retooled and got a lot of sophisticated equipment, and then they wanted to charge more, and what happened was a lot of studios started to discount prices."[114] Studio rates did not increase that much during the sixteen years he managed Bell, from 1958 to 1974.[115] Tape copies and acetate discs cost extra, and many studios used a lot of tape, since this was something they could mark up substantially. At Bell

Sound, they operated on a strict budget. Even though they were one of the most popular studios in New York, they had to meet payroll and pay the steadily increasing rent. That, along with the increased competition meant that by the time Teig left in 1974, they were barely paying the bills. Rising retail rents and real-estate prices in Manhattan were precisely what drove many small studios out of business, forcing many to sell out in the 1970s.[116] Weintraub sold Bell Sound to Viewlex, a large conglomerate that also acquired Buddha Records and several publishing companies, and a board member subsequently sold all the studio assets, ending the era of Bell Sound.[117]

Conclusion

Independent studios and small record labels fueled the recording industry in the postwar period because they addressed a growing demand for music the major record labels did not initially record, and because they provided another conduit through which aspiring performers and songwriters could enter the business. The engineers and studios whose histories have been encapsulated in this chapter are only a small sample of the many independent studios that emerged after World War II, but they exemplify key factors in all: an entrepreneurial spirit, technological enthusiasm, openness to new musical expression, and client service. Like other small businesses, these engineers owned their own businesses and thus identified more closely with their enterprise. They thus differed sharply from the major record label studios not only in the clientele they served but also in the way they served them. Independent studio owners and engineers did not decide what material their clients would record, nor did they attempt to shape their image according to some professional or corporate standard, and in most cases, they did not turn away even the rawest talent. In doing so, like the independent record labels of the early twentieth century, they changed the course of musical history simply by opening the door to creative and talented musicians who may never have gotten even a foot in the door of the major labels.

Although Cleveland Recording Company had its origins long before the others, it did not become a full-time studio until Fred Wolf hired Ken Hamann who would soon devote not only his workday but also much of his free time to building equipment, improving the studio, and experimenting with the latest available technology. His ability to adapt older equipment to meet new demands, to invent and design new components, and his interest in experimenting with sound as young musicians explored new sonic identities made the studio successful, even if on a shoestring budget. Schneider Recording Studio offered a unique combination of services that provided recording opportunities to those who may never have thought it possible to make a record. Because of his musical training, arranging skills, and engineering acumen, Hank Schneider provided services that enabled nonmusicians with song ideas to make records that became national hits, truly a springboard for raw talent. Advertising customers supported the business, and musicians from Cleveland's many different ethnic groups used the studio regularly. Unlike any of the other engineers profiled here, Thomas Boddie struggled against racial prejudice and economic hardship and created opportunities for himself and his clients against tremendous odds. For him, the recording studio business was a way to satisfy both his curiosity about "the way one thing works another," and his aspirations to defy racism and establish a successful business. Like Hank Schneider, his accomplishments relied to a great extent on the help of his wife, who managed the business. Both Kay Schneider and Louise Boddie were working wives who labored alongside their husbands while managing the business as well as the household.

Because of its location in New York, which in the 1960s was still the recording center of America, Bell Sound Studios benefited from the overflowing consumer demand that accompanied the growth of the recording industry and the ample supply of experienced engineers. Very quickly, it established its own clientele, its own sound, and a reputation for being "the hottest studio in the nation." Bell Sound was the first independent studio that gave the major labels real competition, and more than any other, its success grew along with the music it began recording in the early 1950s. Its popularity rested on both technological capabilities and the studio's technicians and diplomats—the mixers who worked with clients and made the kind of records they wanted to make. Seeking to distinguish Bell Sound from the many other studios that began to offer similar services at cheaper rates, Bell engineers built a unique 12-track recorder that ultimately began to undermine the studio's business. In Bell's case, technological uniqueness was not an advantage.

With the availability of affordable, quality sound recording equipment, many more amateur recordists started out, often with nothing more than a consumer tape recorder, and ended up with a significant business. By the early 1970s, there were hundreds of recording studios in the United States, and the number throughout the world was growing rapidly. In 1973, there were so many studios near Washington, DC, that a consulting company hired to undertake a feasibility study for establishing a proposed new studio recommended against it, citing the expense of equipping and staffing a studio and the strong market competition from existing recording facilities as major risk factors of such an investment.[118]

To differing degrees, these studios provided opportunities for both the aspiring artists and the engineers who enjoyed building their own equipment, the technological enthusiasts and tinkerers who, like drag racers, loved tweaking the machinery to improve its performance.[119] For many of these technological enthusiasts—or enthusiastic technologists—the profit motive remained secondary to the satisfaction of chasing sound. If a client recorded a million-selling record, fine, that just brought in more business. Hit records were made at all kinds of studios and did not require the most advanced technology. Yet, ironically, some within the business and artistic community were convinced that having certain equipment was essential in making another hit. In the 1960s, the race for more gear had begun.

Channeling Sound

Technology, Control, and Fixing It in the Mix

During the 1950s, recording engineers balanced, or mixed, the relative volumes of instruments during recording based on what they heard coming out of a single control room monitor. With stereo, engineers had to approach their art in a completely new way. Rather than one or two microphones being fed into a single channel to be reproduced by a single speaker, by 1960 as many as ten or more microphones were fed into three channels which in turn had to be mixed to come out of two speakers. Mixing became more complex as engineers placed instruments on the left, right, center, or anywhere in between using pan pots on the mixing console.[1] Monaural mixing was one dimensional—like listening with one ear. Stereo mixing, by contrast, became a kind of aural architecture as recording engineers worked with sound in three dimensions, envisioning how the overall song should sound and how the instruments and voices should be built into the overall mix.[2] By 1960, the recording engineer was considered "the sound-man artist" who was "fast approaching the importance of the orchestra director in attaining artistic results."[3]

Renowned conductor Leopold Stokowski, who recognized the audio engineer's vital role in the sound of his 1930s radio broadcasts, attributed equally transformative powers to recording engineers: "The first step is to make recorded music exactly like the original," he declared. "The next is to surpass the original."[4] How better to accomplish that feat than through skillful engineering? At the height of the era of "high fidelity" when many professionals felt the pinnacle of recording technique had been reached, faithfulness to an original performance had long since given way to splicing, editing, re-recording, and multiple miking, leading classical music director David Hall to observe, "Of some recordings . . . it is often difficult to say how much credit goes to the performing artist and how much to the combined efforts of the recording director, recording engineers and tape editor."[5] In classical recording, technology made it possible to improve, to polish, and to recast a performance, transforming the age-old quest for fidelity into a modern quest for perfection. Popular music recording, however, exploited the creative potential of new recording technology in a quest for new and unique sounds. As early as 1954, inventor Norman Pickering had observed that "there is no attempt to reproduce anything that may have existed before, but rather to invent new sounds and synthesize new combinations."[6] This trend escalated in the 1960s with the introduction of multi-track recording and the use of studio techniques to shape sound. The new controls at engineers' fingertips set the stage for imaginative uses of studio technology and a reconsideration of the methods and objectives of studio recording. By the mid-1960s, most popular recording had rejected any notion of fidelity to live ensemble performances in favor of studio creations or, what one producer called, "the sound that never was."[7]

What made all of this possible was the rapid adoption of multi-track recording. Beginning with 2-channel stereo in the 1950s and then 3-channel stereo recording in the early 1960s, it became possible to bounce tracks, layering and overdubbing additional parts. With 4-track recording the possibilities increased; they began to multiply, from 4-track to 8-track, 12-track, 16-track, and, by the end of the decade, 24-track recording. More channels required more complex mixing consoles, additional signal-processing devices, and a host of new audio equipment innovations. While multi-track recording simplified and streamlined previous methods of overdubbing, it led to increased time spent at every stage of the process—recording, mixing, and even into the mastering stage—as producers and performers sought to create "artistic triumphs as well as commercial success."[8] Ultimately, multi-track recording changed the way artists, producers, and engineers worked together in the studio and triggered sweeping changes in the recording industry with broader long-term consequences for studios and for the direction of popular music.

Recording in Layers

Asked to describe the most significant change in studio recording since the late 1950s, record producer Phil Ramone said "the biggest change ever is the fact that you could control elements [and] control is the name of the game." He cited the multi-track recorder and its role as "a common partner" in the studio and how this changed the direction of recording because "you start to work with overdubbing, and you build the record in layers, architecturally, it's a totally different thing, it's more like motion pictures."[9] By 1966, recording routinely involved three stages: recording, remixing (or post-mixing), and disc cutting, or mastering. The number of choices available to the engineer in each of these three stages contributed to the length and complexity of the recording chain. Recording could take the signal on a circuitous path from the instrument through microphone, preamplifier, mixer, limiter, equalizer, booster amplifier, program amplifier, recording input amplifier, and finally onto the tape via the recording head. Once the basic tracks were recorded, remixing involved playing the tape on one machine and routing it through the same devices, in slightly different sequence, to record the final mixed master tape. From the master tape, the master disc would be cut, and again, the signal traveled through the same set of devices from tape machine to reach the cutter and then the disc. In many cases, instruments and voices would also be routed through a reverberation device (chamber or electroacoustic plate). Not every one of these forms of signal processing was used for every recording, but these were the possibilities in what was considered a sophisticated recording process of the mid-1960s.[10]

In the disc recording era, before the introduction of magnetic tape, overdubbing, that is, adding or redoing a part on an existing recording, while possible was not common. It was an exacting process that introduced excessive noise as each re-recording of the previous disc took place. In 1941, at the RCA Victor studio in New York, engineer Fred Maisch, recording director John Reid, and recording supervisor Steve Sholes watched as saxophonist Sidney Bechet recorded "The Sheik of Araby" and a blues number by overdubbing each of the six different instrumental parts.[11] It was a nerve-racking experience that succeeded because of Bechet's prodigious talent and willingness to learn additional instrumental parts, but it was not one that would be repeated. Victor publicized Bechet's recording feat, sending out a news release and staged photo showing Bechet as a "One Man Band," holding the clarinet with one hand, playing the piano with the other, right foot on a bass drum pedal, and surrounded by other instruments.[12]

Guitarist Les Paul virtually perfected the practice of disc overdubbing by the late 1940s before switching to using tape recorders. After recording and discarding hundreds of discs, Paul managed to achieve the quality he sought, finally unveiling his "New Sound" in 1948: a multigenerational disc recording of "Lover" and "Brazil," featuring several guitar parts layered in careful order so that the parts he wanted most prominent on the final record were recorded last, enabling them to come through with greatest fidelity.[13] Mitch Miller also used disc overdubbing, recording vocalist Patti Page in much the same way Les Paul had recorded his multiple guitars and Mary Ford's vocal overdubs on their 1951 recording of "How High the Moon."[14] It became a bit of a contest between them when in 1950 both Page and the Paul/Ford duo released different versions of the same songs almost simultaneously, "Tennessee Waltz" and "Mockin' Bird Hill."[15] Miller also overdubbed other parts to some of the records he produced, such as the whip crack on Frankie Laine's "Mule Train" in 1949. Laine recorded the song in Chicago on a Sunday morning, with Miller using his belt to create the sound of a cracking whip, but Miller was dissatisfied with its "squashy sound." After he returned to New York that evening, Miller dubbed a better whip crack using wood blocks at a New York studio the next day. By 1951, producers routinely used overdubbing to fix or improve on original recordings, but Les Paul's use of overdubbing, or sound-on-sound recording, was an intentional production technique, integral to the sound of his records.[16]

In 1956, Les Paul acquired a unique 8-track recorder from Ampex Special Products Section that was equipped with Sel-Sync (selective synchronization), an Ampex innovation that allowed Paul to listen to the previously recorded track while playing along with it yet without having to erase it and without creating the noise of successive copies inherent in sound-on-sound recording.[17] As soon as Atlantic engineer Tom Dowd heard about Les Paul's 8-track recorder, he convinced the owners of Atlantic Records that this was an essential tool for their business and by January 1958 Dowd was recording on Atlantic's new Ampex 8-track machine. Working with songwriters and producers Jerry Leiber and Mike Stoller in the 1950s, Dowd recalled that they often collaborated on how the recording should be accomplished: "I was their engineer, but at the same time, I could give them input, and it was always a two-way street, an open door, move things around, 'What about this?'" When he explained what the 8-track recorder could do, that they could make a record sound bigger by overdubbing voices, Leiber and Stoller immediately embraced the concept, and it influenced the way they composed. Reportedly, they said to Dowd, "'You can do that? Here we go, look out world!'" Very quickly they began writing with overdubbing in mind, Dowd recalled, "to take advantage of the new technology."[18] In Les Paul's case, the idea for creating multiple recordings preceded and in part inspired the development of the technology. He had no interest in hiring another guitarist or in forming a band. Les Paul and Mary Ford were an act with a unique sound, and because Paul was equal parts experimenter and entertainer, using technology to create that sound was very much a part of his creative work. However, Tom Dowd, Lieber and Stoller, and others such as Mitch Miller were in turn inspired by the possibilities that multi-track technology offered to record and to write music in new ways.

In most studios of the early 1960s, mono-track, 2-track, and 3-track recordings were standard, making it possible for instruments to be recorded separately from voices, for vocal parts to be doubled or substituted, and, in some cases, for an instrumental track to be recorded without any vocal, referred to as *tracking*. Because tracking could affect musicians' employment, the union attempted to regulate its use. The American Federation of Musicians (AFM), which had organized the nation's professional musicians around the turn of the century, saw tracking as yet another threat to musicians' employment opportunities and another way in which record companies could use technology to reduce labor costs.[19] Through collective bargaining, AFM leaders had established work rules that musicians and companies had to follow. When a performer such as Sidney Bechet played multiple instruments on a session, he or she was paid for each instrument played. To enforce its rules, the AFM routinely had its representatives observe studio practices and urged its members to report any illegal practices, which by 1950 included tracking. The AFM's attempt to regulate tracking not only created new conflicts in the recording industry but also made the union's leaders appear like modern-day Luddites who opposed technological progress. That image had been propagated by the recording industry during World War II, when AFM president James C. Petrillo called two strikes against the record companies to mitigate the sweeping, deleterious effects of "canned" music on musical careers. The record labels ultimately made major concessions to end the strikes, a powerful victory for Petrillo and the musicians' union. By the 1950s, however, the new possibilities introduced by tape and multi-channel recording made it increasingly difficult for the union to regulate industry practices.[20]

Atlantic Records was "notorious for the 8-track machine," according to Dowd, and musicians frequently checked to make sure he was not using it to "track," or overdub. If he was engineering a date employing ten or twelve strings, Dowd recalled, "in the middle of a take there would always be some string player getting up to look and see if I was recording on eight tracks or just one or two tracks, to make sure that I wasn't doubling the strings [because] if they were gonna be doubled they wanted to get paid double."[21] Under certain circumstances, the union reluctantly sanctioned the recording of an instrumental track without the vocal but demanded proof that there was just cause. In 1964, Dowd and producer Bert Berns were in the recording studio, along with some thirty strings and other musicians, to record the Drifters. The lead singer, Rudy Lewis, was late, and after a frantic search they discovered that "Rudy had expired," as Dowd remembered the date.

> So now Jerry Wexler puts in a call to the union, says, 'Hey we got this band here with thirty strings and this and that, the guy passed away, we're not pulling your leg!' . . . We had the musicians in the studio, we had a contractor, everything was legitimate up to snuff, everything was the way protocol dictated. We could not record without a lead singer singing because then somebody would have said, 'Oh, they're screwing us.' So we had to call the union, advise them that we needed to overdub. We are not doing anything illegal, the guy is dead. The union checks with the police department or something, and they say, yeah, he's expired. . . . Now we have permission to record without a lead singer. But you couldn't say that to the musicians and hold the phone up—they want to see a union man, so here comes the delegate from whatever, and he comes walkin' in and says, "Yeah, it's okay, let 'em record."[22]

As this incident shows, the AFM used discretion in enforcing its rules. In cases where there were extenuating circumstances, the union relaxed rules in ways that benefited all concerned. Yet protocol had to be followed, and companies that disregarded the rules risked penalties. When Frank Sinatra failed to show up for a 1950s recording session, producer Mitch Miller was forced to track the orchestra, only admitting to it decades later, saying he "risked [his] neck with the union" by recording the orchestra alone, pretending to be balancing while actually running tape. Weeks later he snuck into the studio with Sinatra at midnight to finish the recording.[23] By 1952, the trend toward multi-dubbing, or "gimmix waxings," as *Billboard* called it, had raised enough concern within the AFM that the union demanded fees for such recordings. Record companies resisted union efforts to collect the fees, and after many protests by record label A&R departments that stood to lose money on cancelled sessions, the musicians' union began to relax the policy, permitting it on presentation of medical documentation that proved a vocalist was unable to appear on the date contracts were filed for the musicians.[24]

By the 1960s, however, tracking had become common, albeit not always for what the musicians' union would consider legitimate reasons. Sometimes the instrumental track was indeed used for more than one record, but more often tracking was used for convenience and increasingly for creative purposes. However, once the New York local learned that tracking of voices and instruments was being done over "previously recorded and paid for instrumental backgrounds," they sought agreements with individual studios to be informed of all tracking occurring during sessions beginning the first of January 1961, a move that was described by *Billboard* as a way to "police the kind and amount of recording done in studios."[25] The union encouraged musicians to report tracking, and while some willingly did so, others did not. Artie Kaplan, who was a union member, a session musician, and a music contractor responsible for hiring the musicians for recording dates, saw the benefit of tracking. He recalled sessions in which a singer or an entire orchestra would sit quietly in a darkened studio when the union representatives arrived. "In those days, the union rule was that you had to do everything live," Kaplan recalled, "and we'd be taking take after take and putting the pieces together to make, the way one makes a record, and waiting for the delegate to walk in. And he'd ask what was going on, and of course the singer knew to keep his or her mouth shut," while the engineers would say they were mixing. Kaplan thought that it was all "a big ruse anyway" because "the delegates were as hip as anybody. They were all former musicians. They don't wanna break our chops. . . . They were sent to do that by the codes of their job, but they knew," and many simply looked the other way.[26] Kaplan's words again illustrate that the union and its members showed discretion when it came to enforcing union work rules.

The recollections of Brooks Arthur, a recording engineer at Mira Sound in the early 1960s, underscored the challenges facing both recording engineers and the AFM. When producing the Dixie Cups' "Chapel of Love" in 1964, Arthur recalled having to resort to duplicitous tactics to avoid detection: "You weren't supposed to overdub the voices, so we would have the voices sing live in the booth and then these . . . undercover guys would walk in to visit your studio to see if you were still working on the track, because if you were still working on the track and adding voices to it, they knew you were overdubbing so you had to pay the musicians again, so you had to do all kinds of fancy tap dancing."[27]

These new recording methods eventually became standard practice in the industry. By the mid-1960s, engineers typically cut basic tracks with all the musicians and vocalists performing simultaneously. The vocals were recorded on separate tracks so they could be fixed later if need be. Once they completed the instrumental tracks, the producer might dismiss the musicians so as to avoid paying them overtime, often for just sitting in the studio waiting or playing the song again and again while inexperienced singers got their vocal performances right. The producer and engineer could then work on improving the vocals later, adding other parts and overdubbing different voices. On the "Chapel of Love" session, for example, songwriter and label executive Mike Stoller overdubbed the celesta part to mimic the sound of chapel bells, and Arthur recalled that "at the end, when one of the Dixie Cups was to go 'yeah, yeah, yeah, ye-ah-ah'—well, she really couldn't cut it," so at Arthur's suggestion, songwriter Ellie Greenwich overdubbed the part. While the initial session lasted the standard three hours, Arthur, producer Jeff Barry, Greenwich, and Stoller continued working, as they often did, until one or two in the morning.[28] By using this kind of overdubbing to both embellish and improve a song, savvy engineers and producers found ways to overcome the limitations of performers' abilities while keeping their costs for session musicians within the small-label budget. The union eventually accepted such practices as the norm. "It was obvious [AFM leaders] couldn't win," said Kaplan, "The technology was moving too quickly and they knew it."[29]

The AFM's response to tracking had its roots in the 1920s and 1930s, when employers in entertainment industries introduced new sound technologies in ways that slashed labor costs. Throughout those years, movie theaters replaced live orchestras with "talking" movies and new sound systems; radio stations replaced orchestras with pre-recorded transcribed programming and remote broadcasts and places where people danced replaced live bands with jukeboxes. Media scholar Tim Anderson argued that the AFM's efforts to control this "sound revolution," including the union's recording bans of the 1940s, represented a refusal to accept a new postwar music industry based on selling records rather than on live music.[30] Labor historian James Kraft, however, has viewed those efforts as rational attempts to protect musicians' interests and employment as technological change put many musicians out of work. The union's response to tracking, then, could be seen as another effort to protect that employment, but its attempt to control how and when new recording technology was used ultimately failed. The union eventually found itself overpowered by the march of technology, by the trend toward multi-tracking for self-contained rock bands, and an expanding audio industry responding to that demand with a technological boom of multi-track tape machines, sophisticated mixing consoles, and other devices to help streamline the recording process and offer more flexibility in the studio.

Music arranger Alan Lorber believed that the desire to record more tracks existed before the technology permitted it, and he recalled that their efforts to embellish songs with countermelodies and additional vocals led to improvements in studio recording.[31] With Neil Sedaka, Lorber recalled devising how to record within three tracks, creating what he called a road map for recording so that they could layer more than three parts. The 1962 hit song "Breaking Up Is Hard to Do," featured Sedaka harmonizing with himself in addition to the background group, the orchestra, and the countermelody of "down-do-be-do-down-down." Since they only had three tracks to work with, they would leave one track open at all times and record on the other two. Lorber explained how this worked:

> Neil would sing on the one, so-called vocal track, the open vocal track, knowing that we probably will erase that and do it again . . . maybe the percussion or the string track was the first of Neil's counter-lines. Then basically he would do a second counter-line, and I think that he would add a harmony. But I think it was all three, and three, and three, and finally doing the final vocal on that empty track, always leaving an empty track, where I think what they did is bounce from one track to the other.[32]

As Lorber explained, bouncing tracks around to get all the parts desired was painstaking work and required careful planning, but it was the only method available in most small studios. Lorber believed that because arrangers wanted to be able to do more, and because producers and engineers wanted to simplify this process, these goals ultimately led to the popularization of multi-track recording. In Lorber's opinion, "it was the creative process that dictated the growth in technology," an opinion borne out by the fact that some manufacturers recognized they were slow to respond to the creative needs of their clients.[33] Ampex manager Leon Wortman admitted in 1967 that, although "the ability to reproduce original sound has been advanced," there remained "limitations from the creative standpoint" as record companies and studios increasingly requested new advancements.[34]

Even with 3-channel recording, in the early 1960s, most recording still involved musicians playing together along with the vocalist, even if the vocal track would later be erased and re-recorded. Although multiple microphones were used, live recording always meant that instruments leaked into surrounding microphones, so it was necessary to use isolating panels, called "baffles," or "gobos," and vocal booths in which the singer stood barricaded from the musicians to prevent as much of this leakage as possible. Separating instruments eased the task of controlling their respective levels in post-mixing. Although the performance was in real time, and the performers heard one another in real time, this was increasingly accomplished with the use of headphones, introducing yet another layer of separation between musicians. The excitement, the pulse, and what engineer Mike Dorrough described as the "symbiotic" interaction between the instruments, "just the harmonics in space," was a quality that gradually began to disappear from recording sessions by the end of the 1960s.[35]

Les Paul and Tom Dowd were mavericks, ahead of the industry standard, which did not universally adopt 8-track recording until 1968. Each, however, had a different motivation for using this technology, using their devices toward different ends. Les Paul layered guitars and voices to create what Capitol Records promoted as his "New Sound," using multi-track recording as a means to realize the ideas he heard in his head, thus multi-tracking had its origins in the idea of the performer taking more control to execute creative ideas. But Dowd's concept in using the 8-track was more like a data recording device—the very use for which Ampex invented its first multi-track flight data recorders—than a sound manipulating or creative tool. Dowd viewed this tool not as a means to create sounds that were never there but rather to ensure the best recording that could be made from the studio performance that had been captured. As he described it:

> I was not using it recording one instrument at a time, I was recording everybody simultaneously on the doggone machine, and if need be, repair a solo, repair a vocal, take this out, change that, add that and so forth. But I was not recording one instrument at a time. [Les Paul] was the one that started in that tradition. But I was using it to collect as much information in one pass as I could, and then when I sent the musicians home I'd figure out how to salvage what the hell they did and make good out of it.[36]

Just as Brooks Arthur and others tracked after sending musicians home, Dowd's method may well have evolved from his need to use what little time he had most effectively. Les Paul had his 8-track in his home so time was not a factor. Whereas Paul used it to create unique sounds, Dowd used the 8-track to collect enough music on tape so that if he needed to fix it later, the parts would be there. In either case, time was necessary to accomplish the goal.

Multi-tracking: Creative Tool or Crutch?

The different uses to which both Tom Dowd and Les Paul put their respective 8-tracks foreshadowed the way multi-track recorders were used once they reached the market in 1967. As the number of tracks increased, the possibilities and the time spent making records expanded as well. Multi-tracking was universally considered a boon for artists, producers, and engineers because it brought greater control over the instruments in recording and the overall sound in mixing. But soon the downside of multi-tracking became apparent to engineers who had previously worked without it. As multi-tracking gave producer and engineer the ability to alter the mix of instruments and voices after the recording was made, the level of musicianship and precision in performance declined, in part due to the increasing flexibility of recording and the forgiving nature of overdubbing and editing. Many engineers who began in the pre–World War II era expressed utter despair at the effect multi-tracking and the advent of rock 'n' roll had on their work. Columbia Records engineer Frank Laico noticed that musicians began to ignore the dynamics on the scores in front of them so that "instead of having that excitement that the arranger or composer had in his head, you no longer got that." Laico told one producer, "We're wasting time. These guys are never gonna play it any differently, cause you're telling them that they're on separate tracks, and you're gonna be able to fix it later," but dynamics are not something the engineer can create, "it has to be coming from the musicians."[37]

In addition to what he saw as a decline in the skills of some trained musicians, Laico also encountered a new attitude toward recording and the use of the studio by rock musicians. He first approached rock 'n' roll recording as a challenge. Listening to what people told him were hit records, he could tell that most of what he heard had been recorded in someone's basement or garage. "Then," he recalled, "the producers, the people involved would say, 'Duplicate that sound.' Well, it's very difficult to duplicate that sound when it was done in the most meager circumstances and equipment, and here you are in a studio that was built for recording, you have all the modern equipment and all the talent that goes with it, and now you have to go all the way down the ladder and start over."[38] After three decades of working with skilled musicians such as Miles Davis, Tony Bennett, and Frank Sinatra, of seeing firsthand the dramatic advances in sound technology since the end of World War II, what was the point of duplicating a sound he had spent a career trying to improve on? To Laico, this directive to recreate the feel of an amateur recording was a pointless step "down the ladder," a return to a technologically unsophisticated past. Nevertheless, Laico embraced the challenge and even enjoyed doing it for a while, until the increasingly protracted sessions grew tedious. Where experienced engineers such as Laico once recorded four sides in a three-hour session, they now had to exercise tremendous patience with musicians who came into the recording studio with just an idea for a song, frequently going over the same song, or section of a song, repeatedly until they felt they had gotten the right sound or the best performance. One day Laico finally told his boss, "'I can't do it anymore.'

He said, 'Why not?' I said, 'They're driving me crazy. I sit there, day after day and I'm hearing the same bass notes. It's impossible, I can't do it.' He says, 'Well, you got to.' I said, 'Well, maybe it's time we separated, cause I'm not going to.'"[39] For the recording engineer who had spent years refining technique, working with skilled musicians and highly polished bands who either arrived well-rehearsed or were crack sight readers requiring little rehearsal, the prospect of working with untrained musicians was an affront to their profession and tantamount to deskilling.

Capitol's John Palladino went from engineering Sinatra sessions in the 1950s to joining the A&R department as well as engineering in the 1960s and working with such rock groups as Quicksilver Messenger Service, Steve Miller, Sons of Champlain, and Joy of Cooking. Palladino felt his interest wane as technology began to dominate recording sessions. To him, the era of recording live, getting it done, and listening to the playback afterward, "that was it, you know, that was the way to make a record." Recording rock groups who were "for*ever* doing, you know, this overdub, that overdub" was taxing. "I mean, you know, you never got *through* with anything!" He eventually found it hard to maintain focus. "I'd be doing a thing with Steve Miller and he'd say, 'Hey John, wake up!'"[40] Rock musicians may have considered the music of their elders outdated and boring, but the prolonged process of multi-track recording lacked the excitement those older engineers had grown to expect when recording live and thus became for them a source of boredom.

Al Schmitt witnessed the same change in Los Angeles studios during the mid-1960s. As an engineer and later producer with RCA, Schmitt had worked with Sam Cooke, Henry Mancini, and Ray Charles, who made albums in two to three days at a cost of less than two thousand dollars. However, Schmitt started to see this change in 1966, with the introduction of multiple tracks. Schmitt recalled,

> Once multi-tracking came in, as we got more tracks, the expense of albums went up. It took longer to make an album, and once we were able to punch in and fix things, the expense of albums went way up. I went from doing complete albums in ten hours—complete, finished in ten hours—to doing an album in five and a half months. I mean to me that was unbelievable. I mean even two weeks; I'd scratch my head thinking, "God, what are we doing?"[41]

The ability to "punch in"—a term derived from the engineering task of hitting the record button—meant that an artist could replace a recorded part or section of a song as short as a single syllable at any given point on the tape. As

Schmitt explained, this greatly extended the time and expense of recording because it enabled musicians to fix anything that they did not like about their performance. And because multi-tracking isolated instruments each onto its own
channel in the recording, it could be listened to individually, giving musicians
the opportunity to take a magnifying glass to their part to fix every mistake, every slightly off-tempo or off-key note, or even the interpretation. For some musicians, this increased control led to a rise in perfectionism in recording, but the
kind of punching in and fixing Schmitt witnessed was not due to perfectionism
but rather to groups coming to the studio totally unprepared. As he recalled,
"I can't tell you how many times I was in the studio with groups that would be
writing the song on the date. They weren't prepared, the song wasn't written
before. And you'd spend forever . . . they'd write the song, change this, change
that." The fact that they knew they could overdub or fix it later in the mix only
made this easier, but many of these musicians were, as Schmitt graciously put
it, "a lot looser," and he discovered he simply had to accept that with some musicians, "hey, this is as good as I'm gonna get this, and as long as it feels good,
that's what we're gonna go with, rather than trying for the perfection that you
got with great studio musicians."[42]

Laying down tracks separately could also lead to problems at a later stage.
Eddie Smith learned to record at King Records in Cincinnati when everything
was recorded live on one track and mixing was done during the take. Compared
to Columbia's 30th Street Studio where Laico worked, or the Capitol Tower
where Palladino worked, the King Records recording facilities, although professional, were limited.[43] Smith had no more than four microphones to use on
the combined band and vocalist, but for most of the artists who recorded for
King, the Cincinnati studio proved adequate. When Smith moved to New York
and Bell Sound in 1961, he worked initially doing live 2- and 3-track recording
and became a master at mixing large ensemble record dates. Songwriter and
producer Richard Gottehrer regarded Eddie Smith as "a great engineer in the
tradition of what engineers had to be at the time; they had to be musicians as
well." When an arranger gave Smith the score, he was able to anticipate what to
expect in balancing, and this, Gottehrer noted, was part of the skill—not merely
getting the sound because before multi-tracking "there wasn't much you could
do."[44] However, overdubbing precluded mixing during the recording, leading to
unanticipated complications. When he recorded "Hang on Sloopy" by The Mc-
Coys in 1965, Smith said the initial rhythm track sounded great, but when they
overdubbed the voices, suddenly the bass and drums fell out because, Smith explained, they had not blended all the instruments and voices together from the

start. Having learned to mix live, Smith knew the importance of listening to the overall sound, "how loud the bass is, how the drum is, what the feel of the guitar is—those are all things an engineer does, [he doesn't] just sit there and putter, you know!" Although the rhythm track was solid, it was so loud that they could not get the singer's voice to blend. Had they recorded as a band, Smith would have adjusted relative volumes, bringing the bass, drums, or guitar up or down as need be, and he would have used a technique he learned from arranger Claus Ogerman when he worked on the Sammy Davis Jr. recording dates. Ogerman orchestrated the strings high and the other instruments in a lower register than the voice to leave a space for the singer; had he put the strings in the same register as Davis's voice, it would have masked the vocals, which was the case with "Hang on Sloopy."[45] Although engineers and musicians thought that multi-track recording would simplify the process of overdubbing and increase control over the final mix, the separation of the musicians could actually complicate getting a good mix for those accustomed to mixing live. If musicians played together, the engineer could adjust levels in the session so that voices and instruments blended. But recording instruments and voices separately required a certain amount of planning, which a seasoned arranger such as Claus Ogerman and an observant, musically trained mixer such as Eddie Smith understood—by leaving a space in the arrangement to accommodate the vocalists' part the same way he would orchestrate another voicing of an instrument.

The examples of Laico, Palladino, Schmitt, and Smith also suggest that recording engineers assigned different meanings to the technology they used. For Laico and Palladino, who learned their trade in the years before tape when performances had to be well rehearsed and free of mistakes, recording technology represented a means of conveying a polished performance from the artist to the record and ultimately to the listener. Schmitt shared that view but also recognized that some of the younger musicians he worked with may never achieve such polish, even if they wanted to. Eddie Smith, who was both an engineer and a musical arranger, perceived that using multi-track recording technology required more attention to voicing, since it precluded the ability to blend the instruments and voices during recording.[46] As engineers came to rely on multi-tracking, the tacit knowledge of balancing instruments during recording, like other aspects of record engineering that predated the introduction of tape, gradually disappeared.[47]

In all these cases, multi-tracking did not promise to be a time-saving technology, rather by offering the ability to improve, alter, and perfect any recording, and requiring the subsequent step of mixing tracks to a stereo master, multi-tracking could be considered the most decisive technical factor in extending the duration of recording sessions during the 1960s. "The talk of the industry," declared a *Billboard* columnist in September 1967, "is the amount of time spent in the studio—and the astronomical studio costs that have resulted—by the Beatles and the Beach Boys."[48] Having set a new standard for creative record making with *Pet Sounds* (Beach Boys, 1966) and *Sgt. Pepper's Lonely Hearts Club Band* (Beatles, 1967) these artists inspired others to experiment in the studio and to take whatever time was necessary to realize their creative vision.[49] "Records are becoming more of an art form," declared producer Felix Pappalardi, who spent six hours in the studio recording Cream's "Strange Brew" with Atlantic engineer Tom Dowd in April 1967 (see fig. 14). "There's a great deal of thought put into a record before ever going into the studio; then you're constantly fighting in the studio to reach your ideal."[50] And that, too, had broader repercussions—more time making records did not always involve playing music but instead playing *with* the recorded tracks, and this meant greater involvement of the engineer/mixer and the producer in creating the record. Multi-tracking meant that artists need not necessarily be present for the record to be mixed and, in many cases, that was the only way to finish the record since band members notoriously disagreed on the final mix, with one member claiming, "'Oh there's not enough me,' and another one would say, 'Oh, no, there's not enough me,'" recalled Tom Dowd.[51] In addition, if a group felt they collectively played well on part of a recording, but not the rest, the engineer could now solve that problem by editing and splicing. Bill Halverson engineered Crosby, Stills, and Nash's first album in 1969, and according to Dowd, "he catered to their every little whim, to the extent that when they did a song and liked two bars or three bars of the song," but did not think they played it that well anywhere else, "Bill would snip the tape and copy the same three or four bars nine times and then wire the song together for them."[52] It is no wonder that producers preferred mixing with only the engineer present. After recording the Incredible String Band in 1966, producer Joe Boyd recalled the first time he felt impatient to "get the musicians out of the way" so that he and the engineer could start mixing the multi-track tapes into a stereo master.[53]

Fixing It in the Mix, and in the Master: Post-Mixing Music, Postponing Decisions

Before multiple tracks entered the picture, "mixing" referred to the act of balancing the instruments and voices during the take. Yet this was a critical aspect of record making that required experience and skill. Beginning with stereo, mixing became an additional step after the recording and editing of the tape. With multi-tracking, the job grew more complex and mixing became even more critical to the final record, not only because there were so many ways it could be ruined but also because a great mix could make a hit record, leading to the oft-repeated claim that "it's all in the mix." Just as recording sessions grew longer, this second stage of recording, even for a single song, could take hours and even days as producer and engineer experimented with different balances, takes, edits, effects, equalization, and fade endings. As musicians had an array of notes, techniques, and effects pedals to choose from in creating songs and sounds, so the producer and engineer now had a wider palette of choices in mixing. Joe Boyd likened mixing to "an endlessly fascinating jigsaw puzzle" and again employed the photographic metaphor to describe what it felt like to see the music "slowly emerging . . . like watching a print in the developing bath. But with sounds you could control the colour, the contrast and even the positioning."[54] The luxury of choice, however, required the luxury of time in the studio. Several engineers who began in the recording industry before the advent of magnetic tape, when recording to disc or full-track tape required quick decisions and well-rehearsed performances, believed that multi-track recording led to postponing decisions at the time of recording, enabling producers who did not know what they wanted to lay down tracks they might not later use.[55] Al Grundy, who traveled throughout Europe in the late 1950s making live recordings of classical repertoire for the Concert Hall Society and later founded the first school for recording engineers in New York, the Institute of Audio Research, recalled that,

> a great many records were made with the attitude of, "well, that doesn't sound very good, but we'll fix it in the mix" [and this] became a method of postponing decisions. You know, if you were recording as I did with the orchestra, you had to listen and say, 'Okay, that's the way it's gonna sound,' and not say, 'OK, gee, well, we could balance that out later, or we could boost the bass or fix that oboe . . . none of that was possible. Whereas in multi-track, it became a means of postponing decisions, postponing, postponing, and we'll fix it in the mix. But not much could be fixed in the mix. A lot of things were, of course, but many things couldn't be.[56]

The need for quick, on-the-spot decision making, characteristic of recording before tape, declined as engineers and producers realized they had the option of altering decisions made during recording as they mixed. This is one of the few points on which all recording engineers seem to agree: that delayed decisions and less precise playing began with the introduction of tape and escalated with multi-tracking. Frank Laico said that with tape the producers, artists, and vocalists grew lazy, confident that changes could be made in the mix. He found that musicians paid less attention to the dynamics written into the score as producers told them they were on separate tracks that could be adjusted later. What is remarkable is that musicians thus relinquished a certain amount of control over their own performance even as they gained the ability to polish and perfect that performance.

Multi-track recording led not only to postponing decisions for producers and the separation of instruments and musicians from one another in the studio but also to further division of labor that separated the engineer who did the original recording from the mixing of the master. Ray Hall pointed out that this was not always the best artistic decision. At one time, he recalled, RCA recording engineers also mixed down the session, and when they began to separate those tasks, assigning the task of mixing to different engineers, Hall thought it was a big mistake. "Guys who didn't know what you were trying to accomplish would mix down your [recording] and sometimes you would hear the finished product and you'd say, 'Oh, my God!' "[57] Hall's regret was that, in another engineer's hands, his original concept for the sound of the record was not realized. Because mixing involves countless choices, most critically the placement of stereo and relative volume and emphasis of different instruments, the same recorded tracks can potentially yield radically different final mixes. In some cases, however, the mixing engineer faced an arrangement of tracks that only the original recording engineer could have understood. One day in the mid-1960s, Atlantic Records' Ahmet Ertegun, placed an 8-track tape on Tom Dowd's console, stating, " "There's a hit record on here. Let me hear a mix!' "[58] Later that evening Dowd sat down and started taking the tape apart, finding to his dismay that it was, in his words, "a checkerboard square job. I mean here's bass for four bars and then all of a sudden there's nothing, and then it's a drum track, and then it's a background vocal track." In other words, multiple instrumental and vocal parts were recorded on each track to capture performances or to repair parts wherever there was an opening, because all of the available tracks had been assigned. So the engineer who made the tape "would open this track and close this track" with the attitude that he would deal with the details later. He knew what he had done, but Dowd had to listen to the tape several times and create a legend to guide him to make a coherent mix.

The separation of instruments involved in multi-track recording led to further fragmentation of the various steps of making a record. Some welcomed the control afforded by this separation of instruments, musicians, and engineering duties, while others saw it as a change for the worse. There was no question that multi-track recording transformed the way people worked in the studio, with the technology, and with one another and put new demands on recording engineers.

Once the final mix was complete, the third stage of the recording process involved cutting a master disc from the mixed tape. Before magnetic tape became the standard recording medium, this master disc was made at the time of recording, just as mixing was done during the recording. When tape became the primary medium, the final mixed tape became the master and the lacquer master cut from it became the master sent for processing to vinyl LP or single disc. To reduce noise inherent in the process, the mastering engineer boosted high frequencies and might also apply some equalization to bring up the sound of certain instruments, but this was the extent of what the mastering engineer was expected to do. Once multi-tracking introduced the post-mixing stage, some of the postponing that took place in recording spilled over to the mastering stage.

Until 1954, when the Recording Industry Association of America (RIAA) established a standard recording curve, record companies had used a variety of different recording, or equalization curves, in cutting a master disc. Each company established its own curve to best suit playback on its own phonograph. To avoid overcutting and to reduce surface noise inherent in disc recording and playback, the RIAA curve reduced the bass and boosted high frequencies at the time of recording (the recording curve), and modern phonograph preamps included circuitry to reverse those changes at the time of playback (the playback curve). Even after the RIAA standard, Bill Stoddard said, "in the mastering room we did what sounded best!" In the late 1950s, the National Association of Broadcasters (NAB) specified disc-recording reference levels, but most mastering engineers cut with as much level as they could get away with. As Columbia engineer Doug Pomeroy recalled, "It was never hard to find discs which were cut MUCH hotter."[59] Stoddard recalled that in the early 1960s there was one popular record that was "the loudest and hottest record ever. We had a copy of that record and would compare everything we cut to that record for level and apparent loudness."[60] Jack Wiener, mastering engineer at Universal Recording in Chicago, agreed that the level that the RIAA curve designated was considered by the major record labels to be a "hard and fast rule that thou shalt never break," but which he consistently ignored, particularly because of the influence of record producer Randy Wood.[61] Wood, who ran Dot Records, had a jukebox delivered to the mastering room at Universal Recording and instructed the mastering engineers to increase the level until they reached a point where the record would not track in the jukebox; then they would reduce the level just enough so that it would play. According to Stoddard, "Wood said he didn't care if he had to eat some records . . . he just wanted [it] to be the loudest."[62]

Mastering "hot," as it was called, seemed to be the goal of every record company in popular recording to make its records stand out when played on jukeboxes. As soon as Mitch Miller moved to Columbia Records from Mercury, Bill Savory recalled that Miller would come down to the engineering department and say to the engineers, " 'Hey fellas, we'll go to lunch today, this little deli has a jukebox, and I'll bet you anything, you play any record on there and it'll be louder than a Columbia record.' "[63] Miller's purpose was to get the engineering department to come up with a better method of mastering to make Columbia's popular records louder. It could be done but not by adhering to the RIAA curve and usually at the cost of record length because louder passages required wider grooves, thus reducing the number a given disc could accommodate. Making hot masters that could still track on most phonographs (play without skipping) was the goal of every mastering engineer and rarely would the artists or producer be involved at this stage. In 1964, mastering engineer Clair Krepps received a tape from his client United Artists, with the request that he cut a master. As he listened to the tape, Krepps recognized it as a song he had just mastered the previous year, "Do Wah Diddy Diddy," a song by Jeff Barry and Ellie Greenwich that the Exciters recorded in 1963. That record had not done all that well, and since the tape of the new version by the British group Manfred Mann did not sound any better to Krepps, he called the company and asked why they wanted to put it out. Claiming contractual commitments to the band, United Artists told Krepps, " 'Look, do any damn thing you want with that record, we don't care!' " Krepps recalled. "So I started playing around with it. And I got the idea to shock the industry, so I used some equipment that my brother and I designed, and I had undoubtedly, I made the loudest record ever made—45 it was. And pretty soon, other engineers, once it became a hit, they called me and complained, 'What the hell are you doing Clair, you know better than this!' "[64] Because no producer or artist was looking over his shoulder, giving direction on how the record should be mastered, and the record company did not care, Krepps was able to experiment. When the record hit the top of the charts in England, Canada, and the United States, of course, no one complained, and it seemed to reinforce the idea that hot masters made hit records.

Few mastering engineers, however, could enjoy that kind of freedom as artists and producers increasingly stipulated that their masters be cut a certain way. Grundy recalled that some producers would send their tapes to be mastered with explicit instructions like "The first four bars on the left track of song two are okay, but the next four bars require a boost in the high end." The engineer was then expected to make changes during the actual mastering process, sometimes adding reverb or changing equalization. Grundy said that these instructions "began to grow to absurdity," to the point where it ultimately became necessary for studios to put in a duplication of all of the signal-processing devices: the equalizers, limiters, compressors, phasers, all of the things that might be used to affect the sound of the music in the mastering process, and each of these would be available on two different systems, A and B. As Grundy explained, "The left and right channel of the first song on the record would have certain requirements for EQ and whatever. Well those are set up on the A channel, the left and right A; and then the requirements for the second song are on the B, [to avoid] switching instantly during the spiral from [band one to band two]. Then while B is being cut, like a three minute song, you've got to set up A again for the third band, and then switch back to A during the spiral between two and three."[65] So where the mastering engineer once could set his equalization curve, making slight adjustments from one song to the next while inspecting the grooves, he now had to do double-duty to set up one song and then switch to the second channel quickly while the cutter spiraled to the next song. Here, too, the job of the mastering engineer grew more complex, even more creative as a result of multi-tracking and post-mixing, and this led to the rise of independent mastering engineers and studios that did nothing but mastering.[66] And certain mastering engineers became highly valued for their ability to do this effectively. Bob Ludwig and Bernie Grundman in New York made their reputations by being able "to take poorly prepared master tapes and produce a master disc that was amazing in the corrections that they were able to implement," Grundy recalled. "And, of course, everyone wanted Bob to do their record because he could make a silk purse out of a sow's ear."[67]

Some mastering engineers refused to respond to these requests. When Bill Stoddard was a mastering engineer at Fine Recording in the late 1950s, Kapp Records was one of the studio's biggest accounts. Kapp made so many requests in mastering that Bob Fine eventually set a limit on the number they could make. But Stoddard recalled getting instructions like "two more notches of bass on band three," which just made no sense as far as he was concerned, so he would just cut it the way he wanted to. He figured, "The hell with it. Nobody ever said anything. This was a foregone conclusion—with Bob's blessing too—it was just a way to get the thing out. The more sides you cut the more money you make."[68] Stoddard's decision to ignore the client's request was not driven by economic motives but by his conviction that it was the mastering technician's job to use his own good judgment to make a record as close to the tape as possible, and it was the client's job to supply an edited tape ready for mastering. But this trend toward expecting more out of the mastering process only escalated during the 1960s. Eventually, the decline in the number of ready masters and the increase in the amount of those that were passed on to the mastering stage for further adjustment led to the rise of mastering as a specialized field. Since engineers who began after the advent of tape were not required to learn disc cutting to record, the disc-cutting skills required for mastering became the area of recording that retained "the most trade secrets," and as those trained during the era of disc recording began to retire, new engineers could only learn from the technicians whose tacit knowledge and skill had been gleaned over decades.[69] By the 1970s, mastering was no longer simply the transferring of a recording from tape to master disc, it was considered the last creative step and the first manufacturing step in the record-making process.[70]

Multi-tracking and the Demise of Live Recording

Jazz producer Teddy Reig, who had recorded Charlie Parker, Miles Davis, Stan Getz, and Count Basie in the 1950s, returned to the recording business in the mid-1970s, missing much of what had taken place technologically in the recording studio with the advent of multi-tracking. He recalled that, although he was happy to be working again, he was not prepared for the new way of working in the studio. Instead of cutting the rehearsed live performance in one take, now recording took place sequentially, first putting down the rhythm section, adding the horns, fixing mistakes by punching in, and a full twenty-four hours of mixing. For Reig, the process seemed to drag on forever, and with all the new technology employed, it seemed to him "the soul was lost."[71] Record producer John Hammond agreed. After nearly five decades of producing records, one of John Hammond's last sessions involved recording the Bill Watrous band in 1975. Hammond had considerable admiration for this artist, but the experience in the studio was as frustrating as Teddy Reig's. Instead of the band warming up while the engineer set the balance, half of the session involved microphone placement, testing each one for leakage into other channels, and balancing each in relation to the others. This, as Hammond put it, "makes me grind my teeth. Instead of rehearsals and music balancing, we have prayerful inaction, waiting for the high priests of electronics to tell us our circuits are working." Even though the material had all been previously rehearsed, "the performances had to be assembled later in the editing room," Hammond recalled, "and no amount of editing, rebalancing, splicing, and enhancing can make a performance great unless it was great to begin with."[72]

Hammond's admittedly old-fashioned notion that the most important thing in an honest recording was the quality of the players rather than the complexity of the equipment was a view that few would disagree with in theory, but it was a view that was losing ground in the recording studio as multi-tracking, post-mixing, and more complex mastering became accepted practice. Brooks Arthur recalled that, when they had limited tracks to work with, producers and engineers had to plan, make quick decisions, and be committed to making whatever they ended up with on tape work out. "When you committed to a rhythm section sound on 4-track—bass and drums on one track, two guitars on the other, percussion on the next—you lived and died with that mix. You loved it and you worked as hard to get to that level of excellence as you might do now with 48 tracks."[73] Arthur describes a way of working in the studio before so many choices were available. Knowing their technological limitations, engineers had to make judicious and early decisions about how they would organize their tracks. Regardless of what they ended up with, they were committed because they had no means of altering it later and that, for Arthur, is just what made those recordings have what he called "punch," because the signal did not have such a complex path to follow.

As these seasoned engineers and producers found excitement in having a finished product at the end of a recording session and getting through it without endless repetitions, other younger engineers who entered the profession in the era of pop music did not always embrace the concept of studio as instrument. Malcolm Addey began his career at EMI Studios in England, recording popular artists such as Cliff Richard several years before the Beatles and George Martin started working together. Addey came from a later generation but had the same reaction as his predecessors to long-term recording projects. He was not, as he put it, "a very good concept engineer with people like the Beatles . . . you know, when they go on for months and months making an LP." Addey liked variety and preferred to be able to walk out of the studio with a finished product. The idea of spending days "putting on a guitar solo, or some guy punching in odd words on the vocal, and stuff like that" was not the kind of recording that Addey or many other engineers found inspiring.[74]

What punching in and endless repetition of guitar parts signaled, of course, was the demise of live recording. One Columbia engineer likened it to mending, telling his assistant engineer to "get out your needle and thread" at the start of an overdub session.[75] In many instances, musicians would enter the studio at different times to lay down their tracks individually. This could prove convenient from a scheduling standpoint and in some circumstances, was the only way to record. John Palladino and others believed multi-tracking was not always appropriate, and he felt it was too often used indiscriminately at the expense of the performance because musicians had "no playoff against each other." Having worked with Frank Sinatra when "there was no recording the strings later, or let[ting] Frank come in later," he was convinced "that was the spark that made those records [because] Sinatra would *charge* the orchestra."[76] Palladino also felt that the decision to use multi-track recording regardless of whether it was necessary or the best choice for a given session was a wrong turn for recording practice. "It is possible," proposed engineer George Alexandrovich in 1969, "that we are trading in the quality of the recording for the convenience of conducting sessions and extra time charges for re-recording and mixing," adding that multi-track facilities should not be used "to make remix session a more important event than the original take—stripping away the art of mixing."[77] While some studios may have taken advantage of the inflated billing multi-tracking made possible, more often than not it became the expected recording format simply because it was there, an approach that engineer Robert Auld has called the "because we can" syndrome, when "the process becomes more important than the reasons for doing it."[78]

The trend in recording always to use multi-track could be seen as an example of technological overuse or the application of a tool originally intended for accomplishing specific goals to situations that may not require it, simply because it is available. Multi-tracking was seen as the way forward, an improved modern method better than what went before, but some engineers disagreed. Mike Dorrough, who in the 1960s worked for Emperor Productions in Los Angeles, believes that when musicians are working together, eye to eye, there is a tension that makes everybody work harder. In his sessions for Emperor Productions, Dorrough only had six microphones, whether he was recording six people or sixty. Although he had less control over individual instruments, he found that, as he put it, "the more people, the better the sound," and he attributes this to the third dimension that exists when musicians play together. "You take two people and they sing a little harmony together, if you stand in front of them, the sound and the harmonic, that third dimension is there, [but] if you mike them separately, you get their amplitudes, but you didn't get the space, the harmonic between them."[79] In addition to that tension and the harmonic interaction between instruments, Dorrough also discovered that having fewer mikes to work with actually served to make the overall sound better. At the time, many engineers strove for as much separation of instruments as possible. With fewer microphones, the leakage of instruments through multiple microphones reduced engineers' ability to control the sound of each instrument in post-mixing. But Dorrough believes that, even with less control, that leakage made things sound better because it gave the recording "a lower threshold that was fat and full," all from one microphone. "But we didn't realize then, and we couldn't wait to get a multi-track because then you could analyze things and even do more. And that's when it started going downhill."[80] Like other engineers, Dorrough found multi-tracking required a trade-off; what he gained in control and in the ability to analyze things, he sacrificed in the blending of instruments and in the spark of spontaneous performances.

Recording Jazz: Control versus Spontaneity

In 1968, Michael and Christian Blackwood filmed jazz pianist Thelonious Monk over an extended period, documenting his daily and nightly routines, club performances, and a recording session at Columbia's 30th Street Studio with Teo Macero producing. Shot years before the advent of music videos, their film offers a view into the working life of a jazz artist, and a rare glimpse of how a recording session was conducted in a major label studio in the late 1960s. It reveals the frustrations of a musician who sought to capture spontaneity in an arena of technical control and illustrates a certain kind of artist-producer interaction.[81]

Columbia's 30th Street Studio, a former church, had a high ceiling, drop lighting, walls lined with curtains, open here and there to reveal a random assortment of coatracks, ladders, musical instruments, chairs, music stands, mike stands, and baffles. The slightly elevated control room with a long, horizontal window rather ominously dark in contrast to the studio's bright overhead lighting provides the only visual communication between musicians in the studio and the producer/engineer team in the control room. The studio appears starkly functional, rather than comfortable or cozy, a place to work, not to socialize or hang out. The four musicians enter the studio one by one, each greeted by producer Macero, who appears jovial, shaking hands with each one and remarking with a laugh that they are all there "and only a half-hour late." He gets a big kick out of Monk's "invisible glasses" which are empty round wire frames.

In the next scene, Macero is at the piano, looking over some music with Monk pacing nearby. Macero says: "What are these, just little sketches, right?" Monk: "Yeah, four bars or so, you know." These are not songs in the traditional sense, but a basic melody around which the band will play and improvise. Macero keeps things light, interspersing talk about what they are going to record with small talk: "Hey, haven't seen you in months. Since the Vanguard." Monk, dryly: "Time flies." Macero: "I'm gettin' older, you're gettin' younger." Monk replies flatly, "I knew you were going to say that." Macero proceeds to play a few bars of something on the piano. Monk appears very quiet and contemplative, then says, as if wanting to be sure his producer understands his objective for this recording session, "I want it to be as easy as possible for them to dig it, you know? And then, absolutely good, too." The entire mood seems one of light-heartedness on the producer's part and seriousness for the musicians. Since one of the producer's roles was to provide a positive attitude, Macero, himself a musician, probably knew how best to approach a session with this artist, who had a reputation for being difficult.

Positioned fairly close together, facing each other in a circle, the musicians appear dwarfed by the size of the room. None of them wear headphones. Baffles surround the instruments to prevent the sound from traveling around the enormous room, but they do not prevent the musicians from seeing one another. They are playing a composition in the studio, and it sounds good, the musicians appear to be caught up in their performance. By all indications, they are in the midst of recording. The music is interrupted by Macero's voice over the talk-back microphone, saying: "Monk? [the band stops] Let's really do one now, we're all finished." Monk replies, "Huh?" Macero repeats, "Let's do one." Monk seems confused. Still seated at the piano, he turns to face the control room and sweetly but somewhat incredulously says, "We were just getting ready to go. Why you stop us for?" Then mumbles to himself, but loud enough to be heard, "That's really a problem." Monk is perturbed—the problem is that the producer and engineer were not taping when the artist clearly thought that they were. Monk mutters a few more things under his breath and then turns to the band and says, "Now where were we before we were so rudely interrupted?"

The next scene is a later interview with the saxophone player Charlie Rouse, who recalled that Monk preferred to make recordings with as little repetition as possible. They would record a song once, maybe a second time, but never a third because after that first take, "where the feeling is," things began to go downhill. Rouse explained that it was more of a challenge to play it correctly the first or second time because if Monk had his way, a player's mistake would have been preferable to a retake. "If you mess up, well that's it, you know, that's your problem. You have to hear that all the rest of your life." Rouse hints at two key beliefs of musicians of the pre-rock era. First, the initial take is usually best, and second, playing with the knowledge that it is permanent creates the necessary tension to get it right, the challenge that is invigorating. Monk considered any unexpected delay in recording to be a nuisance because it broke what for him was the feeling, the spontaneity and freshness, qualities that he valued over any technical considerations.[82]

Back in the studio, the band is playing again, and they finish the song beautifully. Everyone remains quiet as the last notes fade slowly in the studio's long reverberation time. A voice from the control room comes over the talk back: "Stand by please." Again, it seems, there has been a delay.

MONK: Did you get all that down? I'd like to hear how that sounds.

CONTROL ROOM: "We're still fooling around with the sound in here. Play one and then we'll play it back for you."

MONK [frustrated]: "Why is everybody so unwilling just to do what I asked them to do?

[Monk paces around the room, repeatedly asking to hear the take as if he refuses to accept the fact that it wasn't recorded.]

CHARLIE: I don't think they taped that one though.

MONK: Yes, he did, he said they were gonna tape it.

OTHER VOICE: Yes he did.

MACERO: No, I said we were gonna MAKE one. You said you wanted to rehearse it.

MONK: I didn't say anything . . . I didn't say shit. . . . well, could we hear that one?

CONTROL ROOM: We didn't record it!

[Bassist Larry Gales slowly bows some notes. Monk settles back at the piano, resigned to play it again.]

CONTROL ROOM: Stand by, please.

This fly-on-the-wall view of a recording session in 1968 vividly illustrates how the technical imperatives of the control room could thwart musical spontaneity in the studio. While technical glitches were nothing new, they multiplied with the increased sophistication of recording technology even as that sophistication made other things possible. For some, the cost of gaining control and of exploiting those possibilities was a growing slavishness to the dictates of those controls. The Monk session reveals how much power those who operate those controls could wield over the musicians whose music they recorded. This, too, had always been true, but as the boundary between studio and control room grew more fortified—a necessary development in gaining greater control over the sound—the communication between musicians and those in the control room, conducted over microphones and loudspeakers, became increasingly strained. The Monk session also illustrates a certain kind of working relationship in the studio, where the producer and engineer, seemingly barricaded in the control room, have complete control over the moment recording would begin, and the artist has control over nothing but his own performance. Thelonious Monk was most concerned that the performance be fresh. He neither knew nor cared about the technicalities of the control room and only wanted tape rolling when he was ready to record. Although the viewer does not know what is going on in the control room, or what may have been said leading up to the scenes shown on film, nor for that matter, how much editing of the sequence of events the filmmakers did, nevertheless Monk's reaction, and that of his band, reveal the depth of the communication gulf between the control room and the studio, and the frustration it caused. It was not that the producer and engineer were unsympathetic to the musicians, they simply had to ensure that everything was set up and working optimally, and, as in all recording sessions, having the musicians run through the numbers prior to recording was a necessary step in that process, tedious as it may have been for some musicians.

Not all jazz musicians shared Thelonious Monk's desire to capture the performance with the first take. Miles Davis, for instance, used the recording studio to improve his performance not by electronic intervention but by playing and listening and by honing his interpretation of the music over successive takes until he had achieved what he wanted.[83] In early 1963, Bill Evans recorded *Conversations with Myself*, an album of piano duets and trios, all performed by Evans using overdubbing, holding his own musical conversations, as the album title implies, with himself. Anticipating the response he might get from jazz fans and critics because of the prevailing view that music "which cannot also be produced in natural live performance is a 'gimmick' and therefore should not be considered as a pure musical effort," Evans felt compelled to make a lengthy "statement" following the album's liner notes. Defending the integrity of his choice, Evans maintained that "to the person who uses music as a medium for the expression of ideas, feelings, images, or what have you; anything which facilitates this expression is properly his instrument."[84] With that, Evans articulated a philosophy that became integral to rock recording: any method of chasing sound was acceptable, in fact the more unorthodox, the better.

Recording Rock: The New Cult(ure) of the Studio

In the late 1960s, even the jazz magazine *Down Beat* noted that the willingness of rock musicians "to put advanced technology to work . . . largely accounts for the vitality and excitement that have infused much current rock and which has led to statements about the relative moribundity of jazz."[85] Some wondered, since recording technology made it possible to create records that could not be reproduced in concert, was this really an authentic representation of a group or artist? If they didn't have the technology, what would they sound like? "Without echo chambers and other electronic trappings," asserted critic George Simon, the musical groups of the 1960s "would be completely uncommunicative . . . dependent primarily upon three chords and a smart engineer."[86] Simon praised the skill of big band musicians at a time when rock had eclipsed swing in popularity and artists such as Count Basie were releasing albums of Beatles covers.[87] But you did not have to be a jazz purist to reject technological dependency; even some rock bands avoided becoming overly reliant on studio technology. At the Monterey Pop Festival in July 1967, one group claimed that it shunned gimmicks and lights on stage, and in recording, "we never double-track or use any other instruments. What the four of us can do is the sound we make. That's all."[88] But that "philosophy of sound" was in the minority, according to most accounts, as electronics became increasingly prominent. "Recordings of electronic rock no longer purport to represent live performances," stated an RCA publication. "They are no longer reproductions but *productions*—studio events, self-contained and self-referring."[89] By the late 1960s, the debate over whether recording technology unnaturally enhanced or boosted performances no longer seemed worth waging. "The recording studio," declared Mick Jagger, "with all the things it has in it, is another form of art, of music." Paul McCartney admitted in the late 1960s that "to do these things just a few years ago was a bit immoral. But electronics are no longer immoral."[90] The Beatles were in fact largely responsible for the wider acceptance of electronics and recording tricks after they released *Sgt. Pepper*, an album that from the first sessions in 1966 was intended to be "a new record with new kinds of sounds," one that would preclude any expectation for live performance since the band had decided to give up touring. To do that, of course, the Beatles required a smart engineer and a clever producer willing to experiment to fulfill requests such as John Lennon's charge to make his voice "sound like the Dalai Lama chanting from a mountaintop."[91]

Skilled recording engineers of the 1950s were considered artists in their own right and producers set the stage for creativity, but their ability to alter the sound of a performer or group was relatively limited by both the available technology and the reigning conventions of record production. The idea that the engineer could play a creative role in recording emerged during the 1960s as studio technology became ever more integral to the sonic character of a record and to the sonic identity of a band. "Only their sound man knows for sure," declared an ad in the September 1968 issue of *db: The Sound Engineering Magazine*, over a picture of a rock band. "The Philharmonic they're not. Joyful noisemakers is really more like it. But that's between them, their audio engineer and the lamppost. On tape they sound strictly euphonious." Pitching to the "unsung hero . . . the guy at the controls," B&K Instruments sold its Model 123 Graphic Frequency Equalizer as an essential tool for the engineer to work his magic.[92] Playing on a popular hair color ad of the day, "Only her hairdresser knows for sure," the B&K ad's implicit message was that, like the hairdresser and his hair color to the woman's natural looking tresses, the engineer and his technology were the secret to the band's "euphonious" sound. Some recording engineers believed their role was as a facilitator who set the stage for creativity. "If it was a hit record," Bill Stoddard insisted, "it was a hit record before it ever got into the studio."[93] David Thomas of the group Pere Ubu sees this as an older engineering aesthetic, one that held that the engineer was "part of the machine. His function was not to intrude himself on the process, but to aid the process by being as unobtrusive as possible."[94] This indeed was the message of the B&K ad: engineer as unsung hero working behind the scenes; but with the increasing use of electronics by groups wanting to experiment with sound, the engineer's role expanded to the point where, as producer Jon Landau asserted in 1971, "engineers and remixers play a role as great as any musician in affecting the final sound."[95]

Similarly, producers were no longer only A&R men who matched artist with repertoire and oversaw sessions; they became increasingly important as they had ultimate control over the final mix. In the 1950s, Mitch Miller introduced the idea that popular records should offer something more than simply a reproduction of a live performance; he wanted to "create an image of who was coming out of that disc." Producer George Avakian became a master at editing the many takes and extended performances of jazz recording sessions so that they fit the playing time of an LP, sounding like seamless original performances on the finished albums. But these producers did not see themselves as creators so much as enablers. Mitch Miller's view was "you can't make bullets out of shit," the talent and creativity had to be there and it was his job to bring together the artist, engineer, and studio, and "hopefully you get a performance."[96] Classical producer Wilma Cozart Fine stressed that her job, as she saw it, was to capture the music and the performance and never forget that the most important elements are the artists and the music that they're playing.[97] Producers of the pre-rock era generally share this concept of the producer's role as facilitator rather than as creator, but that concept changed with a new generation of independent producers, such as Phil Spector. Using large ensembles of session musicians to create what came to be known as his "wall of sound," Spector elevated the role of the producer by making his production technique, as Brian Wilson noted, "more integral to the process than the singer's talent."[98]

During the 1960s, artists began to take more control of their recording, leading many to move away from company studios because there staff producers directed the sessions. The Beach Boys became the first major rock group to go outside their label's in-house studio when Brian Wilson chose to use a number of other Los Angeles area studios, including Western, Gold Star, Wally Heider, and Sunset Sound, as well as Capitol, Columbia, and RCA, to record various parts of songs.[99] Capitol engineer John Palladino remembered that Brian "didn't like to come in the studios," which he attributed to Wilson's desire to work with a different group of engineers. Rock musicians "were more involved than with the old thing," recalled Palladino, "they would have their favorite studios or favorite mixers and they'd hang out with them and it would be a lifestyle." The contract artists who preceded rock groups "never," as far as he knew, "ever raised a fuss. They were all delighted to come into Capitol Records because it had all been worked out and it sounded great." But rock groups "wanted their choice [and] that's when all these independents started wooing them," by offering "little amenities . . . and the fact that you cater to them and you could almost live there . . . do whatever you want to get them going, you know, and we couldn't." For some groups, the company studios were both too public and too restrictive. "They didn't want to come into Capitol Records in the midst of all this . . . I mean, you come in the front doors and here's this big thing . . . all [these] people around . . . people hanging out. They didn't want any of that. It's just the feeling of when you're walking into an institution, you know? How you gonna smoke pot in an institution?"[100]

The shift in attitude toward the major recording studios Palladino describes had to do with a number of factors, not only the ability to openly consume recreational drugs. In Brian Wilson's case, choosing independent studios had more to do with his desire to produce his own records, to be able to translate the music he had in his head to the recording without company interference, and to do so, he had to go outside the Capitol Tower. In the studios of Capitol, Columbia, and RCA, staff producers were accustomed to controlling every aspect of the artists' music and staff engineers were unionized and governed by certain rules, including exclusive control over the equipment and designated tasks. While this division of labor made sense from an engineering perspective, some considered it unnecessarily cumbersome. One rock musician described the chain of command at Columbia's New York studio in the mid-1960s as something out of a World War II submarine movie:

> "Take one!" cries Roy. "Tape is rolling," informs the man in charge of starting the 8-track, in an entirely separate room. "Echo on!" finishes the third in a yet more distant room, punching on the 2-track tape delay. After that, you got to start playing . . . all that was missing was the wait to hear if the torpedo had found its mark. . . . Mixing was equally as crazy, as you still had all three to deal with, and you were never—I mean NEVER—allowed to touch a fader, lest you get your fingers slapped. You had to tell each engineer what you wanted and hope he understood.[101]

While this recollection dramatizes the bureaucratic nature of the operations in major label studios, it does not exaggerate the growing communication gulf between artist and engineer, which was difficult even for those who worked together regularly, as demonstrated by the Mitch Miller–Tony Janak session described in chapter 4. When musicians with little or no technical knowledge attempted to explain the sounds they were after to engineers unfamiliar with the new aesthetics of rock, lines of communication were strained even further. Doug Pomeroy began at Columbia in New York in 1969, starting as tape operator on the night shift. Although he understood the restriction on preventing nonengineers from touching the controls—called "jurisdiction," a fundamental principle originally intended to protect union engineer's jobs—he felt that it had become absurd, "since allowing a producer to 'ride a fader' was in no way going to cause anybody to lose his job." Pomeroy allowed producers and artists to touch faders "if they thought it was going to help." As far as he knew, no one ever reported him, but not all of his fellow engineers shared his liberal views.[102]

The preference for outside recording studios had to do with more than their permissive atmosphere about riding faders or taking drugs, it also stemmed from a desire for a looser atmosphere and a shared musical aesthetic. Capitol's attitude toward rock acts, according to one A&R man, was that "as long as you're selling records, you can do what you want."[103] But the dismissive attitude of some label executives who still considered rock 'n' roll a fad, even as sales of Beatles and Beach Boys records combined accounted for more than half of Capitol's revenue in the mid-1960s, may well have contributed to the growing alienation of rock acts from the corporate studios. When The Band went to the West Coast to make their second album, they also avoided the Capitol Tower, choosing instead to set up a studio in the pool house of a rented Hollywood home belonging to Sammy Davis Jr. Making music that in 1969 bucked the trend in psychedelic rock and use of studio technique, The Band recorded as a group with minimal overdubbing or echo effect and the resulting album thus more accurately represented the group's live sound. Contrasting their approach with the increasingly segregated technique employed in multi-track recording sessions, drummer Levon Helm recalled in a later interview that it was "more like makin' music and less like makin' tracks," and guitarist Robbie Robertson characterized the communal atmosphere during their recording sessions as "the clubhouse technique of making music" because they could take their time. "You were not on the clock; you could do whatever you wanted to do. There's nobody in a glass booth saying, 'What was that?' there was none of that."[104] Although the record company provided the equipment and a Capitol engineer was present as required by the label, according to John Palladino, who was responsible for transporting some of Capitol's recording gear to the pool house, "they were doing it themselves."[105]

The stipulation that a staff engineer be present when label artists recorded outside the company studio was a compromise to satisfy the engineers' union contracts with the major labels as well as artists' demands to record where and with whom they chose. Artists often chose an independent producer with a track record of hits who preferred working in studios "and with his own people who 'generate the right vibes.'"[106] But it became contentious and "extremely distasteful" to some staff recording engineers who had to sit and watch while others worked, and as Doug Pomeroy noted, it was "no real solution to the underlying problem" of artists and their producers choosing independent studios, which he and many others believed had more to do with lagging corporate response to rapid technological change and union (International Brotherhood of Electrical Workers) resistance to new trends in engineering practice.[107] Most independent studios had a dedicated drum booth with sets of drums fully miked and ready to go at the start of any session. In the Columbia studios, engineers had to set up and mike a drummer's kit and arrange the necessary baffles at the start of every session, a highly time-consuming activity with an obvious competitive disadvantage.

With their full-time skilled maintenance engineers and staff mixers, the major labels were second to none both technically and professionally, but some independent studios were ahead of the label studios in keeping up with cutting-edge technology and trends in recording, particularly the number of tracks they offered. At Capitol Tower, John Palladino recalled that the evolution from 4- to 8- to 16-track was gradual, and each change required an update to the mixing console both for recording and for playback of a rough mix, noting that "to be able to hear it as if it were a final mix made those mixers very, very complicated." Palladino recalled that there was a "technological lag" once studios reached a certain level of popularity, and he experienced this at both Radio Recorders and Capitol. If a company like Columbia or RCA changed recording equipment in one of its studios, it had to equip the others with the same capabilities to accommodate artists who might record in several different locations if they were touring or doing film work. Columbia engineer Roy Halee explained that this was why big studios appeared to resist new technology; going from 8- to 16-track would mean a significant investment to equip all their studios with the same technology, which was company policy.[108] Independent studios may have had a smaller budget, but they also had more flexibility, no larger infrastructure to worry about, and no long-standing tradition of established production methods or strict job classifications. Where it was once the goal of every rising musical artist to be offered a recording contract and to record in the company studio, considered "the best in the business," by the late 1960s, this was no longer the case. Increasingly, as Cleveland recording engineer Don White recalled, artists were "following the technology," going where the hits were made, which in some cases was attributed to the studio's equipment.[109]

After years of building mastering rooms for major labels, Clair Krepps built his own studio, Mayfair Recording, which in 1965 was the first 8-track studio in New York City outside of Atlantic Records, and because of that, it was in high demand. Krepps recalled, "MGM [Records] booked my first studio, they picked up my schedule book and blocked out 12 noon to 12 midnight ... five days a week, for I think it was six months in advance. And they would pay me whether they used it or not." Tom Wilson, the producer who requested that studio, had acquired a reputation in A&R at Columbia Records, producing several Bob Dylan albums as well as the hit single "Like a Rolling Stone" before moving on to other labels and eventually going independent. Concerned that the record label was not aware that paid-for studio time was going unused, Krepps alerted Val Valentin, director of engineering at MGM Records, who responded, " 'Yeah, that's okay, Clair. Don't worry about it. What Tom wants, that's what we give him.' "[110] In the new era of rock, Tom Wilson was one of a new breed of independent producers who were accorded unprecedented freedom. A *New York Times Magazine* profile called him "the Mitch Miller of tomorrow," and "a psychoanalyst with rhythm" who was successful commercially because he was sensitive to artists as well as to consumer demand, keeping "one ear attuned to the music, the other to 'the audience's proved responses.'"[111] Knowing the demand for multi-track recording, Wilson block-booked Mayfair to "keep out the competition," according to Krepps, a practice that became increasingly common in the hotly competitive popular recording industry. By the end of the 1960s, that competition frequently involved studio technology. "Music is no longer written with just the lyric and performer in mind," declared a *Billboard* reporter, "music is written taking into account the full potential of the studio."[112] A decade earlier, producers competed for exclusives on hit songs, but in the era of multi-tracking and psychedelic rock, working with rock band members who usually wrote their own material, independent producers were more likely to compete for exclusive use of state-of-the-art recording technology.

Conclusion

The introduction of multi-tracking triggered a series of dramatic changes in recording studios and ultimately transformed the art of chasing sound. As studio technology was used more creatively, the importance of engineers and producers in artistic decision making increased. More choice in recording and organizing tracks added time spent at every phase of record making, from recording to mixing to mastering, enabling the postponement of decisions that once had to be made on the spot. For many engineers and producers who had developed their skills in the earlier era, the transition was a mixed blessing. They gained greater control over sound, but that control came at the cost of what for some was a more exciting way of making records—live, face to face, and with no chance of endless overdubs. For other engineers, producers, and musicians who were accustomed to multi-tracking, using the tools of sound recording to shape the record was simply part of the process. In the right hands and with the luxury of time in the studio, these tools yielded magnificent results.

Ironically, the two albums most often cited as inspiring more creative approaches to recording did not have the benefit of much of the equipment that became standard by the end of the 1960s. However, in both cases, the artists did have the luxury of time to experiment. The instrumental tracks for the Beach Boys' *Pet Sounds* were recorded with the full ensemble of studio musicians—L.A.'s Wrecking Crew—playing live on three tracks, which were then mixed down to the fourth track to make room for the Beach Boys' vocal overdubs. According to musician Lyle Ritz, Brian Wilson could spend a whole night getting the musicians to play one sixteen-bar phrase exactly as he wanted it played, yet two of the album's biggest hits, "Wouldn't It Be Nice" and "God Only Knows," were completed in a single, long session without any instrumental overdubs. In addition, Wilson intentionally mixed the album to mono for listeners to hear the record exactly as he intended, "without any interference from the listener's stereo, which could be set up in many different ways that might affect the sound," according to Mark Linett, who remixed the album in stereo in 1996.[113]

In a classic example of creative one-upmanship, the Beatles, whose *Rubber Soul* album had inspired Wilson to make *Pet Sounds*, were in turn inspired by *Pet Sounds* to make *Sgt. Pepper*. At EMI studios, where permission had to be obtained in writing to move a microphone closer than two feet to an instrument, and the maintenance engineers actually wore white laboratory coats, the Beatles were accorded unprecedented freedom and time, and their experimentation signaled a radical break with the rigidly hierarchical structure of the company.[114] They, too, recorded on 4-track, but they made greater use of studio tricks by creative production techniques and clever engineering experiments with microphone placement, combinations of instruments, running tape backward and at different speeds, and various other ingenious methods of achieving unusual sounds, some of which, according to an Abbey Road tape operator, "would be impossible to make, even with today's 48-track equipment and all the microchips imaginable."[115]

The Beatles, in turn, influenced other musicians to approach recording in new ways. Bruce Botnick acquired an advance copy of *Sgt. Pepper* before its June 1967 release while he was engineering the second Doors album, *Strange Days*. He remembered playing it in one of the studio mastering rooms and, "absolutely flipping out. I brought the Doors in—I don't think Jim [Morrison] was there, but I brought them in and we listened, and our jaws dropped. And what it did for us, it said, 'Let's not do it the same way we did before, let's invent new techniques of recording. No holds barred.'"[116] The Doors' first album had been a straight documentary recording of the band as it sounded in live performance, but the second album, according to producer Paul Rothchild, employed more "studio technique. We were trying to be avant-garde, and in so doing, we sacrificed some of the freshness and spontaneity for the drama and staging that you can get with a 'produced' album."[117] They were not alone.

By 1969, the trend toward more highly produced albums escalated to such a point that the man who was arguably most responsible for it, Beatles producer George Martin, voiced his concern over the ascendance of technology to fellow producers at the International Music Industry Conference in Nassau.

"Technical development has made great strides in the past few years," he said, "[and the] electronic and mechanical aids to record producers are immense and complicated." But he asked, "Are we becoming too hung up on production technique?" Martin warned that electronic equipment could be "a trap" and that too much emphasis on technology was undermining artist development. That sentiment was echoed by fellow producer Felix Pappalardi, who said that although the "principle trend in the area of artistic popular music today must be, and is, experimentation," too few independent producers were capable of fostering "artistic incubation" for the acts they worked with in the studio.[118] Artist development had been one of the key investments record companies made toward long-term success for both the artist and the label, but this arrangement changed with the rise of independent producers and the move of rock acts away from a close working relationship with a staff A&R man in the company studios. Although there were several reasons for this, not all technological, new technology increased the desirability of studios where younger producers and engineers were open to experimentation in ways that established studios and engineering staff often were not. Those willing to experiment to "find a novel guitar sound, for instance, that guy became the musician's friend," recalled Phil Ramone.[119]

Like any professional, engineers as well as musicians had to adapt to change, whether to new technology, to new uses for old technology, or to new musical style. Session musicians, like actors, performed to fit the style of the record, and it is fitting that music contractors referred to their work as "casting" a session. When Artie Kaplan was hired to cast musicians for Aldon Music sessions in the 1960s, producer Al Nevins told him, "I want the guys that have their finger on the pulse TODAY—not the old guys, not the people from the other era. I want the people from this era." In New York, Kaplan recalled, most of the older musicians "had a terrible attitude toward the music that we were playing."[120] Kaplan admits he may have had the same attitude, but he had been in the army for two years and when he attended NCO and officers' club parties he "saw what the music did to people." A similar attitude prevailed on the West Coast. In Los Angeles, "rock 'n' roll was still a dirty word" in traditional music circles when the Wrecking Crew session musicians—so named, the story goes, by older musicians who accused them of wrecking the music business—began providing instrumental tracks for nearly all rock records of the 1960s.[121] While some musicians made the transition, those who did not lost work. Just as the generation gap played out in the nuclear family and in society at large, the recording studio became a battleground for competing musical ideologies and differing concepts of the studio's ultimate purpose that were rooted in different musical traditions and technological practices.

Rock music was central to the youth movement that swept across the nation in the 1960s. What gave that music new shape and form was the way in which musicians exploited electronic instruments and effects and the way engineers and producers experimented with recording technology to create new sounds, new styles, new musical expression that combined older musical forms with new technological possibilities. They could only accomplish this by breaking with past practice, by discarding previous formulas for success that had been long in the making. In so doing, they transformed not only musical culture but also recording methods and the working relationship between musician, producer, and engineer.

Conclusion

I n 1970, *Billboard* proclaimed the recording studio "The Crucible of Creativity." No longer a facility for merely transferring an artist's performance to disc, the studio was now "the chief tool of the producer . . . the final catalyst, the crucible wherein the talents of producer, artist, songwriter and musician may be brought together and into the market place and exposed to the ultimate consumer."[1] Aimed at the independent producers who were normally the decision makers in the selection of studios for the artists they produced, the *International Directory of Recording Studios* featured detailed technical profiles for each of the nearly eight hundred studios listed, about three hundred of which were located outside the United States. Just three years earlier *Billboard* had reported that studios were booked solid, and it was so difficult to find openings that labels had to reserve time a month or more in advance.[2] By 1970, that demand had been met and exceeded by an explosion in the number of new studios worldwide to meet the needs of touring rock bands. Clair Krepps's engineer, Gary Kellgren, left Mayfair in 1969 to found the Record Plant, which along with Jimi Hendrix's Electric Lady became New York City's newest independent recording studios and the first to be designed with artist comfort and state-of-the-art technology as dual priorities.

In addition to multi-tracking and its effects on recording, the 1960s instigated sweeping changes in studio design. Since people were now spending more time in studios—weeks and months rather than hours or days—client comfort received more attention. A major complaint against the older studios had been their cold, institutional aesthetic, "high ceilings, bare walls, and fluorescent lights [that] were either off or on, no in between, . . . too bright, too green," recalled folk musician Tony Glover about working at New York's Mastertone Studio, a description that fit nearly every studio built before the 1960s.[3] The austere décor suited the no-nonsense work ethic of three-hour sessions, but with the extended time and open-ended sessions of multi-track recording and mixing, younger musicians wanted surroundings more conducive to creativity. Unlike the days of Liederkranz Hall, the ambience of a studio no longer meant its acoustical perfection, rather it became the "vibe" given off by those who worked there, the "feel" of the surroundings, an exotic or secluded location, hip interior design, amenities such as hot tubs and sleeping accommodations, and access to all kinds of indulgences. Elektra Records founder Jac Holzman moved the label's base of operations from New York to Santa Monica, California, outfitting the studio with Mission furniture, hand-laid distressed wood floors, variable lighting (with mood and color adjustable from the control room), paisley-covered acoustical panels, and "Persian carpets, throw rugs and mounds of fluffy velour cushions, everything to make it seem less like a studio and more like a living room."[4] As the recording studio business became increasingly competitive, artist comfort and hip interior design became selling points along with state-of-the-art technology.

By the early 1970s, there were enough studios throughout the United States that competition for customers in some markets became fierce. Faced with decreased business from its signed artists and the costs of maintaining an engineering staff of twenty-eight, Columbia Records became the first major label to close its Los Angeles studios in 1972.[5] Soon the label sought to attract independent producers to its New York, Nashville, and San Francisco studios by advertising in the *Billboard* directory, "Now you, too, can record at Columbia."[6] Capitol appealed to outside customers as well, inviting them to "Make your hits at our place."[7] While the major label studios traded on their established reputations, independent studios offered privacy, comfort, and a creative environment. Rock stars seeking refuge from urban environments and prying fans, and fired by success and newfound wealth, spent lavishly on home recording studios or relocated to exotic locales and brought studios with them. The Rolling Stones spent the summer of 1971 recording the basic tracks for *Exile on Main Street* in a villa in the south of France, using a remote recording truck parked outside.[8] Seeking to escape the distractions of Los Angeles, the Beach Boys commissioned a 16-track studio to be designed and constructed in Los Angeles. It was then broken down, packed into $50,000 worth of custom transport crates, and shipped to Holland where it was reassembled and built into a custom-modified barn in rural Baambrugge.[9] The Record Plant in Los Angeles was a luxurious live-in palace, where, according to musician Mick Farren, "the facilities for the after hours orgy [we]re as spectacular as the recording equipment."[10] As recording involved more tracks and additional devices, control rooms expanded to accommodate the equipment and consoles resembled the flight decks of spaceships with blinking lights, meters, sliding faders, dials, and switches. The flight metaphor was invoked often in descriptions of studios and engineers; in 1968, Ampex advertised its AG-440 4-track recorder—"For the recording engineer who wants to fly"—depicting an engineer in a pilot's cap and uniform, seated in a control room as if it were a cockpit.[11] By the centennial of the invention of the phonograph, recording studios had become something that the early recordists would have found unrecognizable.

Between 1969 and 1973, the sheer number of studios and audio equipment manufacturers soared along with booming record sales. In 1967, *Billboard* reported record sales of $1 billion in the United States alone, of which $874 million were LPs, and three-quarters of the retail sales comprised "pop, rock, psychedelic music, and comedy."[12] By 1972, *Time* magazine reported that sales of records and tapes hit $2 billion ($3.3 billion worldwide), "making music, for the first measurable time in history, the most popular form of entertainment in America."[13] As recording industry profits soared, many newcomers who had little or no technical background could open a studio. In the 1950s, building a recording studio independently was an expensive undertaking that required getting components from a variety of manufacturers and either building your own console or having it custom built. In the 1970s, MCI, a Florida company, offered complete studio systems for $50,000 and only 10 percent down, and Westlake Audio in Los Angeles offered assistance to turnkey clients by designing and outfitting studios from scratch, giving those who were drawn by the glamour of the recording industry and potential financial reward the opportunity to open a recording studio without knowing a tube from a transistor. Established companies such as 3M in Minnesota, Swiss-based Studer, and Scully, the Bridgeport, Connecticut, makers of the original cutting lathes dating back to the early twentieth century, competed with Ampex in the multi-track recorder market. The audio equipment industry mushroomed with 24- and 48-track tape recorders, more complex mixing consoles, more microphones, studio monitors, headphones, Dolby and other forms of noise reduction, and an array of special electronic equipment: attenuators, compressors, delay systems, graphic equalizers, filters, limiters, oscillators, panners, synchronizing equipment, echo devices, measuring and testing equipment, and in the early 1970s, quadraphonic equipment, the short-lived first attempt at surround sound.[14] More studios added film and video recording for scoring television commercials as advertising moved beyond simple jingle productions to adopt some of the sophistication and style of serious rock in pitching to an ever-expanding youth market.[15]

This technological boom represented exponential growth in the audio industry. In 1958, only twelve manufacturers and distributors exhibited their wares at the Audio Engineering Society's annual convention in New York City; by 1974, there were sixty-six, and the society now had chapters located around the world.[16] The technology that had once been limited to the technically proficient and manufactured by a small number of companies was by the 1970s being produced by dozens of companies, both old and new, and it was accessible to anyone who could afford it. With generous lines of credit available for those who wanted to be part of what seemed to be a limitless growth industry, cost was not a deterrent.

In 1971, engineers gathered in New York for a forum on the state of the independent studio business. The challenges they faced included the large amount of "elaborate equipment the independent studio operator needs in order to keep up with the competition."[17] At the same time, the cost of producing a good multi-track album had skyrocketed, leading some bands to buy used recorders and a bare minimum console and resistive combining network and work out everything at home, then bring their tape to a professional studio for mixdown. Studio recording had always been a group effort, but as multi-tracking made it possible for musicians to lay down tracks separately, days and even months apart, the synergistic quality of group performance and teamwork changed and, in some cases, disappeared altogether. With the rise of the project studio in the 1970s, many musicians took control of producing their own records. The producer-artist-engineer team and the multiple sets of ears that had once offered a much-needed perspective on the process were no longer there to listen. The growing trend toward bands self-producing their albums became a contentious issue among producers and artists. While some musicians thought they knew their sound better than an outside producer, producers argued that musicians "lacked objectivity" and with the increased number of tracks and time spent in the studio were becoming "too analytical."[18] While this may well have been true, implicit in this debate is the growing struggle over who should have control— over artistic expression and over technology.

The sounds we hear on records are the result of creative collaboration between artists and technicians. Teamwork, adaptation, and cooperation were essential features in recording from the very beginning, but the technological changes of the post–World War II era increased the engineer's importance and, eventually, the producer's power, and the musical input of both. While recording stars and record producers enjoyed public recognition, recording engineers often contributed far more to recordings than they received credit for doing, as did session musicians whose playing on some records was attributed to others. The Los Angeles session musicians known as the Wrecking Crew played on records for the Beach Boys, Sonny and Cher, and Phil Spector. Sometimes they read scored arrangements, but in rock music, the emphasis on the feel and the sound meant that they not only followed the producer's directions but also made unique contributions of their own—in their playing style, musical suggestions, ideas for playing a part or setting an amplifier or effect in a certain way, yet their contributions were anonymous. Motown recorded singers and vocal groups, and every record they made used the same group of session musicians, known as the Funk Brothers, who like their Los Angeles counterparts were "standing in the shadows" of the name artists, yet it was their playing that established the record's groove, beat, and melodic structure that are as much a part of the record as the singer's voice.

The designations "engineer," "control man," and "mixer" were used interchangeably through the 1950s, but by the 1970s, there were real differences between engineering and mixing. As recording technology increased in sophistication and in complexity, fewer recording engineers understood how the equipment worked inside and out. In professional studios with a maintenance crew, mixers operated the console on sessions and engineers serviced the equipment and handled technical matters, and even the technicians specialized in different areas. Taking stock of the audio engineering profession on the hundredth anniversary of Edison's invention of the phonograph, DeWitt F. Morris assessed the state of the field for the *Journal of the Audio Engineering Society*. By 1977, he noted, the audio engineering field included men—and now women— working in a variety of industries, environments, and cultures, but the entertainment field and specifically the recording industry had become "the premier audio industry," second only to the telephone industry.[19] Recording at the top professional recording studios required "a battalion of skilled audio engineers," including experts in acoustics and studio design, recording director or *Tonmeister*, console operator/mixer, tape recordist, mixdown mixer, disc recordist, maintenance engineer, and engineers in charge of tape, cartridge, and cassette duplication. With the growth and maturation of audio technology, "the roles of scientists, engineers, technicians, and artists have become progressively more interrelated," and in the fields of entertainment especially, the boundaries between disciplines and responsibilities had grown less distinct. Engineer Doug Pomeroy observed the "vanishing line between engineers and producers" as the meteoric growth of the recording industry since 1968 affected all engineers, but especially recording engineers, who "had to become increasingly involved, creatively, in the use of multi-track techniques, and of the studio itself as a giant musical instrument."[20]

It was also clear by 1978 that an engineer's role was defined by the type of music he or she recorded. In popular music, noted classical recording engineer Carson Taylor, "the engineer is frequently a part of the group being recorded or broadcast as he contributes various electronic controls and processes to create the final effect." Classical recording, however, required that the engineer "always be mindful that it is the music he is attempting to transmit or reproduce, and he must be musical in everything he does in his technical handling of symphonic music." The more perfectly he performs his job, "the less aware the artists and audience will be of his existence."[21]

In the early 1970s, Edward Kealy undertook a sociological investigation of the working environment of recording studios in Chicago, interviewing numerous "cultural workers in record production" all of whom maintained anonymity.[22] Kealy's fieldwork concentrated on the job of sound mixer, which he described as a "technician-artist occupation," because the work involved various levels of collaboration with musicians and producers. Kealy concluded that, as rock musicians assumed a greater role in the technical aspects of recording, the job of the sound mixers became less of an art and more technical and bureaucratic. However, recording engineering has retained elements of art and tacit knowledge that were essential to the work of early recordists, even as their tools grew sophisticated and capable of performing tasks once carried out by human hands. In the 1950s, microphoning exemplified the tacit knowledge of sound engineers, and even today, much of what engineers and mixers do still involves tacit knowledge, like the ability to assist musicians in presenting their ideas. "Sometimes, musicians' ideas—there can be too many, and maybe too much at the same register," noted engineer Steve Gursky. As Eddie Smith recalled about voicing instruments, leaving a space in an arrangement, Gursky pointed out that if voicings are too similar, "they can get in the way of each other sonically. . . . There's no 'theaterical cue' with three of the same voice and the listener will not get the depth of field."[23] A good engineer knows how to introduce enough alteration of this "sameness" to give the background voices a distinct identity. By using a different microphone or with some compression, engineers can place these other voices "behind" the lead vocal, but not all compressors work well on all instruments. Like knowing which microphone to use, good engineers know which compressor works best on which instrument.

∽

From the early years of sound recording to the middle of the twentieth century, improvements in recording technology were aimed at the listener, at improving the fidelity of records. With the introduction of professional magnetic tape recorders, improved microphones, and high-quality vinyl records after World War II, fidelity seemed to have been achieved, but it was a fiction because these same improvements also introduced editing and splicing that altered the original musical performance, begging the question: fidelity to what?

Critics of the high-fidelity craze believed the only way to faithfully convey music on records was through the "natural sound" created by room acoustics, but even the definition of natural sound came into question during the 1950s.[24] With the introduction of stereophonic recording and reproduction, which added depth and dimension to music, the temptation to exploit the potential for further manipulation proved irresistible. Each new technological innovation introduced with the purpose of improving the fidelity of records seemed to invite other uses. The high-fidelity craze, Bill Savory recalled, became "quite an indoor sport and it got to the point where it was high, but not much fidelity." The reason for this was the continued boosting of certain frequencies to make them sound "brilliant" on the recording. Savory knew things had really gotten out of hand when he heard a lecture in which the speaker declared that live music was not high fidelity because "he'd sit in front of the New York Philharmonic, and the strings didn't scream at him, and the brass didn't knock his hat off. So he says 'Live performances are not High Fidelity.' "[25] Technically, of course, he was right, because high fidelity had become something other than what those who forged the path toward achieving better sound had intended. "We all thought we were carrying the torch for Edison," said Clair Krepps, "to advance Edison's art."[26] Yet by the 1960s even Krepps succumbed to what violinist Louis Kaufman called "high *in*fidelity" when he mastered Manfred Mann's version of "Doo-Wah Diddy" so hot that he caught flak from fellow engineers.[27] Rock recording had always emphasized volume, rhythm, and power, not fidelity. Kauffman blamed popular recording for "poisoning" the recording perspective of classical music by "accentuat[ing] things artificially. . . . You get that overwhelming boom, and you get an overemphasis on the highs, and there is a kind of peculiar distortion that they like to call high fidelity." In some classical recording of virtuoso violinists, Kaufman noted, "the better the recording technique gets, the worse their recordings sound. You hear much more surface, much more rosin, and much more extraneous noise," sounds that a live hall would swallow up.[28] Clearly, for some listeners, "high fidelity" did not equate to better sound. On records, some things are better left unheard.

Technological change could, in some cases, result in a return to previously discarded practices, and the changing use of echo and room reverberation illustrates this. In 1956, Bob Fine established Fine Recording in the Great Northern Hotel on West Fifty-Seventh Street in New York City. Located across from Steinway Hall and down the street from Carnegie Hall, the Great Northern had seen better days, but its attraction for Fine was the hotel's ballroom, which became his Studio A and one of the best-sounding rooms in New York City or anywhere (see fig. 10). Throughout the 1950s and 1960s, Fine recorded big band and small combo jazz and popular vocalists such as Dinah Washington. By 1968, Fine claimed that market demand for exaggerated echo effects forced him to abandon natural room reverberation and to use EMT reverb chambers because listeners' ears were now attuned to exaggerated reverb. To obtain the isolation of instruments required in multi-track recording, singers and instruments had to be close-miked. In doing so, Fine said, "you eliminate so much of the natural reverb you must make up for it with EMT's. Unfortunately, when you are recording rock 'n' roll and all these wild, screaming, brassy things . . . a live studio is extremely difficult to handle."[29] The best rooms for rock 'n' roll recording, he found, were low-ceilinged and very dead studios where control of the sound could be maintained. Thus, nearly three decades after the rejection of dead studios in favor of live sounding rooms, technological innovations and new musical styles had completely reversed the trend. Once reverberation or echo could be achieved artificially, and amplified instruments like electric guitars, keyboards, and drums dominated group sounds, the dead room returned.

This study has focused on popular and rock recording because these genres exploited studio technology to the fullest. How did the transformation of studio recording affect jazz, country, folk, and other musical genres? By 1977, it was impossible to escape the impact of multi-tracking and the perfectionism that the increased control in studio recording made possible. By the 1970s, listeners had already begun to expect technological intervention as natural. This phenomenon was the natural extension of a listening culture in which ever greater accomplishments in sound engineering were met with, perhaps exceeded by, ever greater expectations for a more physically stunning quality of sound. The perception of what was natural, acceptable, or even desirable continually changed over the century, as recording and reproduction of sound on disc, on film, over public address, and piped into increasingly noisy public spaces grew exponentially.[30] This continual dialectic in mechanical reproduction—between the goal of authenticity and the achievement of something close but ultimately imitative—is not unique to the work of the recording studio. It has a long history in American culture.[31] The recording has long since ceased to take the live performance as its model, even when it is a so-called live recording. Since the introduction of magnetic tape, such performances have been edited to combine the best of live performances. As stage actors require exaggerated makeup and fashion photographs need digital airbrushing, the recording requires a certain amount of sonic retouching to achieve the natural quality—or whatever quality—it is meant to represent. Is it any less the real thing, any less authentic for the manipulation? In the 1960s, rock musicians shunned natural sound as amplification, feedback, synthesizers, fuzz, and wah-wah became the new means of achieving "the sound" in the studio and on stage.[32] In popular recording, "natural" ceased to have the same meaning it had before. By the end of the twentieth century, as one arts critic declared, "artifice is the new nature."[33]

During the 1960s, the needs of the creative artist had begun to dictate changes in recording technology and how it was used, generating an increasing reliance on tools to manipulate and control different aspects of recording, from the sound of a guitar or voice to the signal processing used in a final mix. The more such tools became useful, the more artists relied on them and on engineers and producers to use them effectively in the studio. The growth of professional audio products manufacturers escalated, introducing new, improved versions and different gadgets, assuming a kind of technological momentum that seemed inevitable, full of potential for creative uses, and thus technology could be used to influence the cultural product. Electronic effects and techniques derived in the studio, by accident or by design, became tools to create new sounds. Higher expectations of sound increased demand for more controlled, precision sound machines, which in turn laid the groundwork for disco music, and later embodiments of technologically driven genres such as rap, electronica, techno, and hip-hop.[34]

Chasing sound began with a technological challenge, to find the solution to problems of sound recording and reproduction and to improve sound quality for listeners. By the 1960s, chasing sound had become an art form, made possible by technologies that were the product of those earlier efforts to streamline recording and improve the listening experience. Yet recording technology was now used in ways older engineers would have rejected—in the service of creative destruction of rules and standards of recording that had been painstakingly developed over decades. In "the search for the sound" in rock music, making records with more depth, more volume, and a more captivating sound required technological support, an engineer and producer willing and knowledgeable about how to achieve that. Anything was fair game to get the sound. Distortion, feedback, and phase-shifting became common effects rock musicians used in recording and on stage, a rejection of anything resembling high fidelity. Looking back on a career in classical recording, Columbia producer Leroy "Sam" Parkins recalled the enormous battery of signal-processing equipment rock musicians deployed in the 1960s, including wah-wah and fuzz tone, and other effects that originated in the studio, such as phase splitters. "Here we're knocking ourselves out in classical music to get everything reasonably in phase," said Parkins, "and these people want to knock it *out* of phase." To Parkins and so many others, classical and rock recording now embodied "quite an opposite set of worlds and aesthetic requirements."[35] Musician, composer, conductor, and music scholar, Gunther Schuller observed that producers of most types of recording no longer even attempted to accurately reproduce a live performance, especially in rock music, where achieving "a certain sound that is new or fashionable or current" made amplification and electronics part of the creative effort, in some instances built into the composition process. Such music, Schuller argued, could never be an accurate, live representation; rather it should be considered a "separate process parallel to live performance."[36] What struck Schuller as remarkable enough to comment on in 1977 has become accepted practice. Today, technological manipulation is no longer considered cheating, it is expected; it is the art of recording.

This book has investigated how technology became intrinsic to a particular art form and, in turn, how objectives and expressions within that art form changed and in the process influenced the development of the technology and the way it was used to shape the sound of records. The recording studio provides an ideal window through which to observe how the give and take of technological and cultural change played out in music. At times, technology has influenced the course of music, as when the weak power of acoustic recording limited the types of music and voices that could be heard on record, effectively determining what record buyers would hear, or when the microphone completely reversed that situation by allowing softer voices to be picked up by the recorder, or when amplified instruments required different studio acoustics. Some argue that social and cultural forces have more to do with how musical style emerges, but even these scholars recognize the significance of technology at different points in the development of recorded music.[37] Musicians have insisted that the creative process drove technological innovations, yet those same innovations have been seen as the inspiration for musicians and composers to do things in new ways. Some believe that without technological intervention certain recording artists of today would not have a career, but engineers and producers have employed skillful manipulation to aid performers "whose popularity exceeds their abilities" for decades, although that manipulation was kept under wraps. Some high-fidelity projects of the 1950s boasted of their technical prowess with lengthy descriptions of the recording process and detailed illustrations inside gatefold album covers, but it was not until the 1960s that record companies regularly credited the engineer.[38]

During the 1970s, rock recording grew more refined along with the technology, advancing the illusion that the record was a bigger, better version of a live performance. In reality it was made by the combined efforts of artists and engineers and could only be performed live with a good deal of sophisticated sound reinforcement. In many cases, prerecorded tracks surreptitiously added to the live mix. In rock concerts, sound engineers became equally critical to what the audience heard coming from the stage, and live sound-mixing consoles adopted the same design and signal-processing capabilities of studio consoles. Companies began marketing equipment that promised to translate the benefits of studio recording to the stage. In 1967, Hohner marketed its Echolette, a portable sound studio that could "do things that no other sound system in the world will do. Like let your group achieve recording studio sound outside the recording studio. At ballparks, dance halls, auditoriums, gymnasiums—you name it." The system offered volume, tone, echo, and reverberation controls. An ad claimed: "The Echolette Tape Loop Reverberation and Echo Unit even allows you to achieve multiple echos [sic] with trick effects."[39] Although it was unlikely that this modest system could fill even a little league ballpark, by the mid-1960s it was clear that recorded sound had far outpaced live sound reinforcement, challenging musicians and engineers to deliver live sound comparable to the record. William Ivey declared in 1977 that even a well-established musician who presumably would be freed by success to create and experiment in a live concert, was being "held prisoner by the demand of his audience that he sound like his studio product," forced to duplicate the "super reality" of recording in a live performance because "the audience spurns creative experiment for the blander perfections of the recording studio." Recording, Ivey noted, had also accelerated the pace of change in music with negative effects on the existence of popular and folk traditions. Styles once immune to trends began to change, leading to the creation of national styles and the disappearance of regional styles. While "technology has always been seductive," said Ivey, "a technological solution is never an unmixed blessing."[40]

Great recorded performances and hit songs emerged from state-of-the-art, high-tech studios and home studios alike, but in the 1960s, certain studios and even certain pieces of equipment began to acquire reputations for working magic in the studio. An enormous audio instrument industry expanded not only to meet demand but also to create new options that in turn created new needs for musicians and technicians. Technology made neither hit records nor memorable performances, but technology did determine the broad outlines of what became possible in the recording studio. In that sense, the growth and development of the recording studio is an example of technological determinism mitigated by user choice, ingenuity, and human aims.[41] The users—engineers, mixers, producers, arrangers, musicians—determined how the technology would and could be used. The irony is that, as engineers got ever closer to the ability to capture the essence of a live performance, they did so only by removing the live performance from the recording studio. In seeking control over all the elements in recording, the separation of instruments and thus of the musicians playing them became increasingly necessary to achieve that control. Many engineers argued that this was not always the best choice and, in most cases, led to the loss of spontaneity and spark that no amount of technological intervention can replace.

With the rise of disco music in the 1970s, which relied on synthesizers, click tracks, drum machines, and repetitive perfectly timed rhythm tracks often created by one individual musician, live concerts usually involved singers performing with prerecorded backing tracks and performances in which "the equipment is the real star."[42] Audio innovators came up with technologies to perfect the sound, fix intonation, and even add psychoacoustic "sparkle" with the Aphex Aural Exciter.[43] The Aphex Effect was so closely associated with disco music that rock musicians distanced themselves from it. Eagles' producer Bill Szymczyk stated on liner notes of *The Long Run*, "This album was NOT mixed through the Aphex Aural Exciter."[44] Alice Echols summed it up well: "Disco did favor the synthetic over the organic, the cut-up over the whole, the producer over the artist, and the record over live performance."[45] At the same time, rock recording sessions extended in length along with the expanded number of tracks and inflated recording budgets. The 1970s became the era of excess, of the "album considered as a Cecil B. Demille epic" that took months and years to finish. By the mid-1970s, there was a reactionary punk movement seeking to return to the "raw power" of an earlier era but with all the technical development and control that recording technology and engineering practice had made possible. Many of these groups may have been untutored musicians, but the studio tools, as well as the instruments, amplifiers, and effects they benefited from gave their rawness a level of technological sophistication well beyond that of the early rock-and-roll records.

Chasing sound meant different things during the twentieth century, from solving the problems posed by technical limitations, to exploring new realms of sound that seemed limitless. The quest for control over sound inspired engineers and producers to devise techniques and devices to gain that control, which, in turn, increased their reliance on technology to maintain it. In music recording, technology offered the way forward, a solution to problems of sound quality, but it eventually became a self-perpetuating creative force. "It is the popularity of certain processes which stimulate a desire for the device that creates it, and for other manufacturers to quickly develop devices with similar characteristics. Specific processes have become commodities in the marketing and application of these devices, and the demand to use them will determine which equipment is purchased and which studios are used." So wrote engineer Tom Lubin about the state of audio engineering in 1996. "The desired sound may come from a state of the art device, or a vintage piece of equipment, since it is 'the sound' which determines the desirability of such a device," and ultimately, the desire for new sounds has turned the devices and the sounds they create into commodities.[46]

That process had been set in motion decades earlier as various technologies of sound recording inspired *audiotechnophilia*, the love of audio gear and technologically driven sounds, as distinct from "audiophile" or "music lover" or "technophile." In 1994, at a gathering of key inventors and innovators in the recording industry, sponsored by the Audio Engineering Society Los Angeles branch, Ham Brosius, who represented audio manufacturers Scully and MCI, warned those in attendance that "we sometimes become overwhelmed with the majesty of technology," and that younger people "confuse talent with a mechanical ability to move faders and work with samplers and things that you can do by rote." He concluded with the admonishment: "Respect technology but revere talent."[47] Many of the audio professionals who began in the pre-digital era have expressed this sentiment, and it speaks to the difference in attitudes toward technology and its relationship to art.

Looking back on his work for Cobra Records during the 1950s, Chicago session musician Phil Upchurch remembered the studio as the backroom of a record store, with a basic, simple setup where the band performed live with no overdubbing and no fixing it in the mix. "[We'd] come in and set up, put mikes on everybody and they'd record performances, as opposed to these days when they perform recordings."[48] Performing recordings was something singer Anna Case knew well from her Edison Tone Test appearances in the 1920s: "Instead of the recording mirroring the artist," Case recalled in 1972, "the artist mimicked the recording and the performances bec[a]me a frozen, unchanging entity to be copied."[49] Although these artists refer to different periods in recording history, Upchurch and Case both express unease with the idea that a live performance should do nothing more than copy a record. Whereas Thomas Edison used the Tone Tests in which Case performed to promote the sound of his Diamond Disc as indistinguishable from the sound of live music, by the 1980s when Upchurch compared his past experience with his present, performing the recording had become a necessity, not to sell the record so much as to meet listeners' expectations that the performance could sound as good as the recording.

This complete reversal of the relationship between recording and live performance came about because of the transformation of musical culture through the changing technology and practice of sound recording. The more we make technology do, the more we expect and the more we rely on it to do things that not too long ago were done solely by human effort. We welcome technological innovation while simultaneously waxing nostalgic about the way things were done—or the way things sounded—in some earlier time. Yet, technological innovation has enabled us to achieve the once inconceivable, such as recovering sounds recorded before Edison invented the phonograph, or resurrecting historic radio performances long thought to have vanished with the airwaves.[50] Whether we like it or not, our musical expectations have been shaped by the technology and the art of studio recording.

INTRODUCTION

1. Frankie Yankovic, *The Polka King: The Life of Frankie Yankovic*, as told to Robert Dolgan (Cleveland, OH: Dillon/Liederbach, 1977), 67.

2. Brian Wilson and Todd Gold, *Wouldn't It Be Nice: My Own Story* (New York: HarperCollins, 1991), 145.

3. John McDonough, "Pop Music at the Crossroads," *High Fidelity* 26 (Apr. 1976): 50.

4. Mark Katz, *Capturing Sound: How Technology Changed Music* (Berkeley: University of California Press, 2004), 101.

5. Eric D. Barry, "High Fidelity Sound as Spectacle and Sublime, 1950–1961," in *Sound in the Age of Mechanical Reproduction*, ed. David Suisman and Susan Strasser (Philadelphia: University of Pennsylvania Press, 2010), 115–138. See also Tim J. Anderson, *Making Easy Listening: Material Culture and Postwar American Recording* (Minneapolis: University of Minnesota Press, 2006); Joseph Lanza, *Elevator Music: A Surreal History of Muzak,® Easy-Listening, and Other Moodsong®* (Ann Arbor: University of Michigan Press, 2004); and V. Vale and Andrea Juno, eds., *Incredibly Strange Music* (San Francisco: Re/Search Publications, 1993–1994), vols. 1, 2.

6. Barry Cleveland, *Creative Music Production: Joe Meek's Bold Techniques* (Vallejo, CA: Mix Books, 2001); John Repsch, *The Legendary Joe Meek: The Telstar Man* (London: Woodford House, 1989). Meek's reputation lived on, inspiring an audio equipment company (Joemeek) dedicated to perpetuating the "Classic Joe Meek Sound" with studio gear and accessories. See www.joemeek.com.

7. See Jeff Chang, *Can't Stop, Won't Stop: A History of the Hip-Hop Generation* (New York: St. Martin's Press, 2005); Mark Katz, *Groove Music: The Art and Culture of the Hip-Hop DJ* (New York: Oxford University Press, 2012); Tricia Rose, *Black Noise: Rap Music and Black Culture in Contemporary America* (Hanover, NH: Wesleyan University Press, 1994); Aram Sinnreich, *Mashed Up: Music, Technology, and the Rise of Configurable Culture* (Amherst: University of Massachusetts Press, 2010).

8. Joanna Demers, *Steal This Music: How Intellectual Property Law Affects Musical Creativity* (Athens: University of Georgia Press, 2006), 4.

9. Paul Théberge, *Any Sound You Can Imagine: Making Music / Consuming Technology* (Hanover, NH: Wesleyan University Press, 1997).

10. David Suisman, *Selling Sounds: The Commercial Revolution in American Music* (Cambridge, MA: Harvard University Press, 2009).

11. Anderson explores this phenomenon in *Making Easy Listening*.

12. Emily Thompson, "Machines, Music, and the Quest for Fidelity: Marketing the Edison Phonograph in America, 1877–1925," *Musical Quarterly* 79 (Spring 1995): 131–171; Susan Schmidt Horning, "Recording: The Search for the Sound," in *The Electric Guitar: A History of an American Icon*, ed. André Millard (Baltimore: Johns Hopkins University Press, 2004): 105–122.

13. Peter Dellheim, "The Fine Art of Recording," *Saturday Review*, Dec. 12, 1953, 35.

14. Geoffrey Payzant, *Glenn Gould: Music & Mind*, rev. ed. (Toronto: Key Porter Books, 1992), 119. Gould's writings are collected in Tim Page, ed., *The Glenn Gould Reader* (New York: Alfred A. Knopf, 1984).

15. Quoted in Charles L. Granata, *Sessions with Sinatra: Frank Sinatra and the Art of Recording* (Chicago: A Cappella Books, 1999), xiii.

16. George Martin, with William Pearson, *With a Little Help from My Friends: The Making of Sgt. Pepper* (Boston: Little, Brown, 1994); Geoff Emerick and Howard Massey, *Here, There and Everywhere: My Life Recording the Music of the Beatles* (New York: Gotham Books, 2006).

17. Quoted in Payzant, *Glenn Gould*, 39, 43.

18. Jonathan Sterne, *The Audible Past: Cultural Origins of Sound Reproduction* (Durham, NC: Duke University Press, 2003), 222–223.

19. For example, Michael Chanan, *Repeated Takes: A Short History of Recording and Its Effects on Music* (London: Verso, 1995); Timothy Day, *A Century of Recorded Music: Listening to Musical History* (New Haven, CT: Yale University Press, 2000); Katz, *Capturing Sound*. Readers will find more sources on technology and music in the bibliographic essay.

20. Nelly Oudshoorn and Trevor Pinch, eds., *How Users Matter: The Co-construction of Users and Technologies* (Cambridge, MA: MIT Press, 2003); Trevor Pinch and Frank Trocco, *Analog Days: The Invention and Impact of the Moog Synthesizer* (Cambridge, MA: Harvard University Press, 2002); and related studies of how users tinkered with design and performance of their cars: Kathleen Franz, *Tinkering: Consumers Reinvent the Early Automobile* (Philadelphia: University of Pennsylvania Press, 2005); and Robert C. Post, *High Performance: The Culture and Technology of Drag Racing, 1950–1990* (Baltimore: Johns Hopkins University Press, 1994).

21. Hans-Joachim Braun, ed., *Music and Technology in the Twentieth Century* (Baltimore: Johns Hopkins University Press, 2002); Veit Erlmann, ed., *Hearing Cultures: Essays on Sound, Listening and Modernity* (Oxford: Berg, 2004); André Millard, *America on Record: A History of Recorded Sound* (Cambridge: Cambridge University Press, 1995); David Morton, *Off the Record: The Technology and Culture of Sound Recording in America* (New Brunswick, NJ: Rutgers University Press, 2000); Robert Philip, *Performing Music in the Age of Recording* (New Haven, CT: Yale University Press, 2004); Alex Ross, *The Rest Is Noise: Listening to the Twentieth Century* (New York: Farrar, Straus and Giroux, 2007).

22. Louise Meintjes, *Sound of Africa! Making Music Zulu in a South African Studio* (Durham, NC: Duke University Press, 2003); Paul D. Greene and Thomas Porcello, eds., *Wired for Sound: Engineering and Technologies in Sonic Cultures* (Middletown, CT: Wesleyan University Press, 2005); Albin J. Zak III, *The Poetics of Rock: Cutting Tracks, Making Records* (Berkeley: University of California Press, 2001); Thomas Gregory Porcello, "Sonic Artistry: Music, Discourse, and Technology in the Sound Recording Studio" (PhD diss., University of Texas at Austin, 1996).

23. For example, G. A. Briggs, *Audio Biographies* (Yorkshire, UK: Wharfedale Wireless Works, 1961); Ted Fox, *In the Groove: The People behind the Music* (New York: St. Martin's Press,

1986); John Harvith and Susan Edwards Harvith, *Edison, Musicians, and the Phonograph: A Century in Retrospect* (New York: Greenwood Press, 1987); and Elisabeth Schwartzkopf, *On and Off the Record: A Memoir of Walter Legge* (New York: Charles Scribner's Sons, 1982).

24. After that first broadcast proved to be a disaster, Stokowski was provided with what amounted to a "dummy" control booth while the actual controls of the broadcast were handled by the audio engineer in "a well-screened cubicle." Robert E. McGinn, "Stokowski and the Bell Telephone Laboratories: Collaboration in the Development of High Fidelity Sound Reproduction," *Technology and Culture* 24 (Jan. 1983): 38–75.

25. William M. Evan, "On the Margin: The Engineering Technician," in *The Human Shape of Work: Studies in the Sociology of Occupations*, ed. Peter L. Berger (New York: Macmillan, 1964), 83–112; Edward R. Kealy, "The Real Rock Revolution: Sound Mixers, Social Inequality, and the Aesthetics of Popular Music Production" (PhD diss., Northwestern University, 1974); Kealy, "From Craft to Art: The Case of Sound Mixers and Popular Music," *Sociology of Work and Occupations* 6, no. 1 (Feb. 1979): 3–29.

26. Dave Teig, telephone conversation with author, July 1996.

27. While black-owned recording businesses date back to 1919 with the Broome Special Phonograph Records mail-order company and Black Swan Records, the first major black-owned record company established in 1921, these operations were short-lived, and it was not until Berry Gordy established Motown Records in Detroit in 1959 that an African American–owned and –operated recording company achieved international and lasting success. Even at Motown, however, the head of engineering was white. See Tim Brooks, *Lost Sounds: Blacks and the Birth of the Recording Industry, 1890–1919* (Champaign: University of Illinois Press, 2004), 464–470; Suisman, *Selling Sounds*, 204–239; D. W. Fostle, "The Audio Interview: Mike McLean—Master of the Motown Sound, Part I," *Audio* 81, no. 11 (Nov. 1997): 56–61; Fostle, "The Audio Interview: Mike McLean, Part II," *Audio* 81, no. 12 (Dec. 1997): 50–56.

28. See James Lastra, *Sound Technology and the American Cinema: Perception, Representation, Modernity* (New York: Columbia University Press, 2000), esp. chap. 5.

29. Carolyn Marvin, *When Old Technologies Were New: Thinking about Electric Communication in the Late Nineteenth Century* (New York: Oxford University Press, 1988).

CHAPTER 1. CAPTURING SOUND IN THE ACOUSTIC ERA

1. Leonard S. Reich, *The Making of American Industrial Research: Science and Business at GE and Bell, 1876–1926* (New York: Cambridge University Press, 1985); Philip Scranton, *Endless Novelty: Specialty Production and American Industrialization, 1865–1920* (Princeton, NJ: Princeton University Press, 1997).

2. Harry A. Gaydon, *The Art and Science of the Gramophone and Electrical Recording up to Date*, 2nd ed. (London: Dunlop & Co., 1928).

3. Nathan Rosenberg, *Exploring the Black Box: Technology, Economics, and History* (Cambridge: Cambridge University Press, 1994), 9–23.

4. Roland Gelatt, *The Fabulous Phonograph: From Edison to Stereo*, rev. ed. (New York: Appleton-Century, 1965), 47–48.

5. Cited in Jas Obrecht, "George W. Johnson: African-American Recording Pioneer," *Tim Gracyk's Phonographs and Old Records*, http://www.gracyk.com/johnson1.shtml, accessed

Mar. 2, 2001. Johnson's career and that of other early black recording artists can be found in Brooks, *Lost Sounds* (see Introduction, n. 27).

6. Andre Millard traces this research and development work in "The Phonograph: A Case Study in Research and Development," in *Edison and the Business of Innovation* (Baltimore: Johns Hopkins University Press, 1990), chap. 4.

7. "Our New York Recording Plant," *Edison Phonograph Monthly* 4, no. 9 (Nov. 1906): 6–8, quote on p. 6.

8. Ibid., 8.

9. One of the reasons Edison built the West Orange lab was "to search for trade secrets that require no patents, and may be sources of profit until some one else discovers them," Edison quoted in Paul Israel, *Edison: A Life of Invention* (New York: John Wiley, 1998), 284. Eldridge R. Johnson's lawyers advised him to delay filing for a patent but to maintain secrecy in his operations, so William H. Nafey worked on the development of Johnson's recording process under lock and key in a rented room. B. L. Aldridge, *Victor Talking Machine Company*, ed. Frederic Bayh (Camden, NJ: RCA Victor, 1964), 22.

10. Quoted in Gelatt, *The Fabulous Phonograph*, 180.

11. Lloyd MacFarlane, *The Phonograph Book* (New York: Rider-Long Company, 1917), 31. All of the early literature on the phonograph and sound recording uses the masculine pronoun exclusively when referring to those involved in the recording process. Available evidence suggests that no women were "recording experts" or recording engineers until the late 1930s, although women surely operated home recording machines and at least one woman ethnographer, Frances Densmore, used the portable cylinder recorder to collect the songs of American Indians for the Smithsonian. I will therefore avoid clumsy efforts at gender neutrality and simply employ the masculine pronoun where appropriate.

12. Henry Seymour, *The Reproduction of Sound: Being a description of the Mechanical Appliances and Technical Processes employed in the art* (London: W. B. Tattersall, 1918), 94–95. Seymour invented the diaphragm "bearing-bridge," which he patented in 1903 and which he claimed was subsequently adopted by most record manufacturers.

13. F[rederick] W[illiam] Gaisberg, *The Music Goes Round* (1942; repr., New York: Arno Press, 1977), 37–38. Gaisberg himself was forced into a similar role when in 1906 he and his brother made the very first recordings of the legendary opera singer Adelina Patti at her castle in Wales. "It was my job," he recalled, "to pull her back when she made those beautiful attacks on the high notes. At first she did not like this and was most indignant, but later when she heard the lovely records she showed her joy just like a child and forgave me my impertinence" (ibid., 91).

14. Austin C. Lescarboura, "At the Other End of the Phonograph," *Scientific American* 31 (Aug. 1918): 178.

15. Joseph Batten, *Joe Batten's Book: The Story of Sound Recording* (London: Rockliff, 1956), 35. Invented by Augustus Stroh, a German inventor and engineer who moved to England in 1851, the Stroh violin was a modified version of the traditional violin, designed to compensate for the violin's lack of sufficient direction or volume. By incorporating elements of the phonograph (horns, diaphragm), Stroh increased the volume and directionality of the instrument. Instead of a wooden body, this instrument had a metal resonator, or diaphragm, connected to

the underside of what was essentially the violin fingerboard, bridge, tailpiece, and shoulder rest. A metal horn attached to the resonator was aimed at the recording horn or into the ear of the singer, and a smaller horn faced the opposite direction, into the ear of the violinist. Stroh must have come up with the idea after building an improved version of Edison's phonograph in 1878, thereby creating probably the first instrument designed specifically for recording. He patented his invention in 1901, and his son Charles manufactured the instruments until 1924, followed by George Evans through 1942. Cynthia A. Hoover, *Music Machines—American Style: A Catalogue of the Exhibition* (Washington, DC: Smithsonian Institution Press, 1971), 80; Stanley Sadie, ed., *The New Grove Dictionary of Musical Instruments* (London: Macmillan Press Limited, 1984), 3:466–467.

16. Schmidt Horning, "Recording," 105–122 (see Introduction, n. 12). I thank Tim Brooks for bringing to my attention the use of the Hawaiian steel guitar in recording during the 1910s.

17. The louder the instrument, the greater the motive force sent to the diaphragm, which in turn caused the wider excursion of the cutting stylus, reducing the space (called "land") between adjacent grooves causing the groove wall to break and the stylus to "jump" across it and ruin the recording.

18. Paul Whiteman and Mary Margaret McBride, *Jazz* (1926; repr., New York: Arno Press, 1974), 225.

19. Raymond B. Sooy, "Memoirs of My Recording and Traveling Experiences for the Victor Talking Machine Company," entry for Nov. 1923. N.d., unpaginated, unpublished photocopy of typescript. Copy in possession of author, courtesy Alexander Magoun. Original photocopy in Hagley Museum and Library Collection 2138, "Files of Nicholas F. Pensiero." Available on-line at the David Sarnoff Library website (www.davidsarnoff.org/soo.html). For a word on the provenance, see "Editorial Notes" (www.davidsarnoff.org/soo-editorialnotes.html), and my Essay on Sources.

20. R. B. Sooy, "Memoirs," entry following Feb. 12, 1931.

21. Harry Sooy, "Memoir of My Career at Victor Talking Machine Company, 1898–1925" available at the David Sarnoff Library website (www.davidsarnoff.org/sooyh.html).

22. Frank W. Hoffman, Dick Carty, and Quentin Riggs, *Billy Murray: The Phonograph Industry's First Great Recording Artist* (Lanham, MD: Scarecrow Press, 1997), 70.

23. Tim Gracyk with Frank Hoffmann, *Popular American Recording Pioneers: 1895–1925* (New York: Haworth Press, 2000), 17–23.

24. Lescarboura, "Other End of the Phonograph," 164, describes the studio anonymously. A fuller, well-illustrated description of the Aeolian operation appeared in "The Aeolian Co. Announces the Vocalion Record," *The Talking Machine World* (May 15, 1918): 68–70.

25. "How Recordings Are Made in the Victor Laboratory," *New Victor Records* (Sept. 1917) unpaginated, cited in Hoffman et al., *Billy Murray*, 76–77, n. 14.

26. "Ernest L. Stevens," interview in Harvith and Harvith, *Edison, Musicians, and the Phonograph*, 29 (see Introduction, n. 23).

27. Lescarboura, "Other End of the Phonograph," 164. Grooves cut at the beginning of the record enabled the home phonograph to come up to speed before the music started, while grooves cut at the end of the record accommodated automatic stopping devices.

28. "Ada Jones Tells Story of Her Career," *The Talking Machine World* (Feb. 15, 1917): 47.

29. "Caruso's Dread of Recording," *The Talking Machine World* (May 15, 1917): 31.

30. R. B. Sooy, "Memoirs," entry following "Symphony Orchestras."

31. R. B. Sooy, "Memoirs," entry under "A Few Memoirs of Temperamental Artists."

32. Gaydon, *Art and Science of the Gramophone*, 192–193.

33. R. B. Sooy, "Memoirs," entry under "A Few Memoirs of Temperamental Artists."

34. "Rosa Ponselle," interview in Harvith and Harvith, *Edison, Musicians, and the Phonograph*, 80.

35. Harvith and Harvith, *Edison, Musicians, and the Phonograph*, 82.

36. Oliver Read and Walter Welch, *From Tin Foil to Stereo*, 2nd ed. (Indianapolis, IN: Howard W. Sams, 1976), 75.

37. Ogilvie Mitchell, *The Talking Machine Industry*, Pitman's Common Commodities and Industries (London: Sir Isaac Pitman & Sons, 1922), 66.

38. Anonymous recording expert, "who had been in the industry since the time of Edison's tinfoil records," quoted in Lescarboura, "Other End of the Phonograph," 164.

39. On Sabine's work, see Emily Thompson, *The Soundscape of Modernity: Architectural Acoustics and the Culture of Listening in America, 1900–1933* (Cambridge, MA: MIT Press, 2002), chaps. 2–3; Robert S. Shankland, "Dayton Clarence Miller: Physics across Fifty Years," *American Journal of Physics* 9 (1941): 273–283; "The Aeolian Co. Announces the Vocalion Record," *The Talking Machine World* (May 15, 1918): 68–70; Andrew F. Inglis, *Behind the Tube: A History of Broadcasting Technology and Business* (Boston: Focal Press, 1990), 21.

40. Gaydon, *Art and Science of the Gramophone*, 190.

41. MacFarlane, *Phonograph Book*, 33–34; Seymour, *Reproduction of Sound*, 80.

42. Raymond Sooy mentioned using up to twelve horns successfully at Victor, but because this reduced the volume on the record, it required a very sensitive diaphragm. R. B. Sooy, "Memoirs," entry following Feb. 12, 1931.

43. Gaydon, *Art and Science of the Gramophone*, 191.

44. Mitchell, *Talking Machine Industry*, 67. The playback apparatus (phonograph or gramophone), like the recording apparatus, had a sound box, a diaphragm, a stylus (not to cut but to play back or trace the sound in the grooves), and a horn.

45. "Improving the Tonal Quality of the Phonograph," *Scientific American* (Feb. 2, 1918): 121.

46. H. Sooy, "Memoir," entry under "1902."

47. R. B. Sooy, "Memoirs," entry from "Requirements Necessary for a Good Recorder."

48. Seymour, *Reproduction of Sound*, 95.

49. Walter G. Vincenti, *What Engineers Know and How They Know It: Analytical Studies from Aeronautical History* (Baltimore: Johns Hopkins University Press, 1990).

50. R. B. Sooy, "Memoirs," entries for Nov. 16, 1903, Feb. 1, 1904, and Sept. 24, 1908. MacDonough also supervised sessions for Victor during the 1910s. Gelatt, *The Fabulous Phonograph*, 71, 136; Gracyk, *Popular American Recording Pioneers*, 22.

51. Russell Sanjek, *American Popular Music and Its Business: The First Four Hundred Years*, vol. 3, *From 1900 to 1984* (New York: Oxford University Press, 1988), 27.

52. Examples of record labels established by furniture and instrument makers were Paramount (Wisconsin Chair Company), Brunswick (Brunswick-Balke-Collender, makers of bowling equipment and pianos), Vocalian (Aeolian Company, manufacturers of pianos and

organs), and Gennett (Starr Piano, discussed below). Read and Welch, *From Tin Foil to Stereo*, 206, 211; Brian Priestley, *Jazz on Record: A History* (New York: Billboard Books, 1991), 2–3.

53. The Unit Construction Company, "The Unico System—The Sales Builder," *The Talking Machine World* (Jan. 15, 1918): 40–41.

54. "The Aeolian Co. Announces the Vocalion Record," *The Talking Machine World* (May 15, 1918): 68; "Heineman 'OkeH' Record Now Ready for the Trade," *The Talking Machine World* (May 15, 1918): 95.

55. Mansel G. Blackford, *A History of Small Business in America*, 2nd ed. (Chapel Hill: University of North Carolina Press, 2003), 3.

56. Pekka Gronow and Ilpo Saunio, *An International History of the Recording Industry*, trans. Christopher Moseley (London: Cassell, 1998); Michael S. Kinnear, *The Gramophone Company's First Indian Recordings, 1899–1908* (Bombay: Popular Prakashan, 1994).

57. William Howland Kenney, *Recorded Music in American Life: The Phonograph and Popular Memory, 1890–1945* (New York: Oxford University Press, 1999), xi.

58. Victor advertisement in *The Talking Machine World* (Oct. 15, 1905): 36–39.

59. The story of this company is recounted in Rick Kennedy, *Jelly Roll, Bix, and Hoagy: Gennett Studios and the Birth of Recorded Jazz* (Bloomington: Indiana University Press, 1994).

60. Laurence Bergreen, *Louis Armstrong: An Extravagant Life* (New York: Broadway Books, 1997), 215–217; Kennedy, *Jelly Roll, Bix, and Hoagy*, 29.

61. Kennedy, *Jelly Roll, Bix, and Hoagy*, 31–32.

62. The first "race series" records began to appear in 1920, and the first commercial country recordings shortly thereafter. On race records, see M. W. Dixon and John Godrich, *Recording the Blues* (New York: Stein and Day, 1970); on country records, see Nolan Porterfield, *Jimmie Rodgers: The Life and Times of America's Blue Yodeler* (Urbana: University of Illinois Press, 1992); and Charles K. Wolfe, "What Ever Happened to Country's First Recording Artist? The Career of Eck Robertson," *Journal of Country Music* 16, no. 1 (1993): 33–41. The open door policy also created some intense disagreement among Gennett employees when the Ku Klux Klan, which was particularly strong in Richmond, decided to make Gennett its recording headquarters. The company pressed private labels, including the "KKK" label replete with burning cross. A picture of this is in Tony Russell, *Blacks, Whites and Blues* (New York: Stein and Day Publishers, 1970), 21; on the divisiveness within Gennett over recording the Klan, see Kennedy, *Jelly Roll, Bix, and Hoagy*, 36–38.

63. Kennedy, *Jelly Roll, Bix, and Hoagy*, 33.

64. Quoted in Frederic Ramsey, Jr., "King Oliver and His Creole Jazz Band," in *Jazzmen*, ed. Frederic Ramsey, Jr., and Charles Edward Smith (New York: Harcourt, Brace, 1939), 73.

65. Ramsey, "King Oliver," 73. The song was "Chimes Blues."

66. Kennedy, *Jelly Roll, Bix, and Hoagy*, 152.

67. Rick Kennedy and Randy McNutt, *Little Labels—Big Sound: Small Record Companies and the Rise of American Music* (Bloomington: Indiana University Press, 1999), 3.

68. On the lawsuit, see Kennedy, *Jelly Roll, Bix and Hoagy*, 14–27; and Dixon and Godrich, *Recording the Blues*, 8.

69. MacFarlane, *The Phonograph Book*, 31.

70. J. Lewis Young, "Musical Phonographs," *The Phonogram*, 1, no. 2 (June 1893): 43.

71. Gracyk, *Popular American Recording Pioneers*, 3.

72. Jody Rosen, "There Once Was a Record of Smut," *New York Times* (July 8, 2007); R. Crumb, "Crazy Music," in *Popular Culture in America*, ed. Paul Buhle (Minneapolis: University of Minnesota Press, 1987), 108.

73. Erika Brady, *A Spiral Way: How the Phonograph Changed Ethnography* (Jackson: University Press of Mississippi, 1999); Alan Lomax, *The Land Where the Blues Began* (New York: The New Press, 1993).

74. Lawrence Cohn, ed., *Nothing but the Blues: The Music and the Musicians* (New York: Abbeville Press, 1993); Paul Oliver, *Barrelhouse Blues: Location Recording and the Early Traditions of the Blues* (New York: BasicCivitas Books, 2009); Charles K. Wolfe and Ted Olson, eds., *The Bristol Sessions: Writings about the Big Bang of Country Music* (Jefferson, NC: McFarland, 2005).

CHAPTER 2. THE STUDIO ELECTRIFIES

1. C. A. Schicke, *Revolution in Sound: A Biography of the Recording Industry* (Boston: Little, Brown, 1974), chaps. 7, 8. *The Evolution of Recordings . . . From Cylinder to Video Disc* (New York: Audio Engineering Society, 1977), 12.

2. Lizabeth Cohen, *Making a New Deal: Industrial Workers in Chicago, 1919–1939* (New York: Cambridge University Press, 1990), 105. See also Richard K. Spottswood's seven-volume *Ethnic Music in America: A Discography of Ethnic Recordings Produced in the United States, 1893–1942* (Champaign: University of Illinois, 1990).

3. Dixon and Godrich, *Recording the Blues*, 12 (see chap. 1, n. 62); Peter Grendysa, *The OKeh Rhythm & Blues Story: 1949–1957*, compact disc set 48912 (Sony Music, 1993), liner notes, 6.

4. John Stanford Magee, "The Music of Fletcher Henderson and His Orchestra in the 1920s" (PhD diss., University of Michigan, 1992), viii.

5. "U.S. Record Sales 1921–1962," *Billboard 1963–1964 International Music-Record Directory*, 13.

6. *Broadcasting—Broadcast Advertising Yearbook* (1935, 1937, 1940, 1945, 1949). A survey of these five editions of the annual listings for radio professionals revealed that between 1935 and 1949 the listings of recording and transcription facilities increased more than tenfold.

7. Reich, *Making of American Industrial Research*, 184 (see chap. 1, n. 1).

8. Millard, *America on Record*, 139–157 (see Introduction, n. 21); Allan Sutton, *Recording the 'Twenties: The Evolution of the American Recording Industry, 1920–29* (Denver, CO: Mainspring Press, 2008), 163–173; Michael J. Biel, "The Making and Use of Recordings in Broadcasting before 1936" (PhD diss., Northwestern University, 1977), 120–128, 303–311; McGinn, "Stokowski and the Bell Telephone Laboratories," 38–75 (see Introduction, n. 24).

9. "Auxetophone Described," *The Talking Machine World* (Aug. 15, 1905): 15; "The Auxetophone Exhibited," *The Talking Machine World* (July 15, 1906): 36; Aldridge, *Victor Talking Machine Company*, 63 (see chap. 1, n. 9); Oliver Read, *The Recording and Reproduction of Sound*, rev. and enl. 2nd ed. (Indianapolis, IN: Howard W. Sams, 1952), 13–14.

10. W. S. Bachman, B. B. Bauer, and P. C. Goldmark, "Disc Recording and Reproduction," *IRE Proceedings* 50 (May 1962): 739. For the historical context, see Frederick V. Hunt, *Electroacoustics: The Analysis of Transduction, and Its Historical Background* (Cambridge, MA: Harvard University Press, 1954), and Thompson, *Soundscape of Modernity* (see chap. 1, n. 39).

11. The idea for Bell Labs to use its considerable knowledge about sound transmission to

improve the making of records and to provide sound for motion pictures had been proposed in a 1916 memorandum from Harold D. Arnold to E. H. Colpitts, in which Arnold, as new director of research, suggested that Bell was "in a position to carry, with very little work, our results and information into a field of great commercial magnitude, and our own knowledge and technique would be greatly improved by the work." He urged it be considered "from a scientific and patent viewpoint." Arnold to Colpitts memo, cited in McGinn, "Stokowski and the Bell Telephone Laboratories," 41–42. On Bell Lab's work in speech, hearing, and sound, see M. D. Fagen, ed., *A History of Engineering and Science in the Bell System: The Early Years (1875–1925)* (n.p.: Bell Telephone Laboratories, 1975), 926–958.

12. Sutton, *Recording the 'Twenties*, 147–161; and Biel, "Making and Use of Recordings," 113–119, 263–311, 366–380.

13. Frank H. Lovette and Stanley Watkins, "Twenty Years of Talking Movies: An Anniversary," *Bell Telephone Magazine* (Summer 1946): 82–100; Thompson, *Soundscape of Modernity*, 99.

14. J. P. Maxfield and H. C. Harrison, "Methods of High Quality Recording and Reproducing of Music and Speech Based on Telephone Research," *Transactions of the American Institute of Electrical Engineers* 45 (Feb. 1926): 334.

15. The recorder was named for the rubber damping element that ran the length of the tonearm. Arthur C. Keller, *Reflections of a Stereo Pioneer* (San Francisco, CA: San Francisco Press, 1986), 9.

16. Joseph P. Maxfield, "Electrical Phonograph Recording," *The Scientific Monthly* 21 (Jan. 1926): 71–79.

17. Joseph P. Maxfield, "Electrical Research Applied to the Phonograph," *Scientific American* 134 (Feb. 1926): 104.

18. Physicist Frederick V. Hunt called Harrison's recorder "[t]he best-known, and perhaps the first, use of electric-network analogs *as a basis for* the *design* of a mechanical system." Hunt, *Electroacoustics*, 67.

19. Maxfield and Harrison, "Methods of High Quality Recording," 337.

20. Maxfield, "Electrical Research Applied," 105.

21. Maxfield and Harrison, "Methods of High Quality Recording," 337.

22. Maxfield, "Electrical Research Applied," 105.

23. Quoted in Gelatt, *The Fabulous Phonograph*, 232. The speaker was no ordinary listener, but a technical writer for *The Gramophone*, one of the first periodicals to regularly review records.

24. The figures are from Biel, "Making and Use of Recordings," 113. Fred Gaisberg put the frequency limits at 164 to 2,088 vibrations per second, which represented an E below middle C and a triple high C, respectively. Gaisberg, *The Music Goes Round*, 86. Joseph Maxfield claimed that there was no accurate method of calibrating the acoustical recording curve but that there was "considerable evidence to indicate that the fundamentals of notes below middle C were not appreciably recorded and that fundamentals or harmonics lying above the middle of the upper octave on the piano also failed to record." Maxfield, "Electrical Phonograph Recording," 79.

25. Biel, "Making and Use of Recordings," 374. They did not electrify the phonograph for several reasons, mainly because they believed that for home use the power of the electrical

records would be adequate. Maxfield and Harrison, "Methods of High Quality Recording," 337.

26. Maxfield, "Electrical Research Applied," 105.

27. Gelatt, *Fabulous Phonograph*, 210–211, 225–226 (see Introduction, n. 4). Gary Marmorstein, *The Label: The Story of Columbia Records* (New York: Thunder's Mouth Press, 2007), 50–51.

28. Sutton, *Recording the 'Twenties*, 175–181; Biel, "Making and Use of Recordings," 374–375.

29. Gelatt, *Fabulous Phonograph*, 226. Marmorstein, *The Label*, 54–55.

30. Paul Whiteman, *Records for the Millions*, ed. David A. Stein (New York: Hermitage Press, 1948), 7.

31. Victor employees Albertis Hewitt and James W. Owen experimented with electrical recording from 1916 to Owen's death in June 1923. Although Hewitt continued on his own, his electrical recorder appears never to have proceeded beyond the experimental stages. See Biel, "Making and Use of Recordings," 275–278.

32. Read and Welch, *From Tin Foil to Stereo*, 240 (see chap. 1, n. 36).

33. Maxfield's privately communicated account of this incident is reported in Hunt, *Electroacoustics*, 68–69.

34. Hunt, *Electroacoustics*, 69.

35. H. Sooy, "Memoir of My Career at Victor Talking Machine Company," entry dated Jan. 2, 1925 (see chap. 1, n. 21).

36. R. B. Sooy, "Memoirs of My Recording and Traveling Experiences for the Victor Talking Machine Company," entries for Jan. 27 and 30, and Feb. 9, 1925 (see chap. 1, n. 19).

37. R. B. Sooy, "Memoirs," entry dated Mar. 11, 1925. Both Harry and Raymond refer to this as the first electrically recorded records for the Victor catalog, "although the Contract for same has not been signed. This work started on permission from the Bell Company. Mme. Samaroff being the first artist to make records for Domestic use." H. Sooy, "Memoir," entry dated Mar. 11, 1925. Among record collectors there has been considerable debate over the "first" electrical recordings, with recent discoveries of discs recorded with the Western Electric system from late 1924. If Victor did not finalize its agreement with Western Electric until 1925, it is unlikely the company would have made commercial releases prior to that. Of course, Western Electric and Bell Labs were actively making test records prior to 1925.

38. "Chorus of 5,000 Sings for Columbia Special Record," *The Talking Machine World* (May 15, 1925): 34; "New Associated Glee Club Record Made by Columbia," *The Talking Machine World* (Nov. 15, 1925): 12; Gelatt, *Fabulous Phonograph*, 229.

39. T. Lindsay Buick, *The Romance of the Gramophone* (Wellington, New Zealand: Ernest Dawson, 1927), 101.

40. Talk by Cyril Francis, Recording Engineer, Association for Recorded Sound Collections Chapter Meeting, Oct. 23, 1996, Mary Pickford Theatre, Library of Congress, Washington, DC. Transcribed notes of this talk courtesy of Howard Sanner of the Library of Congress.

41. Franklyn Baur interview in *Phonograph Monthly Review* (Sept. 1927) cited in Tim Gracyk, *Popular American Recording Pioneers*, 41 (see chap. 1, n. 23).

42. Nathaniel Shilkret interview in *Phonograph Monthly Review* (May 1927), cited in Gracyk, *Popular American Recording Pioneers*, 2.

43. Millard, *America on Record*, 266.

44. Read and Welch mentioned the collection of reflected sound by Columbia-US recordists, *From Tin Foil to Stereo*, 238. Henry Seymour discussed the importance of "wiring of the recording room," also useful for hanging inverted music stands in *Reproduction of Sound*, 60–61 (see chap. 1, n. 12).

45. Seymour, *The Reproduction of Sound*, 59.

46. Edison's pianist Ernest L. Stevens contrasted the two studios: "Columbia Street had a dead sound. The moment you'd walk into the studio it would be so hard to breathe, because there'd be no vibration and hardly any air, while over in New York it was on the eighteenth floor with no padding. It was much better. I used to like to record in New York." "Ernest L. Stevens," interview in Harvith and Harvith, *Edison, Musicians, and the Phonograph*, 26, 30.

47. Harry F. Olson, "Microphones for Recording," *Journal of the Audio Engineering Society* 25 (Oct./Nov. 1977): 676–684, quote on p. 678.

48. Harry F. Olson and Frank Massa, *Applied Acoustics* (Philadelphia: P. Blakiston's Son & Co., 1934), 306.

49. Ibid. For more on Sabine's work, see Thompson, *Soundscape of Modernity*, 33–45.

50. Maxfield and Harrison, "Methods of High Quality Recording," 335.

51. Christopher H. Sterling and John M. Kittross, *Stay Tuned: A Concise History of American Broadcasting*, 2nd ed. (Belmont, CA: Wadsworth Publishing, 1990), 63; Olson and Massa, *Applied Acoustics*, 347.

52. George Avakian, record album liner notes for Bessie Smith, with Joe Smith and Fletcher Henderson's Hot Six, *The Bessie Smith Story*, vol. 3 (Columbia Records, CL 857); Chris Albertson, *Bessie* (New York: Stein and Day, 1972), 91; Studs Terkel, *Giants of Jazz* (New York: Thomas Y. Crowell, 1957), 45. These sources refer to the use of a carbon microphone, but the Western Electric system employed the high-quality condenser microphone developed by Western Electric's E. C. Wente.

53. Similar acoustical treatment with heavy cloths was used at the Vitagraph Studio, the first sound film production facility, just across the East River in Brooklyn, at roughly the same time, perhaps even employing the same Western Electric engineers. Thompson, *Soundscape of Modernity*, 267.

54. Baur interview in *Phonograph Monthly Review* (Sept. 1927), cited in Gracyk and Hoffmann, *Popular American Recording Pioneers*, 41.

55. "Acoustic Features of WCAU's New Studios," *Electronics* (Dec. 1932): 358–360; "WCAU Uses Dead End and Live End Studios," *Broadcasting* 5 (Oct. 1932): 21.

56. Hughson F. Mooney, "Songs, Singers and Society, 1890–1954," *American Quarterly* 6 (Fall 1954): 221–232, quote from p. 228.

57. For example, Douglas Kahn and Gregory Whitehead, eds., *Wireless Imagination: Sound, Radio and the Avant-Garde* (Cambridge, MA: MIT Press, 1992); Frank A. Biocca, "The Pursuit of Sound: Radio, Perception and Utopia in the Early Twentieth Century," *Media Culture & Society* 10 (Jan. 1988): 61–79; and Biocca, "Media and Perceptual Shifts: Early Radio and the Clash of Musical Cultures," *Journal of Popular Culture* 24 (Fall 1990): 1–15.

58. Eddie Condon, with Thomas Sugrue, *We Called It Music: A Generation of Jazz* (1947; repr., New York: Da Capo, 1992), 154.

59. Ibid., 154.

60. Saxophonist Sidney Bechet, for instance, was such a powerful player, as he later re-

called: "When it got to my chorus the needle would jump. I couldn't play the way I wanted to. The engineers would almost go crazy when they saw me coming into the studios. They'd say, 'Here comes trouble itself.'" John Chilton, *Sidney Bechet: The Wizard of Jazz* (New York: Oxford University Press, 1987), 61.

61. Condon, *We Called It Music*, 158.

62. Ibid., 155–156.

63. Condon's book was first published in 1947 by Henry Holt. Tenor saxophonist Bud Freeman also mentioned the date, albeit far less descriptively, in his autobiography. Bud Freeman, *Crazeology: The Autobiography of a Chicago Jazzman*, as told to Robert Wolf (Urbana: University of Illinois Press, 1989), 26–27.

64. Eric Barnouw, *A History of Broadcasting in the United States to 1933*, vol. 1, *A Tower in Babel* (New York: Oxford University Press, 1966), 129.

65. Whiteman, *Records for the Millions*, 3.

66. Ibid., 7; Sooy, "Memoirs," entry for Aug. 9, 1920.

67. Whiteman and McBride, *Jazz*, 227–228 (see chap. 1, n. 18).

68. Ibid.

69. "U.S. Record Sales 1921–1962," *Billboard 1963–1964 International Music Record Directory* (Aug. 3, 1963): 13.

70. Dixon and Godrich, *Recording the Blues*, 76.

71. Sanjek, *American Popular Music and Its Business*, 3:119 (chap. 1, n. 51). Columbia Phonograph had also briefly and disastrously been involved in network broadcasting in 1927. Sterling & Kittross, *Stay Tuned*, 109–110.

72. Dixon and Godrich, *Recording the Blues*, 64–77.

73. On Nov. 1, 1929, Thomas A. Edison, Inc., announced that production of records and phonographs would be discontinued and the company would concentrate on radios and dictating machines. Gelatt, *Fabulous Phonograph*, 248.

74. Aldridge, *Victor Talking Machine Company*, 99–100.

75. Orlando R. Marsh developed an early method of electrical recording and made recordings for his own Autograph label as well as outside clients by the mid-1920s. Martin Bryan, "Orlando R. Marsh: Forgotten Pioneer," *New Amberola Graphic* (Winter, n.d.): 3–14; Sutton, *Recording the 'Twenties*, 157–160.

76. Edward W. Kellogg, "History of Sound Motion Pictures: First Installment," *Journal of the SMPTE* 64 (June 1955), reprinted in Raymond Fielding, ed., *A Technological History of Motion Pictures and Television* (Berkeley: University of California Press, 1967), 179.

77. Harold B. Franklin, *Sound Motion Pictures: From the Laboratory to Their Presentation* (Garden City, NY: Doubleday, Doran & Compant, 1929), 19–20.

78. R. B. Sooy, "Memoirs," entry for June 15, 1926.

79. R. B. Sooy, "Memoirs," entry under June 16, 1928.

80. Edward W. Kellogg, "History of Sound Motion Pictures: Second Installment," *Journal of the SMPTE* 64 (July 1955), reprinted in Fielding, ed., *A Technological History*, 187.

81. James Lastra, "Standards and Practices: Aesthetic Norm and Technological Innovation in the American Cinema," in *The Studio System*, ed. Janet Staiger (New Brunswick, NJ: Rutgers University Press, 1995), 200–225. See also Lastra, *Sound Technology and the American Cinema: Perception, Representation, Modernity* (New York: Columbia University Press, 2000).

82. Sanjek, *American Popular Music and Its Business*, 131; Susan Smulyan, *Selling Radio: The Commercialization of American Broadcasting, 1920–1934* (Washington, DC: Smithsonian Institution Press, 1994), 122.

83. C. Sterling Gleason, "From Microphone to Modulator," *Radio-Craft* (Jan. 1930): 324–325, 347–348.

84. Rodney W. Baum, "Radio Music Transcription Services: Their Development and Decline" (MA thesis, Bowling Green State University, 1964), 6.

85. Electrical Research Products, Inc. (ERPI), a subsidiary formed by Western Electric in 1927 to handle non-telephone interests, developed and distributed recording equipment and sound systems to the major Hollywood studios. Stephen B. Adams and Orville R. Butler, *Manufacturing the Future: A History of Western Electric* (Cambridge: Cambridge University Press, 1999), 138. When the film business shifted from disc to sound-on-film recording, ERPI went into the transcription business with World. Sanjek, *American Popular Music and Its Business*, 131.

86. Commercial records were cut laterally, but Bell Labs and ERPI developed vertically cut discs that were thought to be superior during the 1930s. See H. A. Frederick, "Vertical Sound Records," *Journal of the Society of Motion Picture Engineering* 18 (Feb. 1932): 141–152, reprinted in H. E. Roys, ed., *Disc Recording and Reproduction* (Stroudsburg, PA: Dowden, Hutchinson & Ross, 1978), 38–49.

87. "Electrical Transcriptions Are Being Used by Many Radio Stations," *Electronics* (Nov. 1930): 365.

88. Sterling and Kittross, *Stay Tuned*, 98; Biel, "Making and Use of Recordings," 953–956.

89. "Electrical Musical Instruments," *Electronics* 1 (Dec. 1930): 435; "Electrical Transcriptions for Broadcasting," *Electronics* 1 (Dec. 1930): 438.

90. Invented in 1920, the Theremin, was itself a radio-inspired instrument consisting of a combined radio transmitter-receiver housed in a wooden box with two antennas. Playing the instrument required moving one's hands near the antennae, one of which controlled frequency (pitch), the other volume, thus altering the electromagnetic field surrounding the device and creating what amounted to controlled feedback. Obtaining that control was not easy, and the difficulty of playing this instrument undoubtedly inspired the keyboard version, which resembled a tiny piano topped by a panel with twenty-seven control knobs, similar to those of a broadcast console and later a recording console.

91. For more on Judson and the evolution of CBS, see Barnouw, *A Tower in Babel*, 1:221–224; Sterling and Kittross, *Stay Tuned*, 108–110; and Sanjek, *American Popular Music and Its Business*, 86, 139.

92. *Broadcasting 1935 Yearbook*, 80–81. By the mid-1930s, Gennett was no longer recording music but only sound effects records. Kennedy, *Jelly Roll, Bix, and Hoagy*, 193 (see chap. 1, n. 59).

93. Sanjek, *American Popular Music and Its Business*, 131; Sterling and Kittross, *Stay Tuned*, 164.

94. Jack Alicoate, ed., "Transcriptions," *The 1938 Radio Annual* (Radio Daily, 1938), 483–493.

95. "'Dubbing' or Re-recording for Broadcasting," *Electronics* (June 1931): 692.

96. George A. Blacker, "A History of Home Recording," *Audio* (Apr. 1975): 28–34. Biel of-

fers an exhaustive survey of home recording methods and disc materials in "Making and Use of Recordings," 847–905.

97. Biel, "Making and Use of Recordings," 918.

98. Arthur Heine, "Making Money in Sound Recording," *Radio-Craft* (Nov. 1935): 272, 304.

99. Biel, "Making and Use of Recordings," 932.

100. George J. Saliba, *Home Recording and All about It: A Complete Treatise on Instantaneous Recording, Microphones, Recorders, Amplifiers, Commercial Machines, Servicing, Etc.* (New York: Radcraft Publications, 1932). Much of the information in this 62-page booklet had appeared in a series of eleven articles Saliba published in *Radio-Craft* magazine between June 1931 and Aug. 1932. Reader responses revealed the growing interest in home recording at this time.

101. Biel, "Making and Use of Recordings," 839.

102. Ibid., 953–955.

103. Frank Laico, telephone interview with author, Jan. 13, 1999.

104. "William A. Savory," Audio Engineering Society 87th Convention Awards Banquet Program, Saturday, Oct. 21, 1989. The biographical profile appeared on the occasion of Mr. Savory's receipt of an AES Fellowship. Copy in author's collection courtesy of Bill Savory. In 2010, the National Jazz Museum in Harlem acquired Savory's collection of air-check recordings and began the process of cleaning and digitizing the discs, some of which can be heard on its website, http://www.jazzmuseuminharlem.org/savory/index.php.

105. D[onald] G. F[ink], "Sound-on-Disc," *Electronics* (Oct. 1936): 6–10, 48, 50.

CHAPTER 3. A PASSION FOR SOUND

1. Susan J. Douglas, *Inventing American Broadcasting, 1899–1922* (Baltimore: Johns Hopkins University Press, 1987), 187–215; Daniel Boorstin, *The Americans: The Democratic Experience* (New York: Vintage Books, 1974), 370–390.

2. Les Paul, "Portraits of Invention: An Evening with Les Paul," Wednesday, Nov. 13, 1996, Baird Auditorium, National Museum of American History, Washington, DC.

3. Quote from Ken Pohlmann, Irv Joel (Producer), "AES Golden Gala" videotape of presentation at Audio Engineering Society 103rd Convention, New York City, Sept. 26, 1997. I thank Irv Joel for lending me a copy of his recording of this event. Benedict Anderson, *Imagined Communities: Reflections on the Origin and Spread of Nationalism*, rev. ed. (New York: W. W. Norton, 2006).

4. George A. Blacker, "A History of Home Recording," *Audio* (Apr. 1975): 28–34. The term *records*, which came to be identified with the flat disc after the cessation of cylinder production in the 1920s, refers to both discs and cylinders.

5. National Phonograph Company, *The Phonograph and How to Use It* (New York, 1900).

6. Ibid., 150.

7. "Heard Himself as Others Hear Him," *The Talking Machine World* (Apr. 15, 1905): 15.

8. T. J. Jackson Lears, *No Place of Grace: Antimodernism and the Transformation of American Culture, 1880–1920* (New York: Pantheon, 1981).

9. Patricia R. Zimmerman, *Reel Families: A Social History of Amateur Film* (Bloomington: Indiana University Press, 1995); Steven M. Gelber, *Hobbies: Leisure and the Culture of Work in America* (New York: Columbia University Press, 1999).

10. Zimmerman, *Reel Families*, 6.

11. Fleeming Jenkin and J. A. Ewing, "Helmholtz's Vowel Theory and the Phonograph," *Nature* 17 (Mar. 14, 1878): 384; J. Walter Fewkes, "On the Use of the Phonograph in the Study of the Languages of American Indians," *Science* 15 (May 2, 1890): 267–269. A good study of the long-term effect of the phonograph in ethnographic study, written by a Library of Congress audio engineer, is Erika Brady, *A Spiral Way: How the Phonograph Changed Ethnography* (Jackson: University of Mississippi Press, 1999).

12. Edgar Stillman Kelley, "A Library of Living Melody," *The Outlook* 99 (Sept. 30, 1911): 283–287.

13. R. D. Washburne, "Home Recording of Radio Programs and Speech," *Radio-Craft* 2 (Dec. 1930): 340–342. Susan J. Douglas notes that Apgar was employed by the Marconi Company, thus leaving his status as "amateur" in question, although only with regard to wireless, not recording. Douglas, *Inventing American Broadcasting*, 273–274.

14. Michael J. Biel, "The Making and Use of Recordings in Broadcasting before 1936" (PhD diss., Northwestern University, 1977), 288–294.

15. "How to Sell Victor Home Recording," cited in Biel, "The Making and Use of Recordings in Broadcasting Before 1936," 862. The Radiola 86 was also marketed by other branches of the newly merged RCA Victor conglomerate—General Electric, Graybar, Victor, and Westinghouse—using different model names.

16. Gelber, *Hobbies*, 224–225; Susan Currell, *The March of Spare Time: The Problem and Promise of Leisure in the Great Depression* (Philadelphia: University of Pennsylvania Press, 2005).

17. Harvey Green, "The Promise and Peril of High Technology," in *Craft in the Machine Age, 1920–1945*, ed. Janet Kardon (New York: Harry N. Abrams and the American Craft Museum, 1995), 36–45.

18. Douglas, *Inventing American Broadcasting*, 199–201; Barnouw, *A Tower in Babel*, 28, 30 (see chap. 2, n. 65). Gernsback also sold amateur wireless apparatus through his Electro Importing Company and promoted hobbyist magazines as forums for the technological future; see Joseph J. Corn and Brian Horrigan, *Yesterday's Tomorrows: Past Visions of the American Future* (Baltimore: Johns Hopkins University Press, 1984), 6–7, 9–10, 115. Gernsback's interest in the future and in amateur recording was shared by an illustrator whose work graced Gernsback's magazines and appeared in *Yesterday's Tomorrow's* (p. viii): Julian Krupa, an avid home recordist who built his own amplifiers, turntables, and other gadgets, eventually turning his basement in Argo, Illinois, into a professionally equipped home recording studio. "Home Recording Studio," *Radio & Television News* 42 (Nov. 1949): 57.

19. Hugo Gernsback, "Home Radio-Recording," *Radio-Craft* (Dec. 1930): 327; R. D. Washburne, "Home Recording of Radio Programs and Speech," *Radio-Craft* (Dec. 1930): 340. Gernsback also published *Radio News* (originally *Radio Amateur News*) beginning in 1919.

20. R. D. Washburne, "Home Recording of Radio Programs and Speech," *Radio-Craft* (Dec. 1930): 342.

21. George Basalla, "Keaton and Chaplin: The Silent Film's Response to Technology," in *Technology in America: A History of Individuals and Ideas*, 2nd ed., ed. Carroll W. Pursell, Jr. (Cambridge, MA: MIT Press, 1990), 227–236, quote on p. 232; Amy Bix, *Inventing Ourselves Out of Jobs? America's Debate over Technological Unemployment, 1929–1981* (Baltimore: Johns Hopkins University Press, 2000), chap. 3.

22. George Saliba, "Instantaneous Recording of Sound on Discs: A New Activity for the Radio Technician and Experimenter," *Radio-Craft* (June 1931): 724.

23. George J. Saliba, "Recording Equipment and Its Operation," *Radio-Craft* (Aug. 1931): 80–81, 118–119. Three months later, in the magazine's advice section, six out of nine correspondents had queries relating to recording technique; "Radio-Craft's Information Bureau," *Radio-Craft* (Nov. 1931): 296.

24. Saliba, *Home Recording and All about It* (see chap. 2, n. 100).

25. Ibid., 6.

26. Arthur Heine, "Making Money in Sound Recording," *Radio-Craft* (Nov. 1935): 272, 304.

27. "Directory of Transcription and Recording Producers," *Broadcasting Yearbook* (1935): 80; "Directory of Broadcast Equipment Manufacturers," *Broadcasting Yearbook* (1935): 202–204.

28. "Directory of Transcription, Recording, Program Producing, Script, Talent and Related Services," *Broadcasting Yearbook* (1945): 314, 316, 320, 322, 324, 326, 330, 332; "Directory of Broadcasting Equipment Manufacturers," *Broadcasting Yearbook* (1945): 348, 352, 356, 360, 364, 366, 371. Examination of years 1935, 1937, 1940, and 1949 revealed a fairly steady and continued expansion of these services. The listings are not complete, for not every studio appeared in these listings.

29. For example, F. H. Goldsmith and V. G. Geisel, *Techniques of Recording: A Practical Hand-Book on Recording* (Chicago: Gamble Hinged Music Co., 1939); Audio Devices, *How to Make Good Recordings* (New York: Audio Devices, 1940); Oliver Read, "The Fundamentals of Recording," *Radio News* (Feb. 1940): 16–19, 48–53; "Build Your Own Recording Studio," *Radio News* (Dec. 1940): 11–14, 60; "Build Your Own Recording Studio: Frequency Response," *Radio News* (Feb. 1941): 24–26, 62–63; "Build Your Own Recording Studio," *Radio News* (Aug. 1941): 20–22, 54–55; R. J. Bergemann, Jr., "How to Build a Modern 30/15-Watt P.A.-Radio-Recording Console, Part 1," *Radio-Craft* (May 1941): 652–653; A. C. Shaney, "Modern Circuit Features of a Semi-Professional 10-Watt Recording and Playback Amplifier," *Radio-Craft* (May 1941): 668–671. The first book-length manual was Read, *The Recording and Reproduction of Sound* (Indianapolis, IN: Howard W. Sams, 1949).

30. William Savory, interview with author, Falls Church, Virginia, Nov. 29, 1997.

31. Ibid.

32. Douglas E. Collar, "'Hello Posterity': The Life and Times of G. Robert Vincent, Founder of the National Voice Library" (PhD diss., Michigan State University, 1988), 139–140. Vincent's collection is now in The G. Robert Vincent Voice Library and part of the Special Collections Division of Michigan State University Libraries.

33. Doug Collar, e-mail to author, Apr. 14, 2011. Collar recounted this story from his interviews with Vincent for his doctoral thesis. Remote broadcasts were transmitted via phone lines.

34. "William A. Savory," Audio Engineering Society 87th Convention Awards Banquet Program, Saturday, Oct. 21, 1989. The biographical profile appeared on Mr. Savory's receipt of an AES Fellowship. Copy in author's collection courtesy of William Savory. At the time of its introduction, Dr. Peter Goldmark of CBS was publicly credited with developing the $33\frac{1}{3}$ microgroove LP record but he did not complete the project. In 1998, at a ceremony to honor the fiftieth anniversary of the LP and to award William Bachman posthumous recognition by

the National Academy of Recording Arts and Sciences, former Columbia Records producers and engineers, who worked on the project under Dr. Bachman, joined together to publicly emend the historical record. They revealed that, although Dr. Goldmark did indeed direct the LP project from 1939 to 1946, Dr. Bachman and his team had actually successfully completed the LP's development, although corporate politics left him in a secondary role in the public eye. George Avakian, Mitch Miller, Louis Porrata, William Savory, Howard Scott, "Panel discussion, National Musical Arts Presents: 50th Anniversary of the Long-Play 33^1/$_3$ Record," National Academy of Sciences, Jan. 18, 1998, Washington, DC. Audiotape and transcript of this event in author's collection.

35. The original two-disc vinyl LP set, no longer available, was later re-released as Benny Goodman, *On the Air 1937–1938* (Columbia Legacy, 1993), compact disc.

36. Peter D. Goldsmith, *Making People's Music: Moe Asch and Folkways Records* (Washington, DC: Smithsonian Institution Press, 1998); Richard Carlin, *Worlds of Sound: The Story of Smithsonian Folkways* (New York: Collins / Smithsonian Books, 2008).

37. Frank Laico, telephone interview with author, Jan. 13, 1999.

38. "In Memoriam: Charles Lauda," *Journal of the Audio Engineering Society* 14 (Jan. 1966): 96.

39. Laico interview, Jan. 13, 1999.

40. Ibid.

41. Fagen, *A History of Engineering and Science*, 291–317 (see chap. 2, n. 11). Project X was known to those who worked on it at Bell Labs as "The X System" but to those in the Signal Corps who handled the system, it was known as "Sigsaly" or "Ciphony I."

42. Donald J. Plunkett, interview with author, New York, Feb. 9, 1999.

43. On the origins of RCA, see Barnouw, *A Tower in Babel*, 57–61, and Sterling and Kittross, *Stay Tuned*, 52–53 (see chap. 2, n. 51).

44. Zeh Bouck, "Choosing a Radio School," *Radio News* (Apr. 1937): 586–587.

45. "Record Comeback," *Business Week* (Sept. 13, 1941): 48–49.

46. "Past, Present, and Prophetic," *Radio* (Jan. 1942): 6; "Recording Blues," *Business Week* (May 2, 1942): 28–30.

47. Ad*Access On-Line Project – Ad #R0833. John W. Hartman Center for Sales, Advertising & Marketing History, Duke University Rare Book, Manuscript, and Special Collections Library, http://library.duke.edu/digitalcollections/adaccess/.

48. Ibid., Ad #R0894.

49. Astatic Corporation, "Radio Amateurs Now Going in for Recording as Alternate Hobby," advertisement, *Radio* (Sept. 1942):, 37.

50. Edward R. Murrow and Fred W. Friendly, *I Can Hear It Now . . . : 1933–1945*, Columbia Masterworks MM-800, 78 rpm (5 discs).

51. The Institute of Radio Engineers was formed in 1912 by wireless operators who date back to Marconi's work in 1899, and the Society of Motion Picture Engineers formed in 1916. Dating the beginning of the commercial film industry is more problematic, but it certainly does not predate commercial recording, which began in 1890. For the origins of radio, see Douglas, *Inventing American Broadcasting*. For background on the IRE, see A. Michal McMahon, *The Making of a Profession: A Century of Electrical Engineering in America* (New York: IEEE

Press, 1984). For film, see Raymond Fielding, ed., *A Technological History of Motion Pictures and Television: An Anthology from the Pages of the Journal of the Society of Motion Picture and Television Engineers* (Berkeley: University of California Press, 1967).

52. "Record Rush," *Business Week* (Apr. 20, 1940): 32; "Record Comeback," *Business Week* (Sept. 13, 1941): 48–49. The ban on recording lifted in 1943 for the record labels, beginning with Decca, which signed agreements with the musicians' union. For more on the history of the 1942–44 recording ban and its effect on musicians and the industry, see James P. Kraft, *Stage to Studio: Musicians and the Sound Revolution, 1890–1950* (Baltimore: Johns Hopkins University Press, 1996), 130–161; and for a cultural analysis of the ban and its implications, see Anderson, *Making Easy Listening*, 3–26 (see Introduction, n. 5).

53. Robert J. Callen, "Hollywood Sapphire Group," *Audio Engineering* (Jan. 1948): 17.

54. For an account of the rivalry among competitors in this arena, see Andre Millard, "The Phonograph," in *Edison and the Business of Innovation* (see chap. 1, n. 6).

55. This is mentioned in Callen, "Hollywood Sapphire Group," 17.

56. Whiteman, *Records for the Millions*, 4 (see chap. 2, n. 30).

57. H. Sooy, "Memoir," entry under "1914" (see chap. 1, n. 21).

58. Gaydon, *Art and Science of the Gramophone* (see chap. 1, n. 2), end matter advertisements.

59. Robert Wiebe, *The Search for Order, 1877–1920* (New York: Hill and Wang, 1967).

60. Frank L. Capps invented the burnishing facet on the recording stylus specifically to address the needs of lacquer disc recording. He founded Frank L. Capps & Co. in 1929, which became the most prominent maker of recording styli for the professional trade.

61. Donald J. Plunkett, "Reminiscences of the Founding and Development of the Society," *Journal of the Audio Engineering Society* 46 (Jan./Feb. 1998): 5–6; Donald Plunkett, interview with Author, New York City, Feb. 9, 1999.

62. Plunkett, "Reminiscences," 6.

63. Donald J. Plunkett, "A Brief History of the Audio Engineering Society," undated typescript sent to the author by Louis Manno, Director of the Audio History Library, New York. The sapphire cutting stylus was used more than other materials such as ruby or diamond, not only because of cost considerations but also because it possessed properties that enabled it to be ground to acute angles yet maintain a fine enough edge for groove cutting. Richard Marcucci, "Design and Use of Recording Styli," *Journal of the Audio Engineering Society* 13 (Apr. 1965): 130–133.

64. Callen, "Hollywood Sapphire Group," 17.

65. Since the Sapphire Group was an informal organization with no written records, information on its history, outside of the articles cited, is from oral testimony. William Savory recalled that "there was no attempt by the [New York] Sapphire Club to go beyond a sort of a social group that met periodically." William Savory, telephone interview with author, Oct. 21, 1999.

66. "Sapphire Group Second Anniversary Meeting, March 10, 1948," *Audio Engineering* (June 1948): 14; Callen, "Hollywood Sapphire Group," 17, 39–41; the list is printed on pp. 40–41.

67. Pickering interview featured in Irv Joel (Producer), "AES Golden Gala," (see n. 3 above).

68. John H. Potts, untitled letter to subscribers, *Radio* 31 (Feb./Mar. 1947), inside cover.

69. The board included J. P. Maxfield, who directed the research at Bell Telephone Labora-

tories, which resulted in the development of electrical recording; Howard Chinn, chief audio engineer of CBS, George Nixon of NBC, John D. Colvin of ABC, and C. J. LeBel of Audio Devices, a major manufacturer of blank recording discs.

70. J[ohn] H. P[otts], "Transients," *Audio Engineering* (May 1947): 4.

71. Record companies, in fact, strove to produce the loudest records, at least in the popular field, so that they would be heard above the competition when played in jukeboxes, on the radio, or in the home.

72. Lynne C. Smeby, "Recording and Reproducing Standards," *Proceedings of the IRE* 30, no. 8 (Aug. 1942): 355–356.

73. "NAB Recording Standards Meeting," *Audio Engineering* (Oct. 1947): 32, 44.

74. R. C. Moyer, "Evolution of a Recording Curve," *Audio Engineering* (July 1953): 19–22, 53–54.

75. McMahon, *Making of a Profession*, 129. McMahon's account of the "defection of radio engineers" from the parent organization of power engineering (American Institute of Electrical Engineers, or AIEE) strongly parallels that of the audio engineers from the radio engineers thirty-six years later. Primarily concerned with radio technology, the IRE was formed by members of AIEE, who had grown dissatisfied with how little attention was accorded the growing field of wireless at conferences and in the publications. Just as the radio engineering contingent of the AIEE finally broke ranks and formed their own organization, so the IRE later became the parent organization from whose ranks the sound engineers defected to form the Audio Engineering Society. Some within the IRE feared losing members to the AES (just as the AIEE had feared losing the radio people nearly four decades earlier), but they didn't act swiftly enough to keep them. See ibid., 127–132.

76. Russell Sanjek, *American Popular Music and Its Business*, 3:227 (see chap. 1, n. 51). For Haddy's brief personal account of the story behind "ffrr," see "A. C. Haddy" in G. A. Briggs and 64 Collaborators, *Audio Biographies* (Yorkshire, UK: Wharfedale Wireless Works, 1961), 157–163. During the 1960s, the unit of frequency measurement officially changed to Hertz (Hz) from cycles per second (cps), although many studies continued to use cps, or "frequency in cycles" for some time. See, for example, John H. Bubbers, "A Report on the Proposed NAB Disc Playback Standard," *Journal of the Audio Engineering Society* 12 (Jan. 1964): 51–54.

77. J[ohn] H. P[otts], "Transients," *Radio* (Feb./Mar. 1947): 3.

78. William Savory, telephone interview with author, Oct. 21, 1999. Savory was a charter AES member, as well as a member of the IRE.

79. Julian Reitman, conversation with author, Society for the History of Technology Annual Meeting, Detroit, Michigan, Oct. 9, 1999. Mr. Reitman was himself an audio hobbyist, who on hearing Peter Goldmark's demonstration of the new Columbia LP at a 1948 IRE meeting, went home and promptly built himself a turntable.

80. "The Electron Art: Technical Papers Delivered at National Electronics Conference," *Electronics* (Nov. 1944): 190.

81. C. J. LeBel, "Recent Improvements in Recording," *Electronics* (Sept. 1940): 35.

82. "Letters: Audio Association?," *Audio Engineering* (Dec. 1947): 3. According to more than one source, this letter, signed by Frank E. Sherry, Jr., was "a fake" to get things going and quite possibly was written by C. J. LeBel, but subsequent events proved that the undercurrent of interest existed. No "Frank Sherry" appears on the roster of the first meetings or the

first official membership directory. Jack Hartley, hand-typed list of attendees of Feb. 17, 1948, Organizational Meeting of the Audio Engineering Society, n.d.; Audio Engineering Society, "Directory of Members 1954," *Journal of the Audio Engineering Society* 2, no. 2 (Apr. 1954). Copies of both documents in author's collection, courtesy of Irv Joel, Audio Engineering Society Historical Committee.

83. "Letters: Audio Engineering Society," *Audio Engineering* (Jan. 1948): 5.

84. C. G. McProud, "Audio Engineering Society Celebrates 20th Anniversary," *Audio* (Nov. 1968): 28. The committee consisted of LeBel, John D. Colvin, Norman Pickering, Chester A. Rackey, and C. G. McProud.

85. Hartley, Organizational Meeting of the Audio Engineering Society, n.d. Jack Hartley worked as a recording engineer for the Voice of America and then joined Fisher Radio and was instrumental in Fisher's growth as a major audio manufacturer. Hartley attended the first meetings of the AES and was among eleven founding members honored at the fiftieth anniversary commemoration on Mar. 11, 1998. New York Section of the AES and the Committee for the 50th, AES Goes Gold Program, copy in author's collection, courtesy of Irv Joel.

86. Jerry Minter, "The AES Begins Its Seventh Year: A Report by the President of the Audio Engineering Society," *Journal of the Audio Engineering Society* 2 (Jan. 1954): 1.

87. McProud, "Audio Engineering Society Celebrates 20th Anniversary," 28.

88. C. G. McProud, "Columbia LP Microgroove Records," *Audio Engineering* (Aug. 1948): 24, 32. McProud was a charter member of the AES and managing editor of *Audio Engineering*. Upon the death of John H. Potts, the magazine's founding editor, McProud became editor. The magazine regularly reported news of the AES until the society began publishing its own journal, the *Journal of the Audio Engineering Society*, in Jan. 1953.

89. "The IRE Professional Group System—A Status Report," *Proceedings of the IRE* 36, no. 12 (Dec. 1948): 1507. In 1953, the same year the AES began publishing its journal, the PGA began publishing its *Transactions on Audio*, so named until 1966 when it changed to *Transactions on Audio and Electroacoustics*. In 1963, the IRE merged with the AIEE to form the IEEE.

90. "The IRE Professional Group System—A Status Report," 1507.

91. In 1937, three magazines (*Radio Engineering, Communications & Broadcast Engineering*, and *The Broadcast Engineer*) merged to become *Communications*. As the editors explained in the inaugural issue, "You can't be in communications in this day and age and be indifferent to what the man next to you is doing." "With the Editors," *Communications* 17 (Sept. 1937): 4.

92. James Gleick, *The Information: A History, a Theory, a Flood* (New York: Pantheon, 2011); Mark Donald Bowles, "Crisis in the Information Age? How the Information Explosion Threatened Science, Democracy, the Library, and the Human Body, 1945–1999" (PhD diss., Case Western Reserve University, 1999).

93. William J. Temple, "We Went to the Audio Fair," *Senior Scholastic* (*Teacher's Ed.*) 55 (Dec. 7, 1949): 23T.

94. By 1954, the AES altered the structure and admissions policy of the Audio Fair, allowing members free admission to the professional exhibit and technical sessions, but charging an entrance fee to nonmembers. "New Audio Fair Contract Signed," *Journal of the Audio Engineering Society* 2, no. 2 (Apr. 1954): 112.

95. Canby (b. 1912) exemplified the true Renaissance man. While studying music at Har-

vard he learned the intricacies of high-fidelity audio by trial and error, a journey entertainingly detailed in "How I Fell into Audio," *Audio Engineering* (Mar. 1952): 38–42, and "I Fall Further into Audio," *Audio Engineering* (Apr. 1952): 36–38. After graduating from Harvard with a degree in music, Canby traveled with ethnomusicologist Alan Lomax to record the folk music of Appalachia. He later taught music at Princeton University and Finch College, sang in and directed a choir, and hosted a weekly radio program on WNYC. His interest in audio led not only to a career in technical reviewing but also to co-authoring (with C. G. Burke and Irving Kolodin) *The Saturday Review Home Book of Recorded Music and Sound Production* (New York: Prentice-Hall, 1952). "In Memoriam: Edward Tatnall Canby," *Journal of the Audio Engineering Society* 46 (May 1998): 487.

96. See, for example, Irving Greene, James R. Radcliffe, and Robert Scharff, *Make Music Live: Handbook of Quality Home Sound Reproduction* (New York: Medill McBride Company, 1951); Everett B. Garretson, "A Glossary of High-Fidelity Talk," *New York Times* (Nov. 22, 1953); Meyer Berger, "Hazards and Joys in Audio Shopping," *New York Times* (Mar. 21, 1954) and "Experts Talk about Records," *New York Times* (Nov. 21, 1954); and Charles Sinclair, "The Theory and Practice of Hi-Fi-manship or, How to Be an Audio Expert without Knowing "Harmonic Distortion" from a "Harmonica," *Audio* (May 1954): 20–21, 44–47.

97. "Editor's Report," *Audio Engineering* (Feb 1954): 16. From 1947 to 1952, news of the AES and some papers presented at the annual meetings appeared in *Audio Engineering*. For more on the AES and its history, see (www.aes.org/aeshc/how-the-aes-began.html).

98. "Editor's Report," *Audio* (Apr. 1954): 16.

99. Al Grundy, telephone interview with author, Jan. 8, 2000. For example, between 1951 and 1956, the number of phonographs shipped in the United States rose more than 475 percent and remained steady through 1970 as foreign-made hi-fi equipment dominated. US Bureau of the Census, *Historical Statistics of the United States: Colonial Times to 1970*, Part 2 (Washington, DC, 1976), 695–696. On the postwar rise in hi-fi component manufacture, see David Lander, "Technology Makes Music," *American Heritage of Invention & Technology* 6 (Spring/Summer 1990): 56–63. On the rise of discretionary spending between 1947 and 1961, especially on domestic items, see Elaine Tyler May, *Homeward Bound: American Families in the Cold War Era* (New York: Basic Books, 1988), 164–166. On the high-fidelity "craze," see Keir Keightley, "'Turn It Down!' She Shrieked: Gender, Domestic Space, and High Fidelity, 1948–59," *Popular Music* 15, no. 2 (1996): 149–177.

100. Carolyn Marvin, *When Old Technologies Were New: Thinking about Electric Communication in the Late Nineteenth Century* (New York: Oxford University Press, 1988), 10–11.

101. See Samuel Brylawski, "Armed Forces Radio Service: The Invisible Highway Abroad," in *Wonderful Inventions: Motion Pictures, Broadcasting, and Recorded Sound at the Library of Congress*, ed. Iris Newsom (Washington, DC: The Library, 1985), 333–344; and Richard S. Sears, *V-Discs: A History and Discography* (Westport, CT: Greenwood Press, 1980).

CHAPTER 4. WHEN HIGH FIDELITY WAS NEW

1. William Braid White, "Aspects of Sound Recording 1-The Echo Difficulty," *The Talking Machine World* (July 15, 1920): 159; Paul E. Sabine, "Acoustics of Sound Recording Rooms," *Transactions of the Society of Motion Picture Engineers* 12, no. 35 (1928): 809–822.

2. "Acoustic Features of WCAU's New Studios," *Electronics* (Dec. 1932): 358–360; C. Gordon Jones, "Pioneer 'Live-End, Dead-End' Studios," *Broadcast News* 6 (Jan. 1933): 12–13; "How Echoes are Produced: NBC Engineers Perfect Artificial Sound Reflection" *Broadcast News* 13 (Dec. 1934): 26–27.

3. Emily Thompson, *Soundscape of Modernity* (see chap. 1, n. 39).

4. "Phonograph Boom," *Time* (Sept. 4, 1939): 36; "Phonograph Records," *Fortune* (Sept. 1939): 94; Colin Escott, *Tattooed on Their Tongues: A Journey through the Backrooms of American Music* (New York: Schirmer Books, 1996), 54–61.

5. H. W. Paro, "Modernizing the Phonograph," *Radio News* (Dec. 1938): 28–30.

6. "Music Aids Workers," *Radio News* (Dec. 1938): 44. In 1922, Brigadier General George Owen Squier of the US Army Signal Corps devised a system for transmitting canned music via electronic wire into concert halls, restaurants, and typing pools, ultimately naming it "muzak." On the history of piped-in music, see Lanza, *Elevator Music* (see Introduction, n. 5).

7. "Phonograph Boom," *Time* (Sept. 4, 1939): 36–37. "Phonograph Records," *Fortune* (Sept. 1939): 72–75, 92–102. "U.S. Record Sales 1921–1962," *Billboard International Music-Record Directory* (Aug. 3, 1963): 13.

8. "Record Rush," *Business Week* (Apr. 20, 1940): 32.

9. Howard Taubman, "Black Discs by Millions," *New York Times Magazine* (Jan. 18, 1942): 11; "Record Comeback." *Business Week* (Sept. 13, 1941): 48–49. Copper, used for the master, mother and stamper dies necessary for mass production of records, and aluminum, used as the base for instantaneous recordings, were also required for military preparedness. Shellac, which was imported from India and the most important rosin used in finished discs, had tripled in price from eight cents to twenty-four cents per pound.

10. Gama Gilbert, "Record Renaissance," *New York Times Magazine* (Jan. 7, 1940): 10.

11. "Record Record," *Time* (July 3, 1939): 26; "Record Revival," *Time* (May 20, 1940): 41–42; Gilbert, "Record Renaissance," 10, 17.

12. The earliest and most comprehensive of these was David Hall, *The Record Book: A Music Lover's Guide to the Phonograph* (New York: Smith & Durrell, 1940), with supplements appearing in 1941 and 1943, and subsequent editions in 1950 and 1955.

13. For background on the ban, see Kraft, *Stage to Studio*, 130–161 (see chap. 3, n. 52); and Anderson, *Making Easy Listening*, 3–26 (see Introduction, n. 5).

14. The V-Disc program was initiated and directed by G. Robert Vincent, who had established the National Vocarium in the 1930s (see chap. 3), in an effort to fill demand for current music for the troops. The V-Disc recordings included live bands, studio recordings, and off-air recordings. Artists performed for free, and Petrillo requested no compensation for the union. Richard S. Sears, *V-Discs: A History and Discography* (Westport, CT: Greenwood Press, 1980); Brylawski, "Armed Forces Radio Service," 333–344 (see chap. 3, n. 101). On the spread of American musical culture during the Cold War, see Penny Von Eschen, *Satchmo Blows Up the World: Jazz Ambassadors Play the Cold War* (Cambridge, MA: Harvard University Press, 2004).

15. Smulyan, *Selling Radio* (see chap. 2, n. 82).

16. RCA controlled NBC from its inception in 1926 and purchased Victor Talking Machine in 1929, making it RCA Victor. In 1938, CBS president William Paley purchased the American Record Company, which had acquired in a series of purchases the Columbia, Brunswick,

Okeh, Vocalion, and Melotone labels. Paley reorganized the group as the Columbia Recording Corporation, later renamed Columbia Records, Inc.

17. Robert M. Morris and George M. Nixon, "NBC Studio Design," *Journal of the Acoustical Society of America* (Oct. 1936): 81.

18. "NBC's New Studios in Radio City," *Electronics* (Dec. 1933): 324–326.

19. On the development of acoustical materials, see Thompson, *Soundscape of Modernity*, chap. 5.

20. Morris and Nixon, "NBC Studio Design," 90.

21. "CBS Studio Building, New York City," *Architectural Forum* 73 (Sept. 1940): 202.

22. Ibid., 201.

23. Erik Barnouw, *The Golden Web: A History of Broadcasting in the United States*, vol. 2, *1933 to 1953* (New York: Oxford University Press, 1968), 99–100, 111. In 1933, sightseers paid forty cents apiece to take these tours; "NBC's New Studios," 324.

24. Victor recording engineer Raymond Sooy documented at least five moves between the 1904 Caruso date at Carnegie Hall and the Paul Whiteman recordings of the 1920s. R. B. Sooy, "Memoirs of My Recording and Traveling Experiences for the Victor Talking Machine Company," entries for 1904, 1907, 1909, 1917, 1921. In 1935, the New York studio address was 411 Fifth Avenue, but by 1937, the studio had moved to Twenty-Fourth Street, according to listings in *Broadcasting Yearbook*.

25. Mitch Miller, interview with author, New York City, Jan. 21, 1999.

26. Howard A. Chinn, "Glossary of Disc-Recording Terms," *Proceedings of the I.R.E.* 33 (Nov. 1945): 761, 763; Warren Rex Isom, "Record Materials, Part II: Evolution of the Disc Talking Machine," *Journal of the Audio Engineering Society* 25, no. 10/11 (Oct./Nov. 1977): 718.

27. Olson, "Microphones for Recording," 676–680 (see chap. 2, n. 48).

28. To accompany the album, Avakian wrote an extensive booklet, which is considered a classic within the "minor literary genre" of jazz album liner notes. This album is now hard to find, but Avakian's booklet is reprinted, along with others, in Tom Piazza, ed. *Setting the Tempo: Fifty Years of Great Jazz Liner Notes* (New York: Anchor Books, 1996).

29. George Avakian, telephone interview with author, Dec. 10, 1996.

30. Walter Sear, interview with author, New York City, Jan. 21, 1999.

31. Tom Dowd, telephone interview with author, Mar. 12, 1999.

32. George T. Simon, *The Big Bands* (New York: Macmillan, 1967), 52; Brooks Arthur, telephone interview with author, Apr. 30, 1999.

33. Donald Plunkett, interview with author, New York City, Feb. 9, 1999. Russell Sanjek recalled that the phrase appeared on a placard hanging around the neck of a wooden cigar-store Indian which stood in the office lobby. Sanjek, *American Popular Music*, 3:137 (see chap. 1, n. 51). Kapp was well-known for this attitude, which he maintained long before starting Decca. Guitarist Eddie Condon described a Brunswick session in 1928, during which Earl Hines "was never better at the piano" and Jimmy Noone "was doing impossible things with his clarinet." Listening to the playback, "Kapp looked perplexed. 'What are they doing?' he asked. 'Where is the melody?'" Eddie Condon, with Thomas Sugrue, *We Called It Music: A Generation of Jazz* (New York: Da Capo Press, 1992), 160.

34. John Hammond, with Irving Townsend, *On Record: An Autobiography* (New York: Summit Books, 1977), 210, 217.

35. Sears, *V-Discs*, liv.

36. James W. Bayless, "Innovations in Studio Design and Construction in the Capitol Tower Recording Studios," *Journal of the Audio Engineering Society* (Apr. 1957): 71–76.

37. "In the Groove," *Business Week*, July 21, 1945, 31.

38. Jay Ranellucci, telephone conversation with author, Mar. 25, 1999. One reason for this was that the Armed Forces Radio Service used Radio Recorders for all its processing.

39. Clair Krepps, telephone interview with author, Mar. 23, 1999.

40. "In the Groove," 32. Krepps said that Capitol's owners wanted the mastering to be done as close to the plant as possible, "because the plant would ruin about half of what you sent them," and although they did some pressing on the West Coast, the majority of their records were pressed in Scranton. Krepps interview, Mar. 23, 1999.

41. In 1974, the National Academy of Recording Arts and Sciences established a Hall of Fame to honor recordings released before the inception of the Grammy Awards in 1958. In Jan. 1975 Krepps was informed by NARAS president Bill Lowery that his recording of "The Christmas Song" was among the first five winners, selected by a committee of ninety music historians, academy members, and musicologists, and as one of those who "creatively participated on the winning recording," Krepps received a certificate acknowledging his contribution. Bill Lowery, letter to Clair Krepps, Jan. 28, 1975, copy in author's collection courtesy Clair Krepps.

42. Clair Krepps, telephone interviews with author, Mar. 31, 1999, and Mar. 5, 2001. Irv Joel, telephone interview with author, Mar. 19, 1999.

43. Manny Albam, interview with author, New York City, Feb. 9, 1999; Brooks Arthur, telephone interview with author, Apr. 30, 1999; Meyer Berger, "About New York," *New York Times* (Jan. 20, 1958); Christopher Gray, "Recalling the Days of Knights and Elks," *New York Times* (Aug. 24, 2003).

44. According to the Manhattan Center website, in 1922 the Ancient Accepted Scottish Rite of Free Masonry added the Grand Ballroom on the seventh floor and a new façade, and the Vitaphone motion picture company was located there in the 1920s. History of Manhattan Center, currently an active recording and entertainment venue, http://www.mcstudios.com/about-mc-studios/mc-studio-history.php, accessed Aug. 21, 2009. Webster Hall continues to serve as an entertainment venue, although not a recording facility. Detailed history of Webster Hall in the online report of the Landmarks Preservation Commission, http://www.nyc.gov/html/lpc/downloads/pdf/reports/websterhall.pdf, accessed Aug. 21, 2009.

45. Ray Hall, telephone interview with author, Mar. 21, 1999. Engineers use different types of filters to intensify, to attenuate, or to equalize specific audio frequencies. High-pass filters transmit all frequencies above a certain cutoff frequency and attenuate all lower frequencies. Read, *Recording and Reproduction of Sound*, 370 (see chap. 2, n. 9); Robert Emmett Dolan, *Music in Modern Media: Techniques in Tape, Disc and Film Recording, Motion Picture and Television Scoring and Electronic Music* (New York: G. Schirmer, 1967), 176. When Columbia Records occupied the 799 Seventh Avenue studio, engineers anchored the recording lathes in 500-pound blocks of concrete to counteract the subway rumble. Savory interview, Oct. 21, 1999.

46. Bernhard Behncke, "Liederkranz Hall: The World's Best Recording Studio?," http://vjm.biz/new_page_3.htm, accessed Sept. 9, 2007.

47. Hammond, *On Record*, 217.

48. Avakian interview, Dec. 10, 1996.

49. Andre Kostelanetz, in collaboration with Gloria Hammond, *Echoes: Memoirs of Andre Kostelanetz* (New York: Harcourt Brace Jovanovich, 1981), 82.

50. Song composed by Alan Courtney and Ben Homer, vocal by Betty Bonney, OKeh 6377.

51. Donald Plunkett, interview with author, New York City, Feb. 9, 1999.

52. Lonsdale Green, Jr., and James Y. Dunbar, "Recording Studio Acoustics," *Journal of the Acoustical Society of America* 19 (May 1947): 412–414.

53. G. M. Nixon, "Recording Studio 3A," *Broadcast News* 46 (Sept. 1947): 33.

54. Ibid.

55. Green and Dunbar, "Recording Studio Acoustics," 413. Emphasis in original.

56. Manny Albam, interview with author, New York City, Feb. 9, 1999.

57. Vincent J. Liebler, "A Record Is Born!" *The Columbia Record* 1, no. 6 (Mar 1954): 4.

58. The dimensions are from Liebler, "A Record Is Born!," 4. Columbia engineer Frank Laico estimated the ceiling at one hundred feet, which probably refers to the apex of the central arch. Frank Laico, telephone interview with author, Jan. 13, 1999.

59. Savory interview, Oct. 21, 1999.

60. Parabolic reflectors had been used in film sound recording since 1930. These "microphone concentrators" differed from Savory's upright panels in that they were bowl-shaped reflectors with the microphone secured to the center and were used primarily on location for high-quality sound recording at distances of twenty to forty feet. Carl Dreher, "Microphone Concentrators in Picture Production," *Journal of the Society of Motion Picture Engineers* 16 (Jan. 1931): 23–30. The same kind of device was used by NBC for a broadcast of the opera *Wozzeck* from the Philadelphia Metropolitan Opera House in 1931, where replacement of multiple microphones by the single parabolic reflector condenser microphone reportedly provided "unusually fine results." Read and Welch, *From Tin Foil to Stereo*, 378 (see chap. 1, n. 36).

61. Savory interview, Oct. 21, 1999.

62. Katz, *Capturing Sound*, 26, 32, 80, 85–98 (see Introduction, n. 4).

63. Morris Hastings, album liner notes to Richard Rodgers and Oscar Hammerstein, *South Pacific*, LP 4180 (Columbia Masterworks, 1949); Brooks Atkinson, "At the Theatre," *The New York Times* (Apr. 8, 1949): 30.

64. William Savory, interview with author, Falls Church, VA, Nov. 29, 1997.

65. Mitch Miller interview, Jan. 21, 1999. For more on the 30th Street studio, see Ashley Kahn, *Kind of Blue* (New York: Da Capo Press, 2000), 75–77; and Granata, *Sessions with Sinatra*, 42–45 (see Introduction, n. 15).

66. Ray is pictured peeking between two of the reflectors in "Men behind the Microphones: Makers of Music for Millions," *Newsweek* (Sept. 8, 1952): 56–59.

67. Will Friedwald, *Johnnie Ray: 16 Most Requested Songs*, compact disc CK46095 (Columbia/Legacy, 1991), liner notes.

68. Clair Krepps, telephone interview with author, Mar. 23, 1999.

69. Tom Dowd, telephone interview with author, Mar. 21, 1999.

70. Irv Joel, telephone interview with author, Mar. 19, 1999.

71. "How Echoes Are Produced: NBC Engineers Perfect Artificial Sound Reflection," *Broadcast News* 13 (Dec. 1934): 26–27; Thompson, *Soundscape of Modernity*, 281–284.

72. Les Paul, interview with author, New York City, Jan. 19, 1999.

73. Théberge, *Any Sound You Can Imagine*, 45 (see Introduction, n. 9).

74. Peter Doyle, *Echo and Reverb: Fabricating Space in Popular Music Recording, 1900–1960* (Middletown, CT: Wesleyan University Press, 2005); Daniel J. Levitin, *This Is Your Brain on Music: The Science of Human Obsession* (New York: Dutton, 2006).

75. "How Echoes are Produced," 26–27; "NBC's New Echo Chambers," *Electronics* (July 1935): 227.

76. Dr. S. J. Begun and S. K. Wolf, "On Synthetic Reverberation," in *Magnetic Tape Recording*, ed. Marvin Camras (New York: Van Nostrand Reinhold Company, 1985), 413–414. This chapter, reprinted from *Communications* 18 (1938): 8–9 describes an endless magnetic tape recording and reproducing machine employing steel tape.

77. Music producer Al Brackman recalled the incident in Irwin Chusid, "Raymond Scott," liner notes to *The Music of Raymond Scott: Reckless Nights and Turkish Twilights*, compact disc CK 53028 (Columbia, 1992).

78. Robert Pruter, *Doowop: The Chicago Scene* (Urbana: University of Illinois Press, 1996), 16–17.

79. Savory interview, Nov. 29, 1997.

80. The term *echo chamber* was used pervasively but inaccurately. The effect achieved was actually reverberation, and these were more properly called "reverberation chambers."

81. John Palladino, telephone interview with author, Oct. 15, 1999.

82. Irv Joel, e-mail correspondence, Sept. 4, 2001.

83. The ramped-up floor and angled walls were necessary to eliminate parallel surfaces that could lead to undesirable echo effects, and the hard and smooth walls increased reflectivity. These and other design characteristics for recording studio chambers are discussed in Michael Rettinger, "Reverberation Chambers for Broadcasting and Recording Studios," *Journal of the Audio Engineering Society* 5, no. 1 (Jan. 1957): 18–22.

84. Capitol's New York studio on Forty-Sixth Street always served more outside clients than Capitol artists who usually recorded in the Hollywood studios, with the notable exception of Les Paul and Mary Ford who recorded at home. Irv Joel, telephone interview with author, Mar. 19, 1999.

85. Joel e-mail, Sept. 4, 2001.

86. P. C. Goldmark and J. M. Hollywood, "The Reverbetron," *1957 IRE National Convention Record, Part 7*, pp. 124–127; P. E. Axon, C. L. S. Gilford, and D. E. L. Shorter, "Artificial Reverberation," *Journal of the Audio Engineering Society* 5, no. 4 (Oct. 1957): 218–237; Carlos E. R. A. Moura and Sergio Lara Campos, "Some Notes on Artificial Reverberation," *Journal of the Audio Engineering Society* 5, no. 4 (Oct. 1957): 182–186; Harry F. Olson and John C. Bleazy, "Synthetic Reverberator," *Journal of the Audio Engineering Society* 8, no. 1 (Jan. 1960): 37–41.

87. "Leading Recording Organizations Use Ai Electronic Echo Chambers," Audio Instrument Company advertisement, *Journal of the Audio Engineering Society* (July 1958).

88. Fairchild Recording Equipment Corporation advertisement, *Journal of the Audio Engineering Society* 13 (Oct. 1965): A-234.

89. Karl Otto Bäder, e-mail correspondence, Sept. 4 and 13, 2001. Mr. Bäder was chief engineer of Elektromesstechnik in Lahr, Black Forest, Germany.

90. Phil Ramone, telephone interview with author, Aug. 1, 1999.

91. Slap-back echo could also refer to a room echo, achieved when a signal bounces off a reflective surface in a room.

92. Peter Guralnick, *Last Train to Memphis: The Rise of Elvis Presley* (Boston: Little, Brown, 1994), 237.

93. See Robert E. McGinn, "Stokowski and the Bell Telephone Laboratories: Collaboration in the Development of High-Fidelity Sound Reproduction," *Technology and Culture* 24 (Jan. 1983): 38–75; Abram Chasins, *Leopold Stokowski: A Profile* (New York: Hawthorn Books, 1979), 92–93, 133–136; and Evan Eisenberg, *The Recording Angel: Explorations in Phonography* (New York: McGraw-Hill, 1987), 151.

94. Kostelanetz, *Echoes*, 68.

95. Ibid., 70.

96. One of the "microphone-inspired instruments" Kostelanetz frequently used, a "sub-tone" clarinet, was actually devised by Chester Hazlett, who played with Paul Whiteman during the 1920s. The unique sound was created when the clarinet was played very soft and low and close to the microphone. "When the result was amplified in the control room you could not tell what the instrument was, only that it created a gorgeous 'sub-tone' accompaniment for the rest of the orchestra." Kostelanetz, *Echoes*, 70–71.

97. Kostelanetz, *Echoes*, 71–73.

98. Miller quoted in Ted Fox, *In The Groove* (New York: St. Martin's Press, 1986), 30. In a 1952 *Collier's* profile, writer Gilbert Millstein reported that a songwriter and friend of Miller's had likened his discovery of Guy Mitchell and Tony Bennett to him having "invented" these singers. Gilbert Millstein, "He Calls the Hit Tune," *Collier's*, Jan. 19, 1952, 21.

99. Miller interview, Jan. 21, 1999.

100. Suisman, *Selling Sounds* (see Introduction, n. 10).

101. John McDonough, "Pop Music at the Crossroads," *High Fidelity* 26 (Apr. 1976): 50.

102. Savory interview, Nov. 29, 1997; Laico interview, Jan. 13, 1999.

103. Robert Rice, "Profiles: The Fractured Oboist," *The New Yorker* (June 6, 1953): 43–63, quote on 60, 63.

104. Hammond, *On Record*, 279–281.

105. Ethnomusicologists have studied the contemporary role of studio language in sound recording practice. See, for example, Thomas Porcello, "Speaking of Sound: Language and the Professionalization of Sound-Recording Engineers," *Social Studies of Science* 34, no. 5 (Oct. 2004): 733–758.

106. Miller interview, Jan. 21, 1999.

107. This experiment proved disastrous, and in the end, the conductor received a "placebo" control booth. See Chasins, *Leopold Stokowski*, 133–134.

108. Miller interview, Jan. 21, 1999.

109. John Ball, *The Phonograph Record Industry*, American Industries Series, no. 13 (Boston: Bellman Publishing, 1947), 40–41.

110. Teddy Reig, with Ed Berger, *Reminiscing in Tempo: The Life and Times of a Jazz Hustler* (Metuchen, NJ: Scarecrow Press and the Institute for Jazz Studies, Rutgers University, 1990), 71.

111. A compressor is an amplifier whose gain decreases as its input level is increased, essentially reducing the dynamic range of a recording. Compressors were used in radio broadcasting, but sparingly by recording engineers in the 1940s because they were challenging to

use properly (see chap. 5). John Woram, *The Recording Studio Handbook* (Plainview, NY: Sagamore Publishing, 1976): 449–450.

112. C. A. Tuthill, "The Art of Monitoring," *Transactions of the Society of Motion Picture Engineers* 13 (1929): 173–177.

113. "Men behind the Microphones: Makers of Music for Millions," *Newsweek* (Sept. 8, 1952): 56–59.

CHAPTER 5. CONTROL MEN IN TECHNOLOGICAL TRANSITION

1. "2 Sound Systems on Display Here: New Binaural Methods Among High Fidelity Equipment Shown at New Yorker," *New York Times* (Oct. 30, 1952); Michael Gray, "From the Golden Age: The Birth of Decca/London Stereo," *The Absolute Sound* (July/Aug. 1986): 103–110.

2. Agreement on these new formats was reached only after a protracted competition between Columbia and RCA. This story, along with a comprehensive history of the evolution of the consumer record is in Alexander Boyden Magoun, "Shaping the Sound of Music: The Evolution of the Phonograph Record, 1877–1950" (PhD diss. University of Maryland–College Park, 2000).

3. John Urban, "'Stand By, Please . . .' Recording Session in Progress," *Musical America* 74 (Feb. 15, 1954): 158; James Deane, "Birth of an LP," *High Fidelity* 2 (Sept./Oct. 1952): 20.

4. Warren Rex Isom, "Record Materials, Part II: Evolution of the Disc Talking Machine," *Journal of the Audio Engineering Society* 25 (Oct./Nov. 1977): 719; Joseph C. Ruda, "Record Manufacturing: Making the Sound for Everyone," *Journal of the Audio Engineering Society* 25 (Oct./Nov. 1977): 703, states that by 1938 lacquer had replaced wax. Edward Tatnall Canby, C. G. Burke, and Irving Kolodin, *The Saturday Review Home Book of Recorded Music and Sound Reproduction* (New York: Prentice-Hall, 1952), 32. Because the original formula consisted of cellulose acetate, the term *acetate* came into common usage as a synonym for "instantaneous disc" and continued to be used after it was superseded by the cellulose nitrate lacquer-coated disc. See Michael Jay Biel, "The Making and Use of Recordings in Broadcasting Before 1936" (PhD diss., Northwestern University, 1977), 858, 879–905.

5. "War Business Checklist," *Business Week* (Apr. 18, 1942): 34; "Records Again?," *Business Week* (Jan. 23, 1943): 29. According to a Presto advertisement in late 1944, the restrictions on aluminum had been relaxed enough to allow disc manufacturers to offer it once again, but with the polite request that customers keep their orders at half aluminum and half glass until they filled the first big demand. "Comin' At You! Aluminum," *Electronics* 15 (Nov. 1944): 195.

6. "In the Groove," *Business Week* (July 21, 1945): 28. Vinylite was also a war matériel, used in waterproofing, but the V-Disc program was considered an important part of the war effort.

7. The history of magnetic recording dates back to the nineteenth century, see Mark Henry Clark, "The Magnetic Recording Industry, 1878–1960: An International Study in Business and Technological History" (PhD diss., University of Delaware, 1992), and David Morton, *Off the Record: The Technology and Culture of Sound Recording in America* (New Brunswick, NJ: Rutgers University Press, 2000).

8. Joel Tall, "Tall Tales," *Audio* (Oct. 1978): 20.

9. Howard A. Chinn, "Magnetic Tape Recorders in Broadcasting," *Audio Engineering* (May 1947): 7–10, 48–49.

10. The story behind the development of Ampex and the German AEG Magnetophon recorder is told in John T. Mullin, "Creating the Craft of Tape Recording," *High Fidelity* 26 (Apr. 1976): 62–67; and Clark "The Magnetic Recording Industry," chap. 9.

11. Harold Lindsay, "Magnetic Recording Part I," *db: The Sound Engineering Magazine* 11 (Dec. 1977): 38–44; "Magnetic Recording: Part II," *db: The Sound Engineering Magazine* 12 (Jan. 1978): 40; From September 1947 until the Ampex production models arrived, the Crosby show was taped by Jack Mullin using two modified German Magnetophons. R. F. Bigwood, "Applications of Magnetic Recording in Network Broadcasting," *Audio Engineering* (July 1948): 31. The phrase appeared in Ampex's first advertisement, along with Crosby's smiling face encircled by a loop of tape, Lindsay, "Magnetic Recording: Part I," 44. Lindsay, one of the Model 200's designers, reports that "the key premise in our design philosophy was 'uncompromising quality and unsurpassed reliability,'" but he adds that this led to a product that was "somewhat overdesigned." Lindsay, "Magnetic Recording: Part II," 40.

12. Lawrence A. Ruddell, in collaboration with Gunnar Asklund, "Miracles of Recording," *Etude* 67 (Oct. 1949): 23, 52.

13. "Cap's Plunge into Magnetic Tape Recording Experiments May Face-Lift Waxery Trade," *Billboard* (May 22, 1948): 27; [Baldridge and Ward], "The History of the Development of Ampex and a General Description of its Operation and Facilities," [Dec. 1953] Ampex Corporation Records. Photocopy courtesy of Mark Clark, who conducted research for his doctoral dissertation at the Ampex Museum in Redwood City, California, before it closed. In 2001, the Ampex Museum collection was transferred to Stanford University Library, Department of Special Collections. Ampex Corporation Records, Collection number M1230.

14. J. A. Ford, "Magnetic Memories," *Signal* (July/Aug. 1954): 2021–2023.

15. Making two copies simultaneously had been standard procedure with disc recording, at least in the major label studios, where two recording lathes always ran. Copy A would be sent for processing and Copy B was considered the "safety" in case anything happened to Copy A. At Columbia, engineers had been making sixteen-inch transcription disc backups since 1938.

16. John Palladino, telephone interview with author, Oct. 15, 1999.

17. According to William Savory, Columbia engineers told Ampex they would have to reduce the speed to about half in order for Columbia to be interested. In the meantime, "Fairchild suddenly came along with a tape machine that ran at fifteen inches per second and we started using that. In fact, all of the early tapes I ran through as masters for the LP were run from a Fairchild, not from the Ampex." Savory interview, Oct. 21, 1999. From 1946 to 1948, Savory was a member of the Columbia team led by William Bachman that completed work on the long-playing record.

18. Ampex produced about 30,000 Model 300's through 1969, and many of these models, which became the industry workhorse, remained in use for decades. Clark, "Magnetic Recording Industry," 318–319.

19. The time limit varied, depending on the number of grooves per inch and the overall volume of the record, between about 2:20 (at 88 lines per inch) and 3:00 (at 104 lines per inch). The closer groove spacing (more lines per inch), meant more playing time on the record but with lower overall volume. Wider groove spacing allowed for wider lateral excursion of the needle in the groove which was necessary for louder volume.

20. Reig, *Reminiscing in Tempo*, 71 (see chap. 4, n. 110).

21. Granata, *Sessions with Sinatra*, 40 (see Introduction, n. 15).

22. Harvith and Harvith, *Edison, Musicians, and the Phonograph*, 82 (see Introduction, n. 23).

23. Ralph Berton, *Remembering Bix: A Memoir of the Jazz Age* (1974; repr., New York: Da Capo Press, 2000), 410.

24. Columbia agreed to loan Holiday to Gabler to record the song for Commodore. W. Royal Stokes, *Swing Era New York: The Jazz Photographs of Charles Peterson* (Philadelphia: Temple University Press, 1994), 156–157.

25. "Swing: The Hottest and Best Kind of Jazz Reaches Its Golden Age," *Life* (Aug. 8, 1938): 58.

26. Stokes, *Swing Era New York*, 159.

27. The story of RCA's early long-playing record is recounted in Magoun, *Shaping the Sound of Music*, 326–333.

28. Wallerstein was general manager of RCA Victor in 1933, then president of Columbia Records in 1939. The story of the LP's development at CBS Labs and Columbia Records between 1939 and 1948 is recounted in Edward Wallerstein as told to Ward Botsford, "Creating the LP Record," *High Fidelity* 26 (Apr. 1976): 56–61.

29. For a contemporary account of the ban, see Anders S. Lunde, "The American Federation of Musicians and the Recording Ban," *The Public Opinion Quarterly* (Spring 1948): 45–56; also Kraft, *Stage to Studio*, 178–191; Anderson, *Making Easy Listening*, 27–50.

30. Tom Dowd, interview with author, Mar. 12, 1999; "Woman with a Disk," *Newsweek* 30 (Dec. 29, 1947): 42. Howard was among a very small number of women recordists in an industry dominated by men until the 1970s. Howard was a perfectionist whose air-check recordings of Toscanini broadcasts were among the few recordings the maestro approved. Little documentation of her life and work survives, but a brief interview with her is in Vivian Perlis, *Charles Ives Remembered: An Oral History* (Urbana: University of Illinois Press, 2002), 209–211.

31. The story of how 78s were transferred to LPs is in Howard H. Scott, "The Beginnings of LP," *Gramophone* (July 1998): 112–113; and a more detailed account of the LP's creation was recounted by a panel of former Columbia producers and engineers associated with the early LP at an event honoring William Bachman and the LP in 1998. George Avakian, Mitch Miller, Lou Porrata, William Savory, and Howard Scott, National Musical Arts presents: "50th Anniversary of the Long-Play 33 1/3 Record," Jan. 18, 1998, National Academy of Sciences, Washington, DC, tape and transcript of this event in author's collection.

32. Bill Stoddard, e-mail correspondence with author, Mar. 1999. Shellac pressings were noisy because of the abrasive limestone dust employed as filler and as the means for grinding the steel playback needle to properly fit the groove.

33. Wilma Cozart Fine, telephone interview with author, May 5, 1999. A trained musician, Wilma Cozart became the first female vice president of a record company, Mercury, in 1950 and initiated the highly acclaimed Olympian Series of classical recordings. She later married and worked with recording engineer Bob Fine on the Mercury Living Presence series of recordings,

34. George Avakian, "50th Anniversary of the LP."

35. Lou Porrata, "50th Anniversary of the LP." Peter C. Goldmark, René Snepvangers, and

William S. Bachman, "The Columbia Long-Playing Microgroove Recording System," *Proceedings of the I.R.E.* 37 (1949): 923–927.

36. William S. Bachman, "The Columbia Hot Stylus Recording Technique," *Audio Engineering* (June 1950): 11–13.

37. Harry F. Olson, "Microphones for Recording," *Journal of the Audio Engineering Society* 25 (Oct./Nov. 1977): 678.

38. Ibid. Microphones have different areas of sensitivity to sound pressures, called directional patterns. Omnidirectional microphones are more or less receptive to sound coming from all directions; bidirectional microphones pick up sounds from front and back; and unidirectional cardioid (referring to the heart-shaped appearance of its directivity pattern), pick up sound best from the front and sides. Martin Clifford, *Microphones*, 2nd ed. (Blue Ridge Summit, PA: Tab Books, 1982), 86–96.

39. Owned by AEG (Allgemeine Elektricitäts-Gesellschaft), the same company that developed the Magnetophon tape recorder on which the Ampex recorders were modeled, Telefunken was a marketing company that contracted with small manufacturers, including Neumann and Schoepps. They built the microphones and Telefunken had the worldwide marketing and distribution franchise and sold the microphones with the Telefunken labels. This was why in the 1950s, the Neumann U47 was referred to as the Telefunken mike, or the "Telly," Sinatra's pet name for his favorite recording microphone. Al Grundy, telephone interview with author, Jan. 8, 2000; Edgar M. Villchur, "The Telefunken Mike," *Saturday Review* (Oct. 12, 1952): 69–70; Georg Neumann GmbH—Historical Collection: U47 The Legend, www.neumann.com/history/h2.htm (viewed Mar. 24, 1999).

40. Clair Krepps, telephone interview with author, Mar. 23, 1999.

41. Michael Gray, "The Winged Champion: Mercury Records and the Birth of High Fidelity," *The Absolute Sound* 60 (July/Aug. 1989): 47–58.

42. Edgar M. Villchur, "The Telefunken Mike," *Saturday Review* (Oct. 25, 1952): 70.

43. Clair Krepps, interview with author, Mar. 31, 1999. An equalizer is a signal-processing device used to change the frequency response of the signal passing through it. Woram, *Recording Studio Handbook*, 455 (see chap. 4, n. 111).

44. Walter Sear, interview with author, New York City, Jan. 21, 1999.

45. Krepps interview, Mar. 31, 1999. "W. Oliver Summerlin," *Journal of the Audio Engineering Society* 3 (Jan. 1955): 53.

46. Walter Sear, interview with author, New York City, 21 Jan. 1999. According to Sear, who bought up many Pultecs for next to nothing when studios got rid of their tube based equipment as transistors took over in the late 1960s and early 1970s, Pultec equalizers could fetch $4,000 to $5,000 per unit in 1999.

47. Al Schmitt, interview with author, Apr. 12, 1999.

48. Ibid.

49. Edward Tatnall Canby, "The Sound-Man Artist," *Audio* (June 1956): 44. Mitch Miller argued that the "engineers were the strongest link in the chain," and he was among the first to suggest that they be given album credit, although the company executives balked at this suggestion. Mitch Miller, interview with author, New York City, Jan. 21, 1999.

50. Miller interview, Jan. 21, 1999.

51. Gene Lees, *Singers and the Song* (New York: Oxford University Press, 1987), 107.

52. John Palladino, telephone interview with author, Oct. 15, 1999. Engineers used the term *splay* to refer to anything that would redirect, or diffuse the sound in the studio. The most highly developed of these were the polycylindrical diffusers developed at RCA by the acoustical engineer John Volkmann. Instead of plane surfaces, the studio walls were composed of polycylindrical surfaces of different radii, different thickness and different length, creating rooms with excellent acoustics. Retired RCA engineer Hans O. Dietze, telephone interview with author, July 12, 1996.

53. Ray Hall, telephone interview with author, Mar. 21, 1999.

54. Dowd worked for the Office of Scientific Research and Development during World War II, specifically on the Manhattan Project, considered by audio engineers to be "merely an extension of electronic research." "Transients," *Radio* (Aug. 1945): 8. After that experience, at his first studio job, Dowd felt amply trained, despite no experience in recording: "I was chasing neutrons down a corridor all night long for chrissakes. To me, we were stone-aging records. I thought, it's so simple it's dumb." Tom Dowd, telephone interview with author, Mar. 23, 1999. Krepps designed and built a flying classroom to teach Navy pilots to use radar and radio. "They should have given the job to two MIT graduates," Krepps declared. "I made it so you could put ten student pilots in it and fly it. Each one had a scope in front of him and intercommunication and so forth. It was successful, I flew it." Krepps interview, Mar. 23, 1999.

55. Cardioid microphones such as the 44BX were subject to "proximity effect," the rise in low-frequency response when used at close distances. Woram, *Recording Studio Handbook*, 471. Krepps interview, Mar. 31, 1999.

56. Krepps interview, Mar. 31, 1999.

57. Ray Hall, telephone interview with author, Mar. 21, 1999.

58. Bill Stoddard, telephone interview with author, Mar. 11, 1999.

59. Philip C. Erhorn, "Audio Console Design Notes," *Journal of the Audio Engineering Society* 4 (Apr. 1956): 65.

60. Tom Dowd, telephone interview with author, Mar. 16, 1999.

61. Ibid.

62. Frank Laico, interview with author, Jan. 13, 1999.

63. Ibid.

64. Ibid.

65. Al Schmitt, telephone interview with author, Apr. 12, 1999.

66. Palladino interview, Oct. 15, 1999.

67. Gray, "From the Golden Age, 103–110. Experiments in stereo recording date back to the 1881 Paris Electrical Exposition where a musical performance was reproduced over telephone lines and listeners heard two separate channels through a set of headphones, and in 1931, British recording engineer Alan Blumlein patented a system of stereophonic recording on disc that eventually became the basis for the Westrex 45-45 stereo-cutting system that became the standard in 1958. Oliver Read and Walter L. Welch, *From Tin Foil to Stereo: Evolution of the Phonograph* 2nd ed. (Indianapolis, IN: Howard W. Sams, 1976), 426; John Mosley "Eliminating the Stereo Seat," *JAES* 8 (Jan. 1960): 50. In 1940, Bell Labs presented a demonstration of stereophonic recordings of the Philadelphia Orchestra conducted by Leopold Stokowski, with the Salt Lake City Tabernacle choir and organists and a dramatic presentation by Paul

Robeson at Carnegie Hall. C. W. "Enhanced Stereophonic Recordings Demonstrated by Bell Laboratories," *Electronics* (May 1940): 30–31.

68. Wilma Cozart Fine, interview, May 5, 1999.

69. Dowd interview, Mar. 16, 1999.

70. Emory Cook had invented binaural disc recording in 1953, the system involved a bifurcated tone arm which played two bands on a disc simultaneously, a complex system that only caught on with audiophiles. Emory Cook, "Binaural Disc Recording," *Journal of the Audio Engineering Society* 1 (Jan. 1953): 1–3.

71. William Francis Shea, "Role and Function of Technology in American Popular Music, 1945–1964" (PhD diss. University of Michigan, 1990), 107–108. The story of how they arrived at the standard cutting system is in H. E. Roys, "The Coming of Stereo," *Journal of the Audio Engineering Society* 25 (Oct./Nov. 1977): 824–827.

72. Roys, "The Coming of Stereo," 824–827.

73. Bill Stoddard, e-mail message to author, Feb. 21, 1999. Bob Dietmeier, "1959 Juke Box Operator Poll," in Music Operators of America, *The 1959 Music Machine Guide*, Supplement to *The Billboard* (Apr. 6, 1959): 3.

74. Savory interview, Oct. 21, 1999. Savory became a member of the RIAA committee on stereo standards.

75. James W. Bayless, "Innovations in Studio Design and Construction in the Capitol Tower Recording Studios," *Journal of the Audio Engineering Society* (Apr. 1957): 75–76.

76. William Savory, interview with author, Falls Church, Virginia, Nov. 29, 1997.

77. Palladino interview, Oct. 15, 1999.

78. "New Recording Facilities for Technical Operations," *The Columbia Record* (Apr./May 1960): 6.

79. Vincent J. Liebler, *The Columbia Record* (Dec. 1964): 9.

80. See Steven Levy, *Hackers: Heroes of the Computer Revolution* (New York: Delta, 1984). Thanks to Molly Berger for drawing my attention to this book and the similarities between recording engineers in this era of high fidelity and the computer hackers, the "adventurers, visionaries, risk-takers, artists" who held in common "the philosophy of sharing, openness, decentralization, and getting your hands on machines at any cost" (p. 7).

81. "Electronics Era," *Business Week* (July 29, 1944): 24, 26–30.

82. "Editor's Report," *Audio Engineering* (June 1953): 16.

83. Dowd interview, Mar. 12, 1999.

84. C. A. Tuthill, "The Art of Monitoring," *Transactions of the S.M.P.E.* 13 (1929): 173.

85. On tacit knowledge, see Essay on Sources.

86. Douglas Harper, *Working Knowledge: Skill and Community in a Small Shop* (Chicago: University of Chicago Press, 1987).

87. Eugene Ferguson, *Engineering and the Mind's Eye* (Cambridge, MA: MIT Press, 1992).

88. Ray Hall, telephone interview with author, Mar. 21, 1999.

89. Hall interview, Mar. 21, 1999.

90. John Woram, telephone interview with author, Dec. 6, 1999.

91. Doug Pomeroy, e-mail to author, June 19, 2011.

92. Irv Joel, telephone interview with author, Mar. 19, 1999.

93. Eddison von Ottenfield, "Acoustical Balance in Recording," *Audio Engineering* (Aug.

1951): 23, 58; Fritz A. Kuttner, "Musical Art vs. Technology," *American Record Guide* 26 (Oct. 1959): 84–85, 130–135; John Borwick, "What Is a Tonmeister?, *db: The Sound Engineering Magazine* (Oct. 1973): 26–28; "Mixing Musical Art and Technology," *Journal of the Audio Engineering Society* 22 (Mar. 1974): 112.

94. Palladino interview, Oct. 15, 1999.

95. John G. Frayne and Halley Wolfe, *Elements of Sound Recording* (New York: John Wiley & Sons, 1949). This book was a revised and expanded text version of Frayne and Wolfe's courses.

96. "A University Is Born," *Radio & Television News* 42 (Nov. 1949): 52–53.

97. Woram interview, Dec. 6, 1999.

98. Krepps interview, Mar. 23, 1999.

99. William H. White, Jr., *The Organization Man* (New York: Simon and Schuster, 1956); Vance Packard, *The Status Seekers* (New York: David McKay Company, 1959), chap. 8.

100. "Men behind the Microphones: Makers of Music for Millions," *Newsweek* 8 (Sept. 1952): 59.

101. "Edison Snares Soul of Music," *New York Tribune* (Apr. 29, 1916): 3, cited in Emily Thompson, "Machines, Music, and the Quest for Fidelity: Marketing the Edison Phonograph in America, 1877–1925," *Musical Quarterly* 79 (Spring 1995): 157.

102. Hinton quoted in Ann Douglas, *Terrible Honesty: Mongrel Manhattan in the 1920s* (New York: Farrar, Straus and Giroux, 1995), 421.

103. Manny Albam, interview with author, New York City, Feb. 9, 1999.

104. Milt Hinton and David G. Berger, *Bass Line: The Stories and Photographs of Milt Hinton* (Philadelphia: Temple University Press, 1988), 258.

105. Ibid.

106. Tony Scherman, *Backbeat: Earl Palmer's Story* (Washington, DC: Smithsonian Institution Press, 1999).

107. Brooks Arthur, telephone interview with author, Apr. 30, 1999.

108. Ibid.

109. Emory Cook, "The Man in the Control Booth," *Musical America* 73 (Feb. 1953): 175.

110. Howard Scott, telephone interview with author, May 1, 1998.

111. Frank Bruno, telephone interview with author, Feb. 4, 1999.

112. Don Rayno, *Paul Whiteman: Pioneer in American Music*, vol. 1, 1890–1930. Studies in Jazz, no. 43 (Lanham, MD: Scarecrow Press, 2003), 188.

113. Rose Heylbut, "Back of the Scenes at a Recording Session," *Etude* 72 (Apr. 1954): 57.

114. Alan Lorber, telephone interview with author, Mar. 5, 1999.

115. Palladino interview, Oct. 15, 1999. Emphasis added.

116. Phil Ramone, telephone interview with author, Aug. 1, 1999.

117. Ibid.

118. Ibid.

119. Albam interview, Feb. 9, 1999.

120. Tom Dowd, telephone interview with author, Mar. 21, 1999.

121. Zak, *The Poetics of Rock*, 165 (see Introduction, n. 22).

122. Peter Galison, *Image and Logic: A Material Culture of Microphysics* (Chicago: University of Chicago Press, 1997), title page.

123. Ibid., xxi–xxii.

124. Eric von Hippel, *The Sources of Innovation* (New York: Oxford University Press, 1988), 5–6.

125. Dowd interview, Mar. 16, 1999.

126. Krepps interview, Mar. 23, 1999.

127. Ibid.

128. John W. Rumble, "The Emergence of Nashville as a Recording Center: Logbooks from the Castle Studio, 1952–1953," *Journal of Country Music* 7, no. 3 (1978): 22–41.

CHAPTER 6. THE SEARCH FOR THE SOUND

1. Escott, *Tattooed on Their Tongues*, 53–84 (see chap. 4, n. 4); Charlie Gillett, *The Sound of the City: The Rise of Rock and Roll*, rev. ed. (London: Souvenir Press, 1983); Kennedy and Mc-Nutt, *Little Labels—Big Sound* (chap. 1, n. 67); Donald J. Mabry, "The Rise and Fall of Ace Records: A Case Study in the Independent Record Business," *Business History Review* 64 (Autumn 1990): 411–450; Albin J. Zak III, *"I Don't Sound Like Nobody": Remaking Music in 1950s America* (Ann Arbor: University of Michigan Press, 2010).

2. Blackford, *A History of Small Business in America*, chap. 5 (see chap. 1, n. 55). Beginning in 1942, the trade publication *Billboard* classified all records made by and for blacks as "Harlem Hit Parade," changed in 1945 to "Race Records," and in 1949 to "Rhythm and Blues." Jerry Wexler and David Ritz, *Rhythm and the Blues: A Life in American Music* (New York: Alfred A. Knopf, 1993), 62.

3. A cultural analysis of this transformation is Diane Pecknold, *The Selling Sound: The Rise of the Country Music Industry* (Durham, NC: Duke University Press, 2007), and Joli Jensen, *The Nashville Sound: Authenticity, Commercialization, and Country Music* (Nashville, TN: Country Music Foundation Press & Vanderbilt University Press, 1998), which looks at the development of a distinct Nashville Sound in the 1950s and 1960s. The history of Nashville's recording and musical culture before the major labels arrived has been exhaustively documented in Martin Hawkins, *A Shot in the Dark: Making Records in Nashville, 1945–1955* (Nashville, TN: Vanderbilt University Press & Country Music Foundation Press, 2006).

4. For more on the boycott, see John Ryan, *The Production of Culture in the Music Industry: The ASCAP-BMI Controversy* (Lanham, MD: University Press of America, 1985), and Russell Sanjek, *Pennies from Heaven: The American Popular Music Business in the Twentieth Century* (New York: Da Capo, 1996), 184–211.

5. Bill C. Malone and David Stricklin, *Southern Music / American Music*, rev. ed. (Lexington: University of Kentucky Press, 2003), 93; Kraft, *Stage to Studio*, 155 (see chap. 3, n. 52).

6. Eddie Smith, interview with author, New York City, Jan. 23, 1999; Kennedy and McNutt, *Little Labels—Big Sound*, 56–72; Escott, *Tattooed on Their Tongues*, 63–68; and Steven C. Tracy, *Going to Cincinnati: A History of the Blues in the Queen City* (Urbana: University of Illinois Press, 1993), 114–153. Vertical integration occurs when "a company that initially engages in only one stage of the production and sale of its goods may acquire control of its sources of raw materials and/or the making and sale of its finished products." Mansel G. Blackford and K. Austin Kerr, *Business Enterprise in American History*, 3rd ed. (New York: Houghton Mifflin, 1994), 139.

Although vertical integration was usually applied to big business, it describes owner Syd Nathan's King Records because records were recorded, pressed, packaged, and shipped from the plant in Cincinnati.

7. Suzanne E. Smith, *Dancing in the Street: Motown and the Cultural Politics of Detroit* (Cambridge, MA: Harvard University Press, 1999). An excellent documentary profiling Motown's session musicians, the Funk Brothers, and their contribution to the Motown Sound: Paul Justman, Director, *Standing in the Shadows of Motown,* DVD (Artisan Entertainment, 2002).

8. Chess Records' first studio was actually Sheldon Recording, built in 1958 by an independent recording engineer, Jack Wiener, in partnership with Leonard and Phil Chess. Wiener did not like being restricted to recording only Chess artists and eventually moved his corporation but left the studio at 2120 S. Michigan Avenue, which later became known as the Chess Studio. Jack Wiener, telephone interview with author, Jan. 25, 1997. On Chess Records' history, see Nadine Cohodas, *Spinning Blues into Gold: The Chess Brothers and the Legendary Chess Records* (New York: St. Martin's Press, 2000), and Peter Guralnick, *Feel Like Going Home: Portraits in Blues & Rock 'n' Roll* (New York: Outerbridge & Dienstfrey, 1971), 180–202. On Atlantic Records, see Charlie Gillett, *Making Tracks: Atlantic Records and the Growth of a Multi-Billion-Dollar Industry* (New York: E. P. Dutton, 1974), and Wexler and Ritz, *Rhythm and the Blues.*

9. McDonough, "1951," 52 (see Introduction, n. 3).

10. Colin Escott with Martin Hawkins, *Good Rockin' Tonight: Sun Records and the Birth of Rock 'n' Roll* (New York: St. Martin's Press, 1991); Robert Palmer, *A Tale of Two Cities: Memphis Rock and New Orleans Roll* (Brooklyn: Institute for Studies in American Music, Department of Music, School of Performing Arts, Brooklyn College of the City University of New York, 1979); and Kennedy and McNutt, *Little Labels—Big Sound,* 89–105.

11. Read and Welch claimed that by 1959 there were more than 600 labels in the United States alone, most of which were small independents. Read and Welch, *From Tin Foil to Stereo,* 506 (see chap. 1, n. 36). One can only estimate the number of studios, but one industry reference lists 136 recording studios in Los Angeles (54), Chicago (28), and New York (55) as of 1954; and a 1970 directory lists 470 across the United States and roughly 300 internationally. Stephen F. Keegan, ed., *The Musician's Guide,* First Annual Edition (New York: Music Information Service, 1954), 249–251; *Billboard 1970 International Directory of Recording Studios* (New York: Billboard Publications, 1970).

12. Rodney W. Baum, "Radio Music Transcription Services: Their Development and Decline" (Master's thesis, Bowling Green State University, 1964).

13. "Sam Phillips' Sound," National Public Radio, *Morning Edition,* Nov. 28, 2001.

14. Engineer Ray Hall recalled that in the 1960s, music recording comprised roughly 80–85 percent of the recording work done in RCA's New York studios. The rest included a variety of work, including jingles and advertising. Ray Hall, telephone interview with author, Mar. 21, 1999.

15. Ken Hamann, interview with author, Painesville, Ohio, Nov. 13, 1995.

16. Joe Valencic, "Polkas," in *The Encyclopedia of Cleveland History,* 2nd ed., ed. David D. Van Tassel and John J. Grabowski (Bloomington: Indiana University Press, 1996): 803–805.

17. Victor Greene, *A Passion for Polka: Old-Time Ethnic Music in America* (Berkeley: University of California Press, 1992), 172.

18. Sheldon Henderschott, interview with author, Cleveland, Ohio, May 24, 1996.

19. Gloria Busse Hamann, telephone conversation with author, May 22, 1996. Gloria Busse began working for Fred Wolf shortly after she went to his studio to make a record for her family in late 1943. There she later met and married his chief engineer, Ken Hamann.

20. Greene, *A Passion for Polka*, 233–34; Yankovic, *Polka King*, 47 (see Introduction, n. 1).

21. Yankovic, *Polka King*, 67.

22. Later in his career, Yankovic not only created a blend of ethnic styles but also contributed to a more polished professionalism and perfectionism in polka performance. Charles Keil and Angeliki V. Keil, *Polka Happiness* (Philadelphia: Temple University Press, 1992), 138–140.

23. Yankovic continued to record exclusively for Columbia until 1968. George Avakian, who ran Columbia's International and Popular Albums departments from 1948 to 1958, worked with Yankovic during those ten years. The first Yankovic record he produced, "Blue Skirt Waltz," was Avakian's first million-seller. George Avakian, telephone interview with author, Dec. 10, 1996.

24. Wayne Mack, telephone conversation with author, May 11, 1996.

25. Until the founding of the Audio Engineering Society in 1948, which Hamann joined shortly after being hired by Wolf, the IRE one of the few professional organizations for audio engineers. Hamann remained active in the IRE and subsequently the IEEE and helped form a local chapter of its Professional Group on Audio. See McMahon, *Making of a Profession*, 216 (see chap. 3, n. 51); Ken Hamann, interview with author, Painesville, Ohio, Nov. 21, 1995.

26. Forerunners of the Powerpoint presentation, these stripfilms advanced one frame at a time in synchronization with a recording on a sixteen-inch disc. A sub-sonic tone triggered the advance at prescribed points within the narration. Duquesne Corporation made a machine that would do this automatically so that the operator needed only to set up the projector and player. Hamann interview, Nov. 13, 1995.

27. Hamann interview, Nov. 13, 1995.

28. Ibid.

29. Ken Hamann, Binaural system, US Patent 2,849,540, filed June 17, 1954, and issued Aug. 26, 1958.

30. "Visit Stereo-Land at the Second Annual High Fidelity Fair," special issue, *Fine Music: Weekly Radio and Concert Guide* 4, no. 181 (Nov. 18, 1957): 10–11, copy in author's collection, courtesy of William Hlavin, Jr. Cleveland's High Fidelity Fair was sponsored by WDOK and the *Cleveland Press*.

31. "4-Channel Air Stereo Debuts in Cleveland," *Billboard* (Nov. 10, 1958): 4.

32. For example, Susan J. Douglas, "Oppositional Uses of Technology and Corporate Competition," in *Technological Competitiveness: Contemporary and Historical Perspectives on the Electrical, Electronics, and Computer Industries*, ed. William Aspray (New York: IEEE Press, 1993), 208–219; and Keir Keightley, "'Turn It Down!' She Shrieked: Gender, Domestic Space, and High Fidelity, 1948–59," *Popular Music* 15 (May 1996): 149–177.

33. Hamann interview, Nov. 13, 1995.

34. Columbia and RCA recording engineers used equipment built by the company engineering department that was of the highest quality and dependability because of its construction and regular maintenance, thus they felt no need to radically change the technology they had used with great success for years. In February 1966, Columbia engineer Fred Plaut is pictured at a 16-channel console with rotary faders, "'Take One!' The Making of a Record: 2,"

The Columbia Record (Feb. 1966): 8. Two years later, Langevin advertised its AM4-A modular mixing console as "an audio mixer for the man who doesn't want to do it himself!" Describing it as a "revolutionary concept in audio mixers . . . a fully integrated system that is completely wired, tested and checked out. Just open the box, connect the input/output lines, drop in the modules and you're ready to go. No wiring, no engineering headaches." Langevin's advertisement appeared in *db: The Audio Engineering Magazine* (Apr. 1968): 5. Both examples demonstrate that the evolution of console design was by no means uniform and moved in a staggered progression, with certain components advanced, others remaining traditional.

35. Ken Hamann, interview with author, Painesville, Ohio, Oct. 25, 1996.

36. Anastasia Pantsios, "History of the Cleveland Rock Scene: Part II–The Sixties," *Exit* (Dec. 19, 1974): 9.

37. The Montclairs, one of several groups by that name, were on the Sunburst label. Tom King and the Starfires became The Outsiders, signed a contract with Capitol Records, and recorded three albums at Cleveland Recording. "Time Won't Let Me" has enjoyed a long afterlife in film soundtracks, television commercials, and oldies radio.

38. Hamann interview, Nov. 21, 1995.

39. In 1961, Fred Wolf sold his interest in radio station WDOK to Transcontinent Television Corp. of New York but remained as consultant until 1965. George E. Condon, "WDOK Radio Is Acquired by N.Y. Firm," *Cleveland Plain Dealer* (Nov. 18, 1961).

40. Don White, interview with author, Akron, Ohio, Jan. 23, 1996.

41. Quoted in Claude Hall, "Studio Track," *Billboard* (Mar. 27, 1971): 4.

42. Hamann interview, Nov. 21, 1995.

43. Hamann interview, Nov. 13, 1995.

44. Susan Schmidt Horning, "Recording," 105–122 (see Introduction, n. 12); Mary Alice Shaughnessy, *Les Paul: An American Original* (New York: William Morrow, 1993); Steve Waksman, *Instruments of Desire: The Electric Guitar and the Shaping of Musical Experience* (Cambridge, MA: Harvard University Press, 1999), 36–74.

45. Jimmy Fox, telephone interview with author, Oct. 29, 1996. Art Thompson, *The Stompbox* (San Francisco: Miller Freeman Books, 1997).

46. Hamann interview, Nov. 13, 1995. Evan Eisenberg mentions that conductor Leopold Stokowski ran into this problem when he insisted on mixing his taped performance: "he had to do it, in deference to union rules, by putting his hand over the engineer's." Eisenberg, *Recording Angel*, 152.

47. For more on the art of audio engineering, see Kealy, "From Craft to Art, 3–29 (see Introduction, n. 25).

48. James T. Fox, interview with author, Concord Township, Ohio, May 10, 1996.

49. Cleveland Recording's last gold records were Wild Cherry's single, "Play That Funky Music," and the subsequent album *Wild Cherry*, recorded in 1976 and released on Epic / Sweet City Records.

50. Hamann interview, Nov. 21, 1995.

51. Even after the group dispersed to different parts of the world, Pere Ubu continued to convene at Suma Recording to do nearly every one of their albums, usually with Ken's son Paul Hamann engineering. Ken Hamann passed away in 2003 and Paul Hamann continues to run Suma Recording.

52. "Arranged and Recorded by Schneider," *Audio Record* 7, no. 5 (June/July 1951): 1, 4.

53. Kay Schneider, interview with author, Cleveland, Ohio, June 1, 1996.

54. In late nineteenth century New York, song sharks preyed on aspiring amateur song-writers by charging them to publish their music with the promise of placing it but merely pocketed the money. David Suisman, *Selling Sounds: The Commercial Revolution in American Music* (Cambridge, MA: Harvard University Press, 2009), 51.

55. Between 1946 and 1957, the number of staff musicians in radio steadily declined from 2,433 to 576, with the most precipitous drop in 1957. Kraft, *Stage to Studio,* 198.

56. Schneider's career path exemplifies the fate of musicians caught in the shifting tech-nological landscape of the twentieth-century music industry, a story charted in Kraft, *Stage to Studio*; and Faulkner, *Hollywood Studio Musicians: Their Work and Careers in the Recording Indus-try* (Lanham, MD: University Press of America, 1985).

57. Mark Henry Clark, "The Magnetic Recording Industry, 1878–1960: An International Study in Business and Technological History" (PhD diss., University of Delaware, 1992), 274.

58. Bob Seltzer, "His Work Is for the Record," *Cleveland Press*, Oct. 3, 1967. Brian Telzrow, interview with author, Cleveland, Ohio, June 1, 1996.

59. The transition to tape was neither immediate nor universal. The first syndicated radio program recorded on tape was the Oct. 1, 1947, *Bing Crosby Radio Show*, but the final program was broadcast from disc. Audio Engineering Society, *An Afternoon with John T. "Jack" Mullin*, videotape (Los Angeles: Webster Communications, 1989), copy in author's possession cour-tesy Brian Telzrow.

60. "Arranged and Recorded by Schneider: Unique Combination of Musical Science and Audio Engineering Enables Schneider Recording Studio Lab to Give Clients the 'Full Treat-ment,'" *Audio Record* 7, no. 5 (June/July 1951): 1, 4.

61. Telzrow interview, June 1, 1996.

62. Panned in the press and by disc jockeys, one of whom called it "the worst record I ever heard," the song was a perfect example of a grassroots hit that could only have been made at an independent recording studio. "Music: Mystery Hit," *Time* (Feb. 9, 1953) online at www.time.com/time/magazine/article/0,9171,889646,00.html. This was not the same "Oh Happy Day" later made famous by the gospel choir The Edwin Hawkins Singers, Howard's song was cov-ered by a number of other artists, including The Four Lads, The Three Suns, Lawrence Welk, and Tab Hunter.

63. Schneider interview, June 1, 1996. Until the Sound Recording Act of 1971 allowed sound recordings to be submitted for copyright, songwriters had to submit their music in written form.

64. Schneider interview, June 1, 1996.

65. Arnold Shaw, *The Rockin' '50s: The Decade That Transformed the Pop Music Scene* (New York: Hawthorn Books, 1974), 197. Michael "Doc Rock" Kelly [Michael Bryan Kelly], *Liberty Records: A History of the Recording Company and Its Stars, 1955–1971* (Jefferson, NC: McFarland & Company, 1993).

66. Brian Telzrow, interview with author, Cleveland, Ohio, Jan. 29, 1996.

67. James T. Patterson, *Grand Expectations: The United States, 1945–1974* (New York: Ox-ford University Press, 1996).

68. Thomas Boddie and Louise Boddie, interview with author, Cleveland, Ohio, Feb. 14, 1996.

69. Ibid.

70. Ibid.

71. Patterson, *Grand Expectations*, 381–382.

72. Boddie and Boddie interview, Feb. 14, 1996.

73. Peerce, then tenor with the New York Metropolitan Opera, appeared as guest soloist on the Jewish Singing Society program in 1959. F. Karl Grossman, *History of Music in Cleveland* (Cleveland, OH: Case Western Reserve University, 1972), 52.

74. Boddie and Boddie interview, Feb. 14, 1996.

75. Jack Renner, interview with author, Cleveland, Ohio, Apr. 5, 1996. In 1970, Renner made almost two hundred different custom recordings—school bands, plays, various memento records—and three years later that number had dropped to about fifty.

76. Rob Sevier and Ken Shipley, "Local Customs: Pressed at Boddie," liner notes to *Local Customs: Burned at Boddie*, audio CD, 035 ½ (The Numero Group, 2011). Numero Group, an independent record label based in Chicago, tracked down a catalog of more than a thousand records pressed by the Boddies. Since 2011, Numero Group has been reissuing records both pressed and recorded by the Boddies (www.numerogroup.com).

77. For an account of the riots of 1968, see Louis Masotti and Jerome Corsi, *Shoot-Out in Cleveland: Black Militants and the Police: July 23, 1968*. A Staff Report to the National Commission on the Causes and Prevention of Violence (Washington, DC: Government Printing Office, 1969).

78. Louise Boddie interview, Feb. 14, 1996.

79. Blackford, *A History of Small Business in America* (Chapel Hill: University of North Carolina Press, 2003), 149–150.

80. Tom Boddie interview, Feb. 14, 1996.

81. Ray Hall interview, Mar. 21, 1999.

82. "$600, a Good Telephone Listing and Ambition," *Billboard* (Feb. 24, 1968): B-3.

83. Ibid.

84. Dave Teig, interview with author, New York City, Feb. 9, 1999.

85. "$600, a Good Telephone Listing and Ambition," B-22.

86. "'Take One!': The Making of a Record: 2," *Columbia Record* (Feb. 1966): 6–11.

87. Eddie Smith, interview with author, Jan. 23, 1999.

88. "Bell Sound's 'Button Pushers' at Head of Their Class," *Billboard* (Feb. 24, 1968): B-22.

89. Stoddard, telephone interview with author, Mar. 11, 1999.

90. Stoddard, e-mail to author, Feb. 21, 1999.

91. Stoddard, e-mail to author, June 30, 2000.

92. Ibid.

93. Teig, telephone conversation with author, July 1996.

94. Smith interview, Jan. 23, 1999.

95. Bill Stoddard interview, Mar. 11, 1999.

96. Smith interview, Jan. 23, 1999. Henry Glover was Dinah Washington's producer and an old friend and colleague of Smith's. The two had worked together since both were with the Lucky Millinder Band in the 1940s. Syd Nathan of King Records hired Glover as A&R for

country and rhythm and blues records, and Glover later brought Smith to King Records in 1951 to do the same job for popular records. John W. Rumble, "Roots of Rock & Roll: Henry Glover at King Records," *Journal of Country Music* 14, no. 2 (1990): 30–42.

97. Stoddard e-mail, Feb. 21, 1999.

98. Bob Thiele, as told to Bob Golden, *What a Wonderful World: A Lifetime of Recordings* (New York: Oxford University Press, 1995).

99. Philip Norman, *Rave One: The Biography of Buddy Holly* (New York: Simon & Schuster, 1996), 91.

100. Thiele, *What a Wonderful World*, 55.

101. Smith, e-mail to author, Apr. 25, 2002.

102. Thiele, *What a Wonderful World*, 56.

103. Ibid., 57.

104. Ibid., 56.

105. Anne Phillips, interview with author, Feb. 7, 1999.

106. Smith interview, Jan. 23, 1999.

107. Alan Lorber, telephone interview with author, Mar. 5, 1999.

108. Al Weintraub, "The Role of the Studio Operation in the Recording Industry: Dan Cronin," *Billboard* (Feb. 24, 1968): B-4; Stoddard interview, Mar. 11, 1999. Bell was known widely as the first studio to convert to solid state.

109. Teig interview, Feb. 9, 1999.

110. Ibid.; Matty Polakoff, "Duplicating Tape CARtridge Product," *Billboard* (Feb. 24, 1968): B-14; Morton, *Off the Record*, 159–162 (see Introduction, n. 21).

111. Mort Fujii recalled the development of Bell Sound's 12-track recorder in Dale Manquen (producer), "*A Chronology of American Tape Recording*," Mort Fujii (Los Angeles: Audio Engineering Society, 1994), chap. 2.

112. Phillips interview, Feb. 7, 1999.

113. Teig interview, Feb. 9, 1999.

114. Ibid.

115. Teig recalled that rates were about $35–$40 per hour in the late 1950s and by 1974 were roughly $60 per hour, which he remembered as being the going rate. However, rates varied widely and were based on a number of different variables: the number of tracks used, size of the room used in a given studio, and how much a studio could charge based on its popularity. In 1974, rates in New York City studios ranged between $20 and $45 per hour for 1-track recording (monaural) to between $75 and $160 per hour for a 16-track recording. Figures based on a sampling of listings in *Billboard 1974 International Directory of Recording Studios*, June 22, 1974, RS-3-75.

116. Walter Sear, interview with author, New York City, Jan. 21, 1999.

117. Malcolm Addey, interview with author, New York City, Jan. 19, 1999.

118. At that point, there were twenty-five recording studios, and more audiovisual recording services. "Breakdown of Recording Studios in the Washington, D.C. Area," in Checchi and Company, *Feasibility of Establishing a Recording Studio in the District of Columbia: Final Report*, prepared for the Economic Development Administration, Office of Technical Assistance, US Department of Commerce (Washington: The Company, 1973), 124–125.

119. Post, *High Performance* (see Introduction, n. 20).

CHAPTER 7. CHANNELING SOUND

1. The pan pot (panoramic potentiometer) controls placement of a sound source at any point between the right and left channels. Although it first appeared on studio mixing consoles after the introduction of stereo, the "Panpot" was originally invented by engineers at the Walt Disney Studios to accommodate mixing eight optical sound tracks to three channels for the 1940 film *Fantasia*. Jesse Klapholtz, "Fantasia: Innovations in Sound," *Journal of the Audio Engineering Society* 39, no. 1/2 (Feb. 1991): 66–68, 70.

2. In his study of design engineers, Eugene Ferguson argued that their work involved not only mathematics and scientific principles, but intuitive, nonverbal, *visual* thinking, an activity that takes place in the "mind's eye" of the right brain that governs musical and artistic as well as spatial thinking. I believe an analogous *aural* thinking takes place in the "mind's ear" of recording engineers, becoming an essential component of their craft that was crucial to developing skills in editing, mixing, and envisioning sound. I use the phrase "aural architecture" to describe their work because stereo and multi-tracking enabled them to "build" and "design" sound. Ferguson, *Engineering and the Mind's Eye* (see chap. 5, n. 86).

3. Edward Tatnall Canby, "The Sound-Man Artist," *Audio* (June 1956): 44–45, 60–61; Mort Fuji, George Reklau, John McKnight, and William Miltenburg, "The Multichannel Recording for Mastering Purposes," *Journal of the Audio Engineering Society* 8 (Oct. 1960): 255.

4. Quoted in Frederick Plaut, *The Unguarded Moment: A Photographic Interpretation* (Englewood Cliffs, NJ: Prentice-Hall, 1964), unpaginated, quote appears as caption to photograph of Stokowski taken by Plaut, a Columbia Records engineer and accomplished photographer.

5. David Hall and Abner Levin, *The Disc Book* (New York: Long Player Publications, 1955), 11.

6. Norman Pickering, "A Somewhat Scientific Look at Music," *Musical America* 74 (Feb. 15, 1954): 162.

7. Jack Somer, "Popular Recording, or The Sound That Never Was," *HiFi/Stereo Review* 16 (May 1966): 54–58.

8. Claude Hall, "Long Sessions Required for 'Serious' Pop," *Billboard* (Sept. 2, 1967): 1.

9. Phil Ramone, telephone interview with author, Aug. 1, 1999.

10. Jack Somer, "Behind the Scenes in Classical Recording," *Stereo Review* (Aug. 1966): 55.

11. The recording by Sidney Bechet's One Man Band of "The Sheik" and "Blues of Bechet," featured Bechet on six different instruments "recorded by re-recording 5 times" on Apr. 18, 1941, and released as Victor 27485, Charles Delauney, *New Hot Discography* (New York: Criterion, 1948), 23.

12. Ross Parmenter, "In the Popular Field," *New York Times* (July 13, 1941).

13. Mary Alice Shaughnessy, *Les Paul: An American Original* (New York: William Morrow, 1993), 141.

14. Miller describes the difficulties they encountered doing disc overdubs in Fox, *In the Groove*, 42–43 (see Introduction, n. 23).

15. Paul claimed that "people were walking into stores to ask for Patti Page's 'Tennessee Waltz' and walking out with ours." Shaughnessy, *Les Paul*, 187.

16. Gilbert Millstein, "He Calls the Hit Tune," *Collier's* (Jan. 19, 1952): 35.

17. Ross H. Snyder, "Sel-Sync and the 'Octopus': How Came to Be the First Recorder to Minimize Successive Copying in Overdubs," *ARSC Journal* 34, no. 2 (Fall 2003): 209–213.

18. Quoted in John Tobler and Stuart Grundy, *The Record Producers* (New York: St. Martin's Press, 1982), 25. See also "Leiber and Stoller," in Fox, *In the Groove*, 156–186.

19. Kraft, *Stage to Studio*, chap. 1 (see chap. 3, n. 52). Although the number of AFM union locals had tapered off by the 1960s, national membership (including professional and amateur musicians) was 200,000 and rising in 1950, and the AFM continued to be a powerful force in the entertainment industry. See membership charts in Kraft, 203–204.

20. Kraft, *Stage to Studio*, 152–160, 162–192.

21. Tom Dowd, telephone interview with author, Mar. 17, 1999.

22. The song was "Under the Boardwalk," and Johnny Moore eventually overdubbed the vocal. In his memoirs, Jerry Wexler told a different version of the story. According to Wexler, Lewis had been found dead in his hotel room, a needle in his arm, the night before the session. Wexler tried to cancel the date, but the union would only grant a twenty-four-hour postponement, not enough time to change the key to better suit Moore's range, so he "had to sing in a register lower than his norm." Wexler and Ritz, *Rhythm and the Blues*, 137–138 (see chap. 6, n. 2). It is impossible to reconcile these conflicting recollections. According to Dowd, after the delay in obtaining permission to track, they still managed to complete the session within the time allotted, "Hey, professionals—it took them maybe a half an hour to play the four tracks, if you know what I'm saying. Sight-read the things and they were done, next, next, next—and we did the four songs that way." Tom Dowd interview, Mar. 17, 1999.

23. Mitch Miller quoted in Fox, *In the Groove*, 47.

24. "AFM Would Banish Multi-dub Records," *Billboard* (June 14, 1952): 43; "Disker's Dub-Drubbing by AFM Hurts; Como's Illness a Point," *Billboard* (Jan. 10, 1950): 11; "AFM 5-Yr. Disk Pact Pulls New Switch in Labor Policy," *Billboard* (Jan. 16, 1954): 14.

25. "AFM Moves to Stop Tracking," *Billboard* (Dec. 26, 1960): 3.

26. Artie Kaplan, telephone interview with author, Sept. 12, 2005.

27. Brooks Arthur, telephone interview with author, Apr. 30, 1999.

28. Ibid.

29. Kaplan interview, Sept. 12, 2005. Ultimately, the AFM established a separate wage scale for sessions using multi-tracking.

30. Kraft, *Stage to Studio*, 137–192; Anderson, *Making Easy Listening*, 3–47 (see Introduction, n. 5).

31. Alan Lorber, telephone interview with author, Mar. 5, 1999.

32. Ibid.

33. Ibid.

34. "Tape Co.'s Earn Mark in Industry," *Billboard* (May 6, 1967): SF-36.

35. Mike Dorrough, telephone interview with author, Aug. 19, 1996.

36. Tom Dowd interview, Mar. 17, 1999.

37. Frank Laico, telephone interview with author, Jan. 13, 1999.

38. Ibid.

39. Ibid.

40. John Palladino, telephone interview with author, Oct. 15, 1999.

41. Al Schmitt, telephone interview with author, Apr. 12, 1999.

42. Schmitt, telephone interview with author, June 8, 1999.

43. Rumble, "Roots of Rock & Roll," 30–42 (see chap. 6, n. 96).

44. Richard Gottehrer, interview with author, Cleveland, Ohio, Nov. 7, 2004.

45. Eddie Smith, interview with author, New York City, Jan. 23, 1999. Eventually, they managed to retrieve the bass and drums by experimenting with different levels of equalization.

46. Smith had a background in both electronics and music. He acquired an amateur radio license at age fourteen and studied piano and musical arranging, the latter with Tom Timothy. He played vaudeville shows at New York's Majestic Theater in the 1940s and became an arranger for bandleader Lucky Millinder. In 1951, trumpeter and songwriter Henry Glover, a friend from the Millinder days, called Smith to ask whether he wanted to replace Glover as King Records A&R director. Smith did not expect that he would soon be operating the recording console, but his background in music and electronics served him well.

47. Susan Schmidt Horning, "Engineering the Performance: Recording Engineers, Tacit Knowledge, and the Art of Controlling Sound," *Social Studies of Science* 34, no. 5 (Oct. 2004): 703–731.

48. Hall, "Long Sessions Required," 10.

49. The recording work of both the Beach Boys / Brian Wilson and The Beatles has been exhaustively analyzed and documented. For more detail on the making of these albums, see *The Making of Pet Sounds*, booklet for the CD boxed set *The Beach Boys, The Pet Sounds Sessions*, produced by Brian Wilson, EMI-Capitol, also available online at http://www.albumlinernotes .com/In_The_Studio.html (accessed July 6, 2011); Martin, *With a Little Help* (see Introduction, n. 16); and Emerick and Massey, *Here, There and Everywhere* (see Introduction, n. 16).

50. Pappalardi quoted in Hall, "Long Sessions Required," 10.

51. Tom Dowd, telephone interview with author, Mar. 17, 1999.

52. Ibid. Dowd worked closely with Halverson on several mixing sessions for *Crosby, Stills, and Nash*, released by Atlantic Records in 1969.

53. Joe Boyd, *White Bicycles: Making Music in the 1960s* (London: Serpent's Tail, 2006), 123.

54. Ibid., 124.

55. Stoddard, e-mail to author, Mar. 6, 1999.

56. Al Grundy, telephone interview with author, Jan. 8, 2000.

57. Ray Hall, telephone interview with author, Mar. 21, 1999.

58. Dowd interview, Mar. 17, 1999.

59. Doug Pomeroy, e-mail to author, June 3, 2011.

60. Stoddard, e-mail to author, Apr. 1999. The record was the Isley Brothers' "Twist and Shout," which was mastered at Bell Sound.

61. Jack Wiener, telephone interview with author, Jan. 25, 1997.

62. Stoddard e-mail, Mar. 6, 1999.

63. William Savory, interview with author, Falls Church, Virginia, Nov. 29, 1997.

64. Clair Krepps, telephone interview with author, Mar. 23, 1999. Anyone who heard Manfred Mann's "Do Wah Diddy Diddy" on the radio in the 1960s will probably recall that it stood out as being unusually loud.

65. Grundy, telephone interview with author, Jan. 8, 2000.

66. Bob MacLeod, "Disc Mastering . . . Almost the Point of No Return," *Recording Engineer/Producer* 1, no. 2 (June/July 1970): 21–25.

67. Grundy interview, Jan. 8, 2000.

68. Stoddard interview, Mar. 11, 1999.

69. George Alexandrovich, "The Recording Studio," *db: The Sound Engineering Magazine* 1 (Nov. 1967): 16. For more on the art of disc cutting, see Schmidt Horning, "Engineering the Performance," 716–721.

70. Scott Sutherland, "In a Maine Outpost, a Master of Sound Has Built a Mecca," *New York Times* (Mar. 29, 1998).

71. Reig, *Reminiscing in Tempo*, 71 (see chap. 4, n. 110).

72. John Hammond, with Irving Townsend, *John Hammond on Record: An Autobiography* (New York: Summit Books, 1977), 381–382.

73. Quoted in Blair Jackson, "Classic Tracks: Neil Diamond's 'Kentucky Woman,'" *Mix* (Oct. 1996).

74. Malcolm Addey, interview with author, New York City, Jan. 19, 1999.

75. Pomeroy, e-mail to author, June 19, 2011.

76. Palladino interview, Oct. 15, 1999.

77. George Alexandrovich, "The Audio Engineer's Handbook: Multi-channel Recording—Why?," *db: The Sound Engineering Magazine* (Apr. 1969): 4–6.

78. Robert Auld, "The Art of Recording the Big Band, part 2," www.tiac.net/users/auld wrks/bbrec2.htm (accessed Sept. 20, 1999).

79. Dorrough interview, Aug. 19, 1996.

80. Ibid.

81. The Blackwood's film footage was featured in Charlotte Zwerin and Bruce Ricker (producers), *Thelonious Monk: Straight, No Chaser* (Hollywood: Warner Bros., 1988). Thanks to engineer Jim Anderson for calling my attention to this film. All quotations in this section, unless otherwise noted, are from the film.

82. An interesting study on the importance of spontaneity in American art, literature, and music after World War II is Daniel Belgrad, *The Culture of Spontaneity: Improvisation and the Arts in Postwar America* (Chicago: University of Chicago Press, 1998).

83. Kahn, *Kind of Blue* (see chap. 4, n. 65).

84. Bill Evans, *Conversations with Myself*, Verve V-8526 (1964), engineered by Ray Hall.

85. Pete Welding, "Modern Electronics in the Studio," *Down Beat* (Jan. 23, 1969): 18–19.

86. Simon, *The Big Bands*, 50 (see chap. 4, n. 32).

87. Count Basie and His Orchestra, *Basie's Beatle Bag*, Verve V-8659 (1966).

88. Chuck Beale of the Paupers, quoted in Michael Lydon, "Monterey Pops! An International Pop Festival," in *The Sound and the Fury: A Rock's Backpages Reader, 40 Years of Classic Rock Journalism*, ed. Barney Hoskyns (New York: Bloomsbury, 2003), 257.

89. "Revving up the Rock Revolution," *Tape Recording* 16 (Aug. 1969): 8 (emphasis added). This article was reprinted from RCA's *Electronic Age*.

90. Mick Jagger and Paul McCartney quoted in "Revving up the Rock Revolution," 8.

91. Emerick and Massey, *Here, There and Everywhere*, 132, 8. In Mark Lewisohn, *The Beatles: Recording Sessions* (New York: Harmony Books, 1988), 70, EMI tape operator Jerry Boys observed that "with another producer and another engineer things would have turned out quite differently," and Lewisohn noted that Emerick's youth and inexperience, the absence of "preconceived or irreversible techniques," were an asset.

92. Advertisement for B&K Instruments, Inc., in *db: The Sound Engineering Magazine* 2 (Sept. 1968): back page.

93. Stoddard interview, Mar. 11, 1999.

94. David Thomas, interview with author, Cleveland, Ohio, Aug. 12, 1997.

95. Jon Landau, "Engineering: What You Hear Is What You Get," in *It's Too Late to Stop Now: A Rock and Roll Journal* (San Francisco: Straight Arrow Books, 1972), 137.

96. Mitch Miller, interview with author, New York City, Jan. 21, 1999.

97. Wilma Cozart Fine, telephone interview with author, May 5, 1999.

98. Wilson, *Wouldn't It Be Nice* 77 (see Introduction, n. 2); Mark Ribowsky, *He's a Rebel: Phil Spector, Rock and Roll's Legendary Producer* (New York: Cooper Square Press, 2000); Dave Thompson, *Wall of Pain: The Biography of Phil Spector* (London: Sanctuary, 2003).

99. Paul Grein, *Capitol Records: Fiftieth Anniversary, 1942–1992* (Hollywood, CA: Capitol Records, 1992), 98; Wilson, *Wouldn't It Be Nice*, 79ff., 147.

100. Palladino interview, Oct. 15, 1999.

101. John Townley, "Laid-Back Multitrack: The Downtown Birth of Modern Recording Studio Style," http://www.astrococktail.com/Apostolic.html (accessed July 2009). This personal reminiscence by the founder of Apostolic Recording Studio in Manhattan was originally published in *EQ Magazine* in 1988.

102. Pomeroy e-mail, June 19, 2011.

103. Quoted in Grein, *Capitol Records*, 98.

104. Robertson and Helm statements from the documentary by Terry Shand and Geoff Kempin (producers), *The Band*. The Band, Classic Album Series (London: Eagle Rock Entertainment, 2005) DVD. Greil Marcus neatly summed up how the Band fit in the context of their time: "Against a cult of youth they felt for a continuity of generations. . . . Against the pop scene, all flux and novelty, they set themselves: a band with years behind it, and meant to last." Greil Marcus, "The Band: Pilgrim's Progress," in *Mystery Train: Images of America in Rock 'n' Roll Music*, 4th rev. ed. (New York: Plume, 1997), 39–64, quote on p. 45.

105. The Band, *The Band*, Capitol STAO-132. The original album release gave no production credit on the album jacket, but the compact disc release in 1987 credits the Band as producer and John Simon, who engineered as well as played various instruments, as co-producer, Capitol CDP 546493.

106. Eliot Tiegel, "Unions' Engineer Stipulation Irks Independent Producers," *Billboard* (Apr. 1, 1972): 1.

107. Pomeroy e-mail, June 19, 2011; Tiegel, "Unions' Engineer Stipulation Irks Independent Producers," 1, 14; Bill Williams, "Col Settles Studio Beef with Nashville IBEW," *Billboard* (Apr. 1, 1972): 14.

108. Blair Jackson, "The Mix Interview: Roy Halee," *Mix* (Oct. 2001): 30–42, quote p. 36.

109. Don White, interview with author, Akron, Ohio, Jan. 23, 1996.

110. Clair Krepps, telephone interview with author, Mar. 31, 1999.

111. Ann Geracimos, "A Record Producer Is a Psychoanalyst with Rhythm," *New York Times Magazine* (Sept. 29, 1968): 32–33.

112. Hank Fox, "Disadvantages of a Large Studio Operation: No Buck Passing," *Billboard* (Feb. 24, 1968): B-3.

113. Lyle Ritz and Mark Linett quotes in *The Making of Pet Sounds*, 126-page booklet accompanying the Beach Boys, *The Pet Sounds Sessions*, 4-CD Box Set (Capitol Records, 1996).

114. Geoff Emerick and Howard Massey, "From Songwriters to Soundmen" interview series, Rock and Roll Hall of Fame and Museum, Cleveland, Ohio, Apr. 19, 2006; Emerick also refers to the EMI hierarchy throughout his memoir, Emerick and Masey, *Here, There and Everywhere*.

115. Jerry Boys, Abbey Road tape operator, quoted in Lewisohn, *The Beatles: Recording Sessions*, 114.

116. Bruce Botnick quoted in Jac Holzman and Gavan Dawes, *Follow the Music: The Life and High Times of Elektra Records in the Great Years of American Popular Culture* (Santa Monica, CA: FirstMedia Books, 1998), 204.

117. Paul Laurence, "A Production Analysis of the Doors Strange Days: An In-Depth Interview with Paul Rothchild, Bruce Botnick, Robby Krieger," *Recording Engineer/Producer* 6, no. 4 (Aug. 1975): 27–35. Quote on p. 35.

118. "Technical Strides May Not Be in Step with the Artist: Martin," *Billboard* (May 10, 1969): 61; "A Metamorphosis Ahead for Electronic Music: Pappalardi," *Billboard* (May 10, 1969): 61–62.

119. Quoted in Jack McDonough, "The Dawn of Rock," *Mix* 6 (Aug. 1982): 34.

120. Artie Kaplan, telephone interview with author, Sept. 12, 2005.

121. The session musicians who played on Phil Spector–produced records, then on the Beach Boys records and many television and film soundtracks, acquired the name because older session players believed "these dungareed, Ray-Banned upstarts were 'wrecking' the business." Chuck Cristafuli, "Endless Summers: The Studio Heavyweights behind the Beach Boys Sound," *Guitar Player* 27, no. 12 (Dec. 1993): 91–98, quote on p. 92. Also see the *The Wrecking Crew* documentary trailer at http://www.wreckingcrewfilm.com (accessed June 9, 2011).

CONCLUSION

1. *Billboard 1970 International Directory of Recording Studios* (May 9, 1970), 6.

2. Hall, "Long Sessions Required," 10 (see chap. 7, n. 8).

3. Tony Glover quoted in Jac Holzman and Gavan Dawes, *Follow the Music: The Life and High Times of Elektra Records in the Great Years of American Popular Culture* (Santa Monica, CA: FirstMedia Books, 1998), 77–78.

4. Holzman and Dawes, *Follow the Music*, 188.

5. Eliot Tiegel, "Studios: Payoffs in Nashville & Col Closes in LA," *Billboard* (Sept. 23, 1972): 1.

6. *Billboard 1974 International Directory of Recording Studios* (June 22, 1974), RS-45.

7. *Recording Engineer/Producer* 1, no. 4 (Nov./Dec. 1970): inside cover.

8. Robert Greenfield, *Exile on Main Street: A Season in Hell with the Rolling Stones* (New York: Da Capo Press, 2006).

9. *The Making of the Album*, booklet accompanying the original LP, the Beach Boys, *Holland* (Brother/Reprise, 1972). The album jacket summarizes the equipment used: "Recorded in Baambrugge, The Netherlands, using a new Clover Systems custom quadraphonic console;

30 input channels with 16 output busses; 1000 position patch bay; 20 Dolby noise reduction units. Among microphones used: Neumann, Sony, AKG, Shure and EV. Custom monitoring systems utilized ME-4 and JBL-4310 speakers. All equipment was designed specially [*sic*] for this project by Brother Records in Los Angeles, then flown to Holland for this recording. Recorded in stereophonic sound." The booklet offers an extended essay by Shelley Benoit, detailing the entire undertaking, from travel and housing for the band and families, to additional technology and the complications of installing it and getting it running.

10. Mick Farren and George Snow, *The Rock 'n' Roll Circus: The Illustrated Rock Concert* (New York: A&W Visual Library, 1978), 52.

11. Ampex advertisement, *db: The Sound Engineering Magazine* 2, no. 6 (June 1968): 18–19.

12. Figures cited in Ann Geracimos, "A Record Producer Is a Psychoanalyst with Rhythm," *New York Times Magazine* (Sept. 29, 1968): 32.

13. "Pop Records: Moguls, Money & Monsters," *Time* (Feb. 12, 1973), reprinted in Frederic Rissover and David C. Birch, eds., *Mass Media and the Popular Arts*, 2nd ed. (New York: McGraw Hill, 1977), 416–426.

14. Quadrophonic sound, or 4-channel stereo, was introduced to consumers in 1969 and heralded as the next big thing in audio but did not catch on because of a number of factors, from conflict over the technique of recording and playback to a weak economic climate. James Cunningham, "Tetraphonic Sound," *db: The Sound Engineering Magazine* (Dec. 1969): 21–23; John Eargle, "Four-Channel Stereo," *db: The Sound Engineering Magazine* (Aug. 1970): 23–25; John M. Woram, "The Sync Track," *db: The Sound Engineering Magazine* (Sept. 1971): 10–11; The Editors of *Electronics*, *an Age of Innovation: The World of Electronics, 1930–2000* (New York: McGraw-Hill, 1981), 161–162.

15. "Witch Doctors of the Jingle Jungle," *Recording Engineer/Producer* 2, no. 5 (Sept./Oct. 1971): 15–21. Anne Phillips did vocal arrangements for Revlon and Pepsi commercials in the styles of the popular recording artists who sang them, including the Shangri-Las, the Four Tops, Wilson Pickett, and Jackie DeShannon, among others. Anne Phillips, interview with author, Feb. 7, 1999. Audio excerpts of these commercials can be heard on Phillips's website, www.annephillips.com/history_1.html#BringingRockIntoCommercials. See also Thomas Frank, *The Conquest of Cool: Business Culture, Counterculture, and the Rise of Hip Consumerism* (Chicago: University of Chicago Press, 1997).

16. "AES News," *Journal of the Audio Engineering Society* 7, no. 1 (Jan. 1959): 57; "Always a New Audience for Audio," *Journal of the Audio Engineering Society* 22 (Oct. 1974): 706–716.

17. "Independent Recording Studios—a *db* Forum," *db: The Sound Engineering Magazine* 5, no. 11 (Nov. 1971): 21–24.

18. Pete Senoff, "The Trend toward Self-Production," *Recording Engineer/Producer* 1, no. 1 (Apr./May 1970): 12, 14.

19. DeWitt F. Morris, "The Audio Engineer—circa 1977: What Does He (or She) Do?," *Journal of the Audio Engineering Society* 25 (Oct./Nov. 1977): 864–872.

20. Doug Pomeroy, "*db* Letters," *db: The Sound Engineering Magazine* (Oct. 1978): 2, 4.

21. Carson C. Taylor, "Practical and Aesthetic Microphone Technique for Recording or Broadcasting Symphonic Music," *Journal of the Audio Engineering Society* 27 (Sept. 1979): 677, 679.

22. Kealy, "The Real Rock Revolution," (see Introduction, n. 25).

23. Steve Gursky, telephone interview with author, June 3, 2003.

24. Record producer John Hammond was an outspoken critic of the overuse of echo. John Hammond, "Jazz Records Deficient in Tone-Quality," *New York Times* (Nov. 22, 1953): 50; and "The Talk of the Town: Natural Sound," *New Yorker* (July 17, 1954): 17–18.

25. William Savory, telephone interview with author, Oct. 21, 1999.

26. Clair Krepps, telephone interview with author, Mar. 23, 1999.

27. Louis Kaufman interview in John Harvith and Susan Edwards Harvith, eds., *Edison, Musicians, and the Phonograph: A Century in Retrospect* (Westport, CT: Greenwood Press, 1987), 117.

28. Ibid.

29. Bert Whyte, "Behind the Scenes: The Professional Viewpoint," *Audio* (Nov. 1968): 10.

30. Thompson, *Soundscape of Modernity* (see chap. 1, n. 39); Karin Bijsterveld, *Mechanical Sound: Technology, Culture, and Public Problems of Noise in the Twentieth Century* (Cambridge, MA: MIT Press, 2008).

31. Daniel J. Boorstin, *The Image: A Guide to Pseudo-events in America* (New York: Harper Colophon Books, 1961); Miles Orvell, *The Real Thing: Imitation and Authenticity in American Culture, 1880–1940* (Chapel Hill: University of North Carolina Press, 1989); Hillel Schwartz, *The Culture of the Copy: Striking Likenesses, Unreasonable Facsimiles* (New York: Zone Books, 1996).

32. Michael Hicks, *Sixties Rock: Garage, Psychedelic, and Other Satisfactions* (Urbana: University of Illinois Press, 1999); Art Thompson, *The Stompbox* (San Francisco: Miller Freeman Books, 1997).

33. Edward Rothstein, "Pursuing 'Real' Sound, with Artifice as the Ideal," *New York Times* (Mar. 11, 2000): A17.

34. Jon Pareles, "Disco Lives! Actually, It Never Died," *New York Times* (Oct. 17, 1999): sec. 2.

35. Harvith and Harvith, *Edison, Musicians, and the Phonograph*, 382.

36. Quoted in Ibid., 399.

37. Mooney, "Songs, Singers and Society," 221–232 (see chap. 2, n. 57); Kenney, *Recorded Music in American Life* (see chap. 1, n. 57); Simon Frith, "Technology and Authority," in *Performing Rites: On the Value of Popular Music* (Cambridge, MA: Harvard University Press, 1996).

38. One example is Les Brown and Vic Schoen, *Double Exposure: Suite for Two Bands*, Medallion MS-7523 (1959), engineered by Bob Fine and recorded in the ballroom of Fine Recording Studios. The genesis of the idea for this recording, a live performance of two big bands set up on opposite sides of the studio with a shared rhythm section, as well as the preparation, studio setup, and engineering, is described in detail on the album jacket. Fine's skill in capturing such a dynamic live performance was an extraordinary feat of engineering. I thank Bill Stoddard for telling me about this record. Other examples, also engineered by Fine, are Mercury Record's Perfect Presence Sound and f:35ᴰ recordings, so named because they were recorded on 35 mm magnetic film, a sound recording technique Fine developed.

39. Ad: "Hohner Echolette," *Guitar Player* 1, no. 2 (1967): 37.

40. William Ivey, "Recordings and the Audience for the Regional and Ethnic Musics of the United States," in *The Phonograph and Our Musical Life. Proceedings of a Centennial Conference*

7–10 December 1977, ed. H. Wiley Hitchcock, I.S.A.M. Monographs: No. 14 (Brooklyn, NY: Institute for Studies in American Music, 1980): 7–13; quotes from pp. 8, 10.

41. Merritt Roe Smith and Leo Marx, eds., *Does Technology Drive History? The Dilemma of Technological Determinism* (Cambridge, MA: MIT Press, 1994). Oudshoorn and Pinch, *How Users Matter* (see Introduction, n. 20).

42. "The World of Sound Will Be Heard at the New Hi-Fi Show," *New York Times* (Nov. 6, 1977).

43. The Aphex Aural Exciter, used in recording live concerts, TV, and movies, added presence and excited certain aural functions, creating a spatial illusion. It did not introduce new audio stimuli, only "heightened" and "stimulated" hearing by a system of encoding, which decodes only through hearing using the sum and difference principle inherent in human hearing. Howard Cummings, "Aphex Processed," *Recording Engineer/Producer* 8, no. 4 (1977): 67–73.

44. The Eagles, *The Long Run*, Asylum Records (1969). I thank Bruce Hensal for bringing this to my attention.

45. Alice Echols, *Hot Stuff: Disco and the Remaking of American Culture* (New York: W. W. Norton, 2010).

46. Tom Lubin, "The Sounds of Science: The Development of the Recording Studio as Instrument," *NARAS Journal* 7, no. 1 (Summer/Fall 1996): 41–99, quote p. 42.

47. Dale Manquen (producer), "A Chronology of American Tape Recording: Chapter Three" (Los Angeles: Audio Engineering Society, 1994).

48. Phil Upchurch quoted in Willie Dixon with Don Snowden, *I Am the Blues: The Willie Dixon Story* (New York: Da Capo Press, 1989), 106.

49. Anna Case quoted in Harvith and Harvith, *Edison, Musicians, and the Phonograph*, 13.

50. In 2007, a research team deployed modern technology to reproduce for the first time the recordings made in 1860 by Édouard-Léon Scott de Martinville, who etched soundwaves onto paper with blackened smoke. Jody Rosen, "Researchers Play Tune Recorded before Edison," *New York Times* (Mar. 7, 2008), and see www.firstsounds.org. In 2009, the National Jazz Museum in Harlem acquired William Savory's collection of air-check recordings from the 1930s and 1940s and through a laborious process of disc cleaning and digital restoration, unique performances by major jazz artists like Billie Holiday and Coleman Hawkins have been recovered. Larry Rohter, "Museum Acquires Storied Trove of Performances by Jazz Greats," *New York Times* (Aug. 16, 2010), and see www.jazzmuseuminharlem.org/savory/index.php.

Essay on Sources

When I began my research in 1995, there were only a few studies in the history of the phonograph, fewer still in the history of sound recording, and nothing that traced the history of studio recording or sound engineering from the acoustic period through to the advent of digital sound. Edward R. Kealy's dissertation, "The Real Rock Revolution: Sound Mixers, Social Inequality, and the Aesthetics of Popular Music Production" (PhD diss., Northwestern University, 1974), was the only scholarly study of studios and engineers, but it had limitations. Since that time, literature on sound recording and reproduction has exploded. This bibliographic essay does not include every primary or secondary source used in the preparation of *Chasing Sound*. Rather it adds to the sources documented in the chapter notes and is intended to point readers to the range of sources consulted and to significant works in the many disciplines that informed my thinking on the subject.

PRIMARY SOURCES

The value of oral history for historians of twentieth-century topics cannot be overstated. Much of the type of evidence historians tend to rely on—textual sources, archives of documents, company files, logbooks and the like for the recording industry are rare or not publicly accessible. While some companies such as EMI in the United Kingdom maintain an archive, the recording industry grew so rapidly in the post–World War era, with companies changing hands, coming and going, that documenting the activities of recording studios was apparently not a priority. Who cared about what engineers were doing? The music was all that mattered. As Eric Clapton reminisced in the wonderful documentary *Tom Dowd and the Language of Music*, "To be perfectly frank, I wasn't interested in people like that." Fortunately, that sentiment has changed and a great deal of interest in what recording engineers do, and in the history of sound recording and its impact on listening has grown in recent decades, with surveys, monographs, memoirs, beautifully illustrated coffee-table books, documentary films, and websites devoted to the era of analog sound. Still, archives devoted to the history of sound recording and audio engineering remain scarce. The major "archive" used in the creation of this book is the collection of oral interviews I began under National Science Foundation Dissertation Improvement Grant No. 9711127 and continued after completing my dissertation. Comprising some 135 hours and more than a thousand transcribed pages, the interviews form the backbone of the research for this book and led me to many of the documents and publications that helped back up and expand on the oral testimony.

The most substantial interviews, the tapes and transcripts of which remain in my possession, were with Malcolm Addey, Manny Albam, Jim Anderson, Brooks Arthur, George Avakian, Thomas Boddie, Louise Boddie, Frank Bruno, Dave Budin, Dick Dale, Jim Davison, Mike Dorrough, Tom Dowd, Steve Epstein, Wilma Cozart Fine, Jim Fox, Bruce Gigax, Richard Gottehrer, Al Grundy, Steve Gursky, Ray Hall, Gloria Busse Hamann, Ken Hamann, Paul Hamann, Pete Hammar, Sheldon Henderschott, Bruce Hensal, Irv Joel, Leslie Ann Jones, Artie Kaplan, Clair Krepps, Frank Laico, Alan Lorber, Wayne Mack, Vlad Maleckar, Mitch Miller, Don Ososke, John Palladino, Les Paul, Anne Phillips, Don Plunkett, Doug Pomeroy, Phil Ramone, Jack Renner, David Sarser, Bill Savory, Phil Schaap, Al Schmitt, Kay Schneider, Howard Scott, Walter Sear, Art Shifrin, Eddie Smith, Ross Snyder, Bill Stoddard, Dave Teig, Brian Telzrow, David Thomas, Mike Thorne, Don White, Jack Wiener, and John Woram.

In addition, those listed below offered invaluable guidance and helpful information through telephone or personal conversation, e-mail, and other correspondence: Larry Appelbaum, Karl Otto Bäder, Matt Barton, George Bowley, Ed Cherney, Douglas Collar, Adrian Cosentini, Earle H. Davis, Hans O. Dietze, John Dixon, Tom Fine, Bud Ford, Dan Gaydos, Kevin Gore, Mike Gray, Joyce Halasa, Laura Helper, Shelley Herman, Bruce Leslie, Juan Marquez, Jack Miller, Tony Noe, Jim Prohaska, Wayne Puntel, Don Ososke, Tim Ramsey, Jay Ranellucci, John Richmond, John Rumble, David Sanjek, Howard Sanner, Gene Savory, Tony Scherman, Steve Smolian, Russ Wapensky, Don Was, and Seth Winner. Only two of my requests were denied: Lee Hazlewood wrote, "I don't do interviews!," and Rudy Van Gelder politely declined citing a reluctance to talk about proprietary technology he still uses. Who said secrecy in the recording industry died?

Traditional archives and libraries that proved useful included the Bowling Green State University Browne Popular Culture Library and the Music Library and Sound Recording Archive; the Library of Congress, Motion Picture, Broadcasting and Recorded Sound Division, Recorded Sound Section; the New York Public Library, Rodgers and Hammerstein Archives of Recorded Sound; the Paley Center for Media in New York, and the Smithsonian Institution's National Museum of American History, Archives Center, which has the Jazz Oral History Project Collection and the Milt Gabler Papers and Joel Dorn papers. Thanks to Mark Clark, who shared important documents he acquired during his research at the Ampex Museum in Redwood City, California, I had copies of Ampex archival material then locked in storage. Fortunately, the Ampex Corporation records, collection M1230, are now held at the Stanford Silicon Valley Archive and a preliminary guide to the collection is currently available online. In the 1990s, former Capitol Records engineer Irv Joel and other members of the Audio Engineering Society Historical Committee began conducting oral interviews with important individuals in audio engineering, and these are accessible through the AES website, albeit for a fee. However, in addition to society and historical committee activities, the AESHC website also offers a wealth of free online sources pertaining to audio engineering history such as historic articles, transcripts of lectures, technical papers, links to archives and museums, and more. See www.aes.org/aeshc.

I also relied on published oral histories, many of which are unfortunately out of print, including Ted Fox, *In the Groove: The People behind the Music* (New York: St. Martin's Press, 1986); Ira Gitler, *Swing to Bop: An Oral History of the Transition in Jazz in the 1940s* (New York: Oxford University Press, 1985); John Harvith and Susan Edwards Harvith, eds., *Edison, Musicians, and the Phonograph: A Century in Retrospect* (New York: Greenwood Press, 1987); Sam Moore and Dave Marsh, *For the Record: Sam and Dave: An Oral History* (New York: Avalon Books, 1998); and John Tobler and Stuart Grundy, *The Record Producers* (New York: St. Martin's Press, 1982). Still available is the entertaining oral history by Jac Holzman and Gavan Daws, *Follow the Music: The Life and High Times of Elektra Records in the Great Years of American Popular Culture* (Santa Monica, CA: FirstMedia Books, 1998).

Documentation of the day-to-day work of recording engineers is scarce. Only one engineer I interviewed recalled keeping a log of his studio setups, which he did not save. Session track sheets, if available, tell only the most rudimentary details of recording sessions. In some cases, even these have been found to disagree with the final lineup of musicians on a session. Raymond Sooy, "Memoirs of My Recording and Traveling Experiences for the Victor Talking Machine Company," and Harry O. Sooy, "Memoir of My Career at Victor Talking Machine Company, 1989–1925," are among the few existing logbook-type sources that document the day-to-day work routine of early recordists. While there are discrepancies between the two memoirs on the dates of certain events, they provide an invaluable view into general information about the nature of their work, the environment of the Victor operations, interactions with co-workers and artists, as well as technical information. Digitized copies of both memoirs as well as the corporate history to 1930 by B. L. Aldridge, *The Victor Talking Machine Company*, edited by Frederic Bayh (RCA Sales Corporation, 1964) can be found on the David Sarnoff Library website (www.davidsarnoff .org/onlinetexts.html). The Sarnoff Library closed in 2009 and transferred its archives to the Hagley Library in Wilmington, Delaware, but the website continues to be maintained.

Trade literature, city directories, newspapers, and other professional and popular periodicals shed light on technical developments in recording, the shifting locations of studios, and information about the recording industry. *Talking Machine World* and *Scientific American* proved useful for the period from the late nineteenth century through the mid-1920s, when radio and electronics periodicals became the richest sources of technical information on recording. In 1947, *Radio* changed its name and focus to *Audio Engineering*, and after the *Journal of the Audio Engineering Society* began publication, it became the more consumer-oriented *Audio* in 1954. Also useful were the *Proceedings of the I.R.E.*, *Radio Annual*, *Radio-Craft*, *Radio News*, *Radio & Television News*, *Billboard*, *Broadcasting Yearbook*, *Business Week*, *Communications*, and *Electronics*. After World War II, audiophile literature flourishes. Beginning in the early 1950s, *High Fidelity*, *Hi-Fi/Stereo Review*, regular columns in the *New York Times* and technical record reviews in *Saturday Review* offer a view into the serious listener. By the late 1960s, the boom in independent recording studios gave rise to semiprofessional periodicals such as *db: The Sound Engineering Magazine*, *Recording engineer/producer*, and by the late 1970s, *Mix Magazine*. Researchers should also consult *The Absolute Sound* and the *ARSC Journal*, which offers in-depth and up-to-date scholarship on recording history, as well as *Tape Op: The Creative Music Recording Magazine*, which often features interviews with recording professionals.

Visual documentation of recording sessions before the era of music videos ushered in by MTV is rare. I was fortunate to gain access to some unique video recordings from my interview subjects. Bill Stoddard shared his 16-mm "home movie" of Duke Ellington and Joni James sessions at Universal Recording in Chicago around 1957, and Howard Scott shared the Columbia Records promotional film *A View from the Microphone: A Day at Columbia Records,* and *Glenn Gould: Off the Record / On the Record* (Toronto: CBC, 1959). The Gould film is commercially available, as are other films, through the Audio Engineering Society, including *An Afternoon with John T. "Jack" Mullin* (Los Angeles: Webster Communications, 1989), and Dale Manquen (producer), *A Chronology of American Tape Recording* (Los Angeles: Audio Engineering Society, 1994), a multivolume videorecording of presentations by pioneers in sound recording, including engineers and inventors such as Marvin Camras, Ray Dolby, and Thomas Stockham. RCA Victor produced marketing films in the 1940s and 1950s that visually document the complex process of recording, mastering, and manufacturing records, and the technicalities of sound reproduction. Three of the best are *Command Performance* (1942), which shows how shellac 78s were made, from studio to manufacturing; *The Sound and the Story* (1956) shows how vinyl records were made; and *Living Stereo* (1958), which explains the "miracle of stereo" through graphics and sound demonstrations. These are viewable at the Prelinger Archive, another rich online source, available at www.archive.org. A wealth of primary sources pertaining to early sound recording are available through the Library of Congress American Memory Collection, in particular "Inventing Entertainment: The Motion Picture and Sound Recordings of the Edison Companies," and "Emile Berliner and the Birth of the Recording Industry," where researchers can read documents, view films, and listen to recordings from the acoustical era. See http://memory.loc.gov/ammem. In the IEEE Oral History Collections, I found useful interviews with RCA engineers Harry F. Olson and Irving Wolff. The collection includes interviews dating back to the 1960s, grouped into collections and searchable, available at the IEEE Global History Network at www.ieeeghn.org.

SECONDARY SOURCES

Several dissertations and theses proved to be rich resources, in particular Michael Jay Biel's exhaustively researched "The Making and Use of Recordings in Broadcasting Before 1936" (PhD diss., Northwestern University, 1977); Rodney W. Baum, "Radio Music Transcription Services: Their Development and Decline" (Master's thesis, Bowling Green State University, 1964); Mark Henry Clark, "The Magnetic Recording Industry, 1878–1960: An International Study in Business and Technological History" (PhD diss., University of Delaware, 1992); Douglas E. Collar, "'Hello Posterity': The Life and Times of G. Robert Vincent, Founder of the National Voice Library" (PhD diss., Michigan State University, 1988); Richard A. Gradone, "Enoch Light (1905–1978): His Contributions to the Music Recording Industry" (PhD diss., New York University, 1980); Alexander Boyden Magoun, "Shaping the Sound of Music: The Evolution of the Phonograph Record,

1877–1950" (PhD diss., University of Maryland, 2000); Thomas Gregory Porcello, "Sonic Artistry: Music, Discourse, and Technology in the Sound Recording Studio" (PhD diss., University of Texas at Austin, 1996); William Francis Shea, "The Role and Function of Technology in American Popular Music: 1945–1964" (PhD diss., University of Michigan, 1990); and John W. Rumble, "Fred Rose and the Development of the Nashville Music Industry, 1942–1954" (PhD diss., Vanderbilt University, 1980).

Discographies offer useful data, particularly if they include historical background on the collection. Richard S. Sears, *V-Discs: A History and Discography*, and *V-Discs: First Supplement* (Westport, CT: Greenwood Press, 1980, 1986) are the best source on the V-Disc program that ran from October 1943 to May 1949, during war, peace, and two different recording bans. Ross Laird, *Brunswick Records: A Discography of Recordings, 1916–1931*, 4 vols. (Westport CT: Greenwood Press, 2001), highlights the important contributions of a label so long overshadowed by Victor and Edison. Similarly, Tim Brooks with Merle Sprinzen, *Little Wonder Records and Bubble Books: An Illustrated History and Discography* (Denver, CO: Mainspring Press, 2011), documents other lesser-known labels notable for their affordable records in the early 1900s; and Richard K. Spottswood, *Ethnic Music on Records: A Discography of Ethnic Recordings Produced in the United States, 1893–1942*, 7 vols. (Urbana: University of Illinois Press, 1990), catalogs the popular music of America's many ethnic groups from the beginning of the recording industry through the mid-twentieth century. There is a vast literature aimed at serious record collectors that chronicles in detail and often lavish illustrations the histories of particular labels and eras of recording, such as Allan Sutton, *Recording the Twenties: The Evolution of the American Recording Industry 1920–1929* (Denver, CO: Mainspring Press, 2008). Another essential, but out-of-print source is Cynthia A. Hoover, *Music Machines—American Style: A Catalogue of the Exhibition* (Washington, DC: Smithsonian Institution Press, 1971).

Biographies, memoirs, and eyewitness accounts offered rich insight and enjoyable reading. Ralph Berton's *Remembering Bix: A Memoir of the Jazz Age* (1974; repr., New York: Da Capo Press, 2000) is an eyewitness account of the world of jazz musicians in the 1920s and 1930s. Hoagy Carmichael, *The Stardust Road* (New York: Rinehart and Company, 1946) is a brief poetic memoir by the composer and actor about his early days. Benny Goodman and Irving Kolodin, *The Kingdom of Swing* (1939; repr., New York: Frederick Ungar Publishing Company, 1961); and John Hammond, with Irving Townsend, *John Hammond on Record: An Autobiography* (New York: Summit Books, 1977) are two memoirs that chronicle the growth years of the recording industry by two of its biggest names. The somewhat stream-of-consciousness style memoir by Perry Bradford, *Born with the Blues* (New York: Oak Publications, 1965), and Eddie Condon with Thomas Sugrue, *We Called It Music: A Generation of Jazz* (New York: Da Capo Press, 1992), are colorful and rich musician autobiographies as is Bob Dylan, *Chronicles, Volume One* (New York: Simon & Schuster, 2004) which offers insight into Dylan's development as an artist and fascinating details about some of his later recording sessions. Joe Boyd, *White Bicycles: Making Music in the 1960s* (London: Serpent's Tail, 2006) is a great read by one who was involved in the rock music world from its blues roots to the psychedelic era, and Tony Scherman, *Backbeat: Earl Palmer's Story* (Washington, DC: Smithsonian Institution Press, 1999) tells the story of one of the most important session drummers of all time. Jerry Wexler and

David Ritz, *Rhythm and the Blues: A Life in American Music* (New York: Knopf, 1993) and Brian Wilson and Todd Gold, *Wouldn't It Be Nice: My Own Story* (New York: HarperCollins, 1991) are revealing memoirs with rich detail about recording.

The history of radio and recording since the 1920s are inseparable. Important works are Hugh G. J. Aitken, *Syntony and Spark: The Origins of Radio* (New York: John Wiley & Sons, 1976) and *The Continuous Wave: Technology and American Radio, 1900–1932* (Princeton, NJ: Princeton University Press, 1985); Eric Barnouw, *The Tower of Babel: A History of Broadcasting in the United States to 1933*, vol. 1 (New York: Oxford University Press, 1966) and *The Golden Web: A History of Broadcasting in the United States 1933 to 1953*, vol. 2 (New York: Oxford University Press, 1968). Susan J. Douglas, *Inventing American Broadcasting, 1899–1922* (Baltimore: Johns Hopkins University Press, 1987), *Listening In: Radio and the American Imagination* (Minneapolis: University of Minnesota Press, 2004), and *Where the Girls Are: Growing Up Female with the Mass Media* (New York: Times Books, 1994) are essential for understanding the evolution of radio and the culture of listening to music over the airwaves, whereas Susan Smulyan, *Selling Radio: The Commercialization of American Broadcasting, 1920–1934* (Washington, DC: Smithsonian Institution Press, 1994) chronicles how broadcasting became a commercial medium. Also useful are J. Fred MacDonald, *Don't Touch That Dial: Radio Programming in American Life, 1920–1960* (Chicago: Nelson-Hall, 1979); Andrew F. Inglis, *Behind the Tube: A History of Broadcasting Technology and Business* (Boston: Focal Press, 1990); and Christopher H. Sterling and John M. Kittross, *Stay Tuned: A History of American Broadcasting*, 3rd ed. (New York: Routledge, 2001) is the standard text.

Record company histories helped to fill out information gleaned in my interviews, especially Charlie Gillett, *Making Tracks: Atlantic Records and the Growth of a Multi-Billion-Dollar Industry* (New York: E. P. Dutton, 1974), and *The Sound of the City: The Rise of Rock and Roll*, rev. ed. (London: Souvenir Press, 1983); Paul Grein, *Capitol Records: Fiftieth Anniversary, 1942–1992* (Hollywood, CA: Capitol Records, 1992); Michael "Doc Rock" Kelly, *Liberty Records: A History of the Recording Company and Its Stars, 1955–1971* (Jefferson, NC: McFarland & Company, 1993); Donald J. Mabry, "The Rise and Fall of Ace Records: A Case Study in the Independent Record Business," *Business History Review* 64 (Autumn 1990): 411–450; and Gary Marmorstein, *The Label: The Story of Columbia Records* (New York: Thunder's Mouth Press, 2007) are a few examples. Record industry surveys essential for anyone studying this period are Russell Sanjek, *American Popular Music and Its Business: The First Four Hundred Years: From 1900 to 1984*, vol. 3 (New York: Oxford University Press, 1988) and the expanded version, updated by his son, David Sanjek, *Pennies from Heaven: The American Popular Music Business in the Twentieth Century* (New York: Da Capo Press, 1996); R. Serge Denisoff, *Solid Gold: The Popular Record Industry* (New Brunswick, NJ: Transaction Books, 1975); Steve Chapple and Reebee Garofalo, *Rock 'n' Roll Is Here to Pay: The History and Politics of the Music Industry* (Chicago: Nelson-Hall, 1977); and Frederic Dannen, *Hit Men: Power Brokers and Fast Money Inside the Music Business* (New York: Times Books, 1990). H. Wiley Hitchcock, ed., *The Phonograph and Our Musical Life. Proceedings of a Centennial Conference 7–10 December 1977. I.S.A.M. Monographs: Number 14* (Brooklyn, NY: Institute for Studies in American Music, 1980), along with the Centennial issue of the *Journal of the Audio Engineering Society* 25 (Oct./Nov. 1977) reflect on the state of the recording industry and technological progress on the one hundredth anniversary of Edison's phonograph.

Recent literature on sound recording history has built on the foundational but somewhat biased accounts of the earlier standard histories, namely, Fred Gaisberg, *The Music Goes Round* (1942; repr.; New York: Arno Press, 1977), Roland Gelatt, *The Fabulous Phonograph: From Edison to Stereo* (rev. ed. New York: Appleton-Century, 1965), and Oliver Read and Walter L. Welch, *From Tin Foil to Stereo: Evolution of the Phonograph*, 2nd ed. (Indianapolis: Howard W. Sams, 1976). Newer studies include Tim Brooks, *Lost Sounds: Blacks and the Birth of the Recording Industry, 1890–1919* (Champaign: University of Illinois Press, 2004); Michael Chanan, *Repeated Takes: A Short History of Recording and Its Effects on Music* (London: Verso, 1995); Timothy Day, *A Century of Recorded Music: Listening to Musical History* (New Haven, CT: Yale University Press, 2000); Andre Millard, *America on Record: A History of Recorded Sound*, 2nd ed. (New York: Cambridge University Press, 2005); and David Suisman, *Selling Sounds: The Commercial Revolution in American Music* (Cambridge, MA: Harvard University Press, 2009), to name only a few of the best. One survey taking a global approach is Pekka Gronow and Ilpo Saunio, *An International History of the Recording Industry*, trans. Christopher Moseley (London: Cassell, 1998), but it is necessarily quite selective.

Also important to this study were the beautifully illustrated "coffee table" books that document recording history through image and text, such as Frank Driggs and Harris Lewine, *Black Beauty, White Heat: A Pictorial History of Classic Jazz, 1920–1950* (New York: William Morrow and Company, 1982); Orrin Keepnews and Bill Grauer, Jr., *A Pictorial History of Jazz: People and Places from New Orleans to Modern Jazz* (New York: Crown Publishers, 1955); Leo Walker, *The Wonderful Era of the Great Dance Bands* (Berkeley, CA: Howell-North Books, 1964); and W. Royal Stokes, *Swing Era New York: The Jazz Photographs of Charles Peterson* (Philadelphia: Temple University Press, 1994). The newest addition, which falls somewhere between illustrated and academic history, is Sean Wilentz, *360 Sound: The Columbia Records Story* (San Francisco: Chronicle Books, 2012). Several well-illustrated tribute studies offering insight into studio history and detailed reconstructions of the making of particular albums include Charles Granata, *Sessions with Sinatra: Frank Sinatra and the Art of Recording* (Chicago: A Cappella Books, 1999); Ashley Kahn, *Kind of Blue* (New York: Da Capo Press, 2000); and Jim Cogan and William Clark, *Temples of Sound: Inside the Great Recording Studios* (San Francisco: Chronicle Books, 2003).

Increasingly, music and technology have been the focus of studies from many disciplines. Sociologists and ethnomusicologists were among the first to write about music technology, especially Simon Frith, *Sound Effects: Youth, Leisure, and the Politics of Rock 'n' Roll* (New York: Pantheon, 1981), and *Performing Rites: On the Value of Popular Music* (Cambridge, MA: Harvard University Press, 1996). Paul Greene and Tom Porcello, eds., *Wired for Sound: Engineering and Technologies in Sonic Cultures* (Middletown, CT: Wesleyan University Press, 2004), offers a wide-ranging anthology on the role of sound engineering technologies in contemporary global music. Two early articles that broke new ground in the study of how musical culture and technology grew inseparable with the phonograph are Marsha Seifert, "Aesthetics, Technology, and the Capitalization of Culture: How the Talking Machine Became a Musical Instrument," *Science in Context* 8, no.

2 (1995): 417–449; and Emily Thompson, "Machines, Music, and the Quest for Fidelity: Marketing the Edison Phonograph in America, 1877–1925," *Musical Quarterly* 79 (Spring 1995): 131–171. Other books that helped me think about the connection between music and technology are Mark Katz, *Capturing Sound: How Technology Has Changed Music* (Berkeley: University of California Press, 2004); Michael Lydon, *Boogie Lightning* (New York: Dial Press, 1974); Paul Théberge, *Any Sound You Can Imagine: Making Music/Consuming Technology* (Hanover, NH: Wesleyan University Press and the University Press of New England, 1997), and Albin J. Zak III, *The Poetics of Rock: Cutting Track, Making Records* (Berkeley: University of California Press, 2001). A recent collection of papers that cover historical, theoretical, and case study approaches to the study of record production is Simon Frith and Simon Zagorski-Thomas, eds., *The Art of Record Production: An Introductory Reader for a New Academic Field* (Surrey, UK: Ashgate, 2012).

The interdisciplinary field of sound studies has grown exponentially in recent years. Some of the most useful works are Karin Bijsterveld, *Mechanical Sound: Technology, Culture, and Public Problems of Noise in the Twentieth Century* (Cambridge, MA: MIT Press, 2008); Trevor Pinch and Frank Trocco, *Analog Days: The Invention and Impact of the Moog Synthesizer* (Cambridge, MA: Harvard University Press, 2002); Jonathan Sterne, *The Audible Past: Cultural Origins of Sound Reproduction* (Durham, NC: Duke University Press, 2003); and Emily Thompson, *The Soundscape of Modernity: Architectural Acoustics and the Culture of Listening in America, 1900–1933* (Cambridge, MA: MIT Press, 2002). Collections of essays on music and technology have appeared more frequently in the past decade. One of the earliest, originally published in Germany in 2000, is Hans-Joachim Braun, ed., *Music and Technology in the Twentieth Century* (Baltimore: Johns Hopkins University Press, 2004); and numerous others have appeared since then, including Veit Erlmann, ed., *Hearing Cultures: Essays on Sound, Listening, and Modernity* (Oxford: Berg, 2004); Trevor Pinch and Karin Bijsterveld, *The Oxford Handbook of Sound Studies* (New York: Oxford University Press, 2011); Jonathan Sterne, ed., *The Sound Studies Reader* (Routledge, 2012); and David Suisman and Susan Strasser, eds., *Sound in the Age of Mechanical Reproduction* (Philadelphia: University of Pennsylvania Press, 2010).

This study is grounded in the history of technology, the field most influential in my thinking about the history of recording. In understanding the mental map of recording engineers and the social aspect of their work, I draw especially on the works of Eugene S. Ferguson, *Engineering and the Mind's Eye* (Cambridge, MA: MIT Press, 1992); Walter G. Vincenti, *What Engineers Know and How They Know It: Analytical Studies From Aeronautical History* (Baltimore: Johns Hopkins University Press, 1990); and Louis L. Bucciarelli, *Designing Engineers* (Cambridge, MA: MIT Press, 1994). If there is one trait shared by recording engineers, it is technological enthusiasm, and the two works that best develop this theme are Thomas Hughes, *American Genesis: A Century of Invention and Technological Enthusiasm* (New York: Penguin, 1989), and Robert C. Post, *High Performance: The Culture and Technology of Drag Racing, 1950–1990* (Baltimore: Johns Hopkins University Press, 1994). Other essential works are Carolyn Marvin, *When Old technologies Were New: Thinking about Electric Communication in the Late Nineteenth Century* (New York: Oxford University Press, 1988), particularly her chapter on the emerging technological literacy of electrical "experts." Carroll Pursell, *The Machine in America* (Baltimore: Johns Hopkins University Press, 2007); David E. Nye, *American Technological Sublime* (Cambridge, MA: MIT Press, 1994); and Merritt Roe Smith and Leo Marx, eds., *Does Technology Drive History? The Dilemma of Technological Determinism* (Cambridge, MA: MIT Press, 1994) are a few of the many books in this field that have helped me think about technology as a cultural construct.

Studies in business and labor history shed light on the small business culture of independent studios and the work of both recording engineers and musicians. Particularly useful were Mansel G. Blackford, *A History of Small Business in America*, 2nd ed. (Chapel Hill: University of North Carolina Press, 2003), Mansel G. Blackford and K. Austin Kerr, *Business Enterprise in American History*, 3rd ed. (Boston: Houghton Mifflin, 1995), and Philip Scranton, *Endless Novelty: Specialty Production and American Industrialization, 1865–1925* (Princeton, NJ: Princeton University Press, 1997). My well-worn copy of James P. Kraft, *Stage to Studio: Musicians and the Sound Revolution, 1890–1950* (Baltimore: Johns Hopkins University Press, 1996) was an essential source for the impact of new sound technologies on the careers of professional musicians and is a must-read for any scholar of recording history. Although Kraft considers himself a labor historian, his study is a rich technological history that benefits from his insight as a musician. Also useful was Robert R. Faulkner, *Hollywood Studio Musicians: Their Work and Careers in the Recording Industry* (Lanham, MD: University Press of America, 1985).

The literature on tacit knowledge ranges across disciplines. Works that helped me think about the skill set of recording engineers included Michael Polanyi, *Personal Knowledge: Towards a Post-Critical Philosophy* (Chicago: University of Chicago Press, 1958), and *The Tacit Dimension* (Garden City, NY: Doubleday & Company, 1966); Robert J. Sternberg et al., *Tacit Knowledge in the Workplace*, Technical Report 1093 (Alexandria, VA: U.S. Army Research Institute for the Behavioral and Social Sciences, March 1999); Stephen R. Barley and Julian E. Orr, eds., *Between Craft and Science: Technical Work in U.S. Settings* (Ithaca, NY: ILR Press, 1997); Jerome Ravetz, "Science as Craftsman's Work," in *Scientific Knowledge and Its Social Problems* (London: Oxford University Press, 1971); and Douglas Harper, *Working Knowledge: Skill and Community in a Small Shop* (Chicago: University of Chicago Press, 1987).

Literature on amateurs and amateurism provided insight into the motivation of recording enthusiasts, including Steven M. Gelber, *Hobbies: Leisure and the Culture of Work in America* (New York: Columbia University Press, 1999); Kristin Haring, *Ham Radio's Technical Culture* (Cambridge: MIT Press, 2006); Robert A. Stebbins, *Amateurs: On the Margin between Work and Leisure* (Beverly Hills, CA: Sage Publications, 1979), and his subsequent *Amateurs, Professionals, and Serious Leisure* (Montreal: McGill-Queen's University Press, 1992); Patricia R. Zimmerman, *Reel Families: A Social History of Amateur Film* (Bloomington: Indiana University Press, 1995); and Karin Bijsterveld, "'What Do I Do with My Tape Recorder . . . ?': Sound Hunting and the Sounds of Everyday Dutch Life in the 1950s and 1960s," *Historical Journal of Film, Radio and Television* 24, no. 4 (2004): 613–634.

Index